Third Edition

MODERN TECHNICAL WRITING

THEODORE A. SHERMAN
University of Idaho

SIMON S. JOHNSON
Oregon State University

PRENTICE-HALL, INC., *Englewood Cliffs, New Jersey*

Library of Congress Cataloging in Publication Data

Sherman, Theodore Allison.
  Modern technical writing.

  Bibliography: p.
  Includes index.
  1. Technical writing.   I. Johnson, Simon S.,
1940–   joint author. II. Title.
T11.S52 1975   808′.066′6021   74-28155
ISBN   0-13-598763-6

10   9   8   7   6   5   4

PRENTICE-HALL INTERNATIONAL, INC., London
PRENTICE-HALL OF AUSTRALIA, PTY., LTD., Sydney
PRENTICE-HALL OF CANADA, LTD., Toronto
PRENTICE-HALL OF INDIA PRIVATE LIMITED, New Delhi
PRENTICE-HALL OF JAPAN, INC., Tokyo

CONTENTS

iii

Part III
BUSINESS CORRESPONDENCE

PREFACE

The third edition of Modern Technical Writing consists of four parts: *Technical Writing in General; Reports, Proposals, and Oral Presentation; Business Correspondence;* and a *Handbook of Fundamentals.* This is the same overall plan that was used in the first two editions except that the material on oral presentation is entirely new and much of the rest has been updated or rewritten to keep abreast of current practices of the better writers in government, education, science, and industry.

The logic of this arrangement is that some of the various skills needed for technical writing can best be acquired if first studied separately and then put to use in the writing of reports and other longer pieces. The parts are written, however, in a manner that permits an instructor to take them up in a different order if he wishes.

Part I starts with a brief introductory chapter and then takes up Style and Organization, covering them in that order because skill or lack of skill in style affects all writing, while much technical writing is so short as to offer no problems of organization. These chapters are followed by one on Mechanics, covering matters of form that are peculiar to technical writing or else crop up in it with abnormal frequency.

The chapter on Special Problems, which follows, performs a dual function. It provides writing assignments that may be used while the study of style, organization, and mechanics is still under way, and it explains ways of handling certain problems that may arise during the writing of reports, proposals, and other longer forms. We have also expanded the treatment of technical articles—recognizing the potential contribution of article writing to the career of the writer and the value of the article to science and technology.

In Part II, a change of emphasis at one point is reflected in the new title for Chapter 8, Nonformal Reports—Their Variation in Form and Purpose, which was formerly called Special Types of Reports. Though certain special types of reports are still discussed, additional emphasis is given to the fact that there does not exist any universally accepted set of types, under which all reports can be classified.

Two other extensive changes have been made in Part II: The chapter on Proposals, which first appeared in the second edition, has been rewritten and substantially expanded so as to cover that important subject more thoroughly. Also, an entirely new chapter, Oral Presentation of Technical Information, has been added. Though a study of this chapter is no substitute for training in public speaking, we believe that its recommendations can nevertheless be of substantial assistance to those who use this book on the numerous occasions when they will be called upon to present their ideas in person before a small group or a large audience.

Part III, *Business Correspondence*, remains largely as it was. It is important because, though not all letters, strictly speaking, are technical writing, those who do technical writing will find it necessary to write more and more letters as their careers continue. Consequently, they should be made aware of the difference between the skills called for in letter writing and those that they need in the other writing that their work involves. Of special interest to the student is the section on letters concerning employment, which has been altered to take into account a rapidly changing job market.

The *Handbook of Fundamentals* has been updated—usually by being liberalized. It is intended primarily for purposes of reference when a writer is doubtful about some matter that concerns correctness and effectiveness of writing in general. Its attitude, like that of the authors, is liberal; but in fairness to those who use it, some of the traditional taboos are pointed out because they may still be taken seriously by people whom those who use this book will address.

In all parts of the book new ideas have been added and changes have been made wherever close scrutiny revealed an opportunity to make the book more effective. The exercises are of the same kind used before, but consist of entirely new material except on a few occasions where something in the second edition brought out a point so well that no satisfactory replacement could be obtained or devised. Also, the subjects for writing assignments are new except where they are broad enough to allow the student considerable freedom in utilizing material from his own special interests.

The specimens of various kinds of technical writing are almost entirely new, and in every case have been written recently enough to be in line with current practice. Particular pains were taken in obtaining reports and proposals to use as specimens. All of those included were written for actual use in government or industry, and are shown in facsimile to let students see an actual cross section of material produced by professional workers in fields they themselves may enter. No specimen is the product of professional technical writers and artists. On the contrary, they are the work of professional men who must write as part of their jobs but whose main occupation is not writing. Most of those who use this book will fall into that category.

These specimens were obtained from many sources, and those that are presented were chosen because they seemed best to represent what is being produced in the field. They are of various types and concern various fields. Geographically, they come from places as far apart as Alaska, Hawaii, and Brooklyn. They comprise only a small percentage of those received. Examples not included were also valuable in that they too were examined with care so that what the book says about reports and proposals would correspond with current practices. Also, they were a source of subject matter in devising exercises and assignments that would be most like what the student will be doing when he is on the job.

Throughout the book—in the text proper and in assignments, exercises, and specimens—material will be found that concerns subjects which have emerged as vital in recent years, not only to students, but to Americans generally. Ecology, pollution control, and energy sources are typical examples of these interests.

One further change that is directly apparent only on the title page but which is indirectly reflected elsewhere through the book is the addition of a second author. This addition permitted us to contact a great many more sources for examination and allowed us to state with greater confidence that the suggestions we provide the student to aid his writing are indeed current and enduring in the field.

In this respect, the number of persons who have been of service while revision was under way has been so great that listing all of them would be impractical, but a particular debt of gratitude is owed each of the following: Allen S. Janssen, Dean (Emeritus) of the College of Engineering, University of Idaho (who helped us to make the contacts that enabled us to obtain much of our material from outside sources); John Banks, President of the National Steel and Shipbuilding Company of

San Diego; Ronald Hurlbutt, Traffic and Engineering Department Manager, Oakland, California; Eugene Lunty, Executive Vice President of the Brooklyn Union Gas Company; William Smith, City Engineer, Moscow, Idaho; Jerry Whiting, Mining Services Manager, First Bank of Montreal, Canada; Nancy G. Baker, Oregon State University Speech Department; Diane Kireilis and Elbert G. Melcher, Rocky Mountain Center on Environment; Robert Larson, Ken Seaman, and Harry Sloffkiss, National Assessment of Educational Progress; and Edmund E. Tylutki, Associate Professor of Biology, University of Idaho.

In addition, the corporations and governmental agencies listed below were extremely cooperative: The Army Corps of Engineers; The Asphalt Institute, College Park, Maryland; Boeing Company; Brooklyn Union Gas; Chevron Chemical Company; Douglas United Nuclear Industries, Inc.; DeKalb Agrisearch, Inc.; Earl V. Miller Engineers, Phoenix, Arizona; General Electric Company; International Snowmobile Industry Association; Montgomery County Maryland Division of Transportation Engineering; National Assessment of Educational Progress; National Steel and Shipbuilding Company; the City of Oakland, California, Traffic Engineering Department; The Potlatch Corporation; Rocky Mountain Center on Environment; 3M Company; T.R.W. Corporation; U.S. Gypsum; United States Environmental Protection Agency; United States Forest Service; University of Washington College of Engineering; and Washington State University College of Engineering Extension Service.

The faculties of the University of Idaho Colleges of Agriculture, Engineering, Forestry, and Mines and the faculty of the Oregon State University College of Science have given encouragement and cooperation at all times.

Theodore A. Sherman
Simon S. Johnson

PART I

TECHNICAL WRITING IN GENERAL

1

INTRODUCTION

In order to profit by a study of technical writing you will find it necessary to recognize one fact at the outset: Writing will be part of your work in the jobs that can do most to advance your career. Skill in writing must therefore be regarded as a professional tool, to be ranked on a par with the other knowledge and skills that will comprise your professional qualifications. If you can think of writing as something you will do on the job rather than only in the classroom—something you will use to convey information for practical use by an employer or client rather than just to demonstrate academic proficiency—you will have a motivation for improvement. And if you are motivated to make a genuine effort to profit from a technical writing course, you have every reason to expect that the time you devote to the course will contribute to a successful career.

No one who has been exposed to the comments of your future employers can doubt that the preceding statements are justified. In articles, public addresses, and personal remarks, high-ranking executives in industry and government have long been stressing the value of good writing in the technical professions. For example, in a personal letter one industrial executive has said, "To rise to an executive position one must be able to present his ideas effectively, first on paper and then orally. If the ideas are not presented well on paper, the chances are that he will never

3

get a chance to present them orally." Another executive wrote, "Unless an engineer can write a good report, he might as well reconcile himself to something a little better than a drafting job for the rest of his life." And the chief engineer of a nationwide company has commented that if its field engineers could write more clearly, the company's long distance telephone bills would be reduced by about 25 percent.

Why do men in positions of authority consider skill in writing so important? The reason is clear to anyone familiar with the postgraduation demands made on a worker with technical training. In engineering, agriculture, forestry, bacteriology, mining, metallurgy, public health, or any other technical field, writing is essential in almost every stage of every important project. Before work is started, the possibilities of a project must be stated and evaluated, and plans must be made in writing. When the work is under way, memoranda and letters must be exchanged as problems arise, and written reports must be made so that those in authority can learn how the work is progressing. When the project has been completed, the results must be recorded. Often, indeed, the only tangible result of a long and expensive project will be a written report; and if that report is ineffective, the benefits of all the time and energy expended may be jeopardized.

Since your writing will be important to your employers, it will also be important to you. Certainly it is one part of your work that is sure to reach the attention of your superiors on every level. If you write badly, not only will your weakness in writing be recorded against you, but your ability in other aspects of your work may be obscured. On the other hand, if you write effectively, you will not only receive credit for being a good writer, but will also find it easier to gain recognition for your other abilities. Many a person first catches the attention of those upon whom promotion depends by means of well-written reports; and unfortunately, many a person entering a technical profession finds progress more difficult because he had not realized the necessity of learning to use language correctly and effectively.

In addition, the skills you acquire in writing can go beyond helping you in your job or your employer in his private pursuits. While personal ambition is not the only motive that influences conduct, a desire for professional advancement is not incompatible with a desire to contribute to the general welfare. The problems plaguing the modern world are far too complicated to be solved by crusading zeal alone. Every conceivable profession will be involved in their solution. As a professional-level employee, you will have access to the latest developments and knowledge in your technical field. It will be these developments and knowledge that will solve the problems of a modern world. The skills in technical writing, finally, are skills in communication, and before society can effect changes, it must have communicated to it the knowledge necessary to make the changes. You, as an individual, can contribute more toward solving these problems and will have more influence on others if you achieve a recognized professional standing than if you remain in an obscure position. Thus in urging you to achieve skill in writing in order to advance your career, we are urging you to reach a position from which whatever efforts you wish to make toward the improvement of the human condition will be more effective.

So much for the reasons that technical writing is important. Another question that may come to mind concerns the relationship of technical writing to writing in general.

Certainly the qualities cultivated in general writing are important in technical writing also. As a technical writer you will continue to work toward correctness in grammar, punctuation, and spelling; and whatever skills you have acquired in making your sentences and paragraphs effective will improve your technical writing. Sound organization, desirable in writing of any sort, is a matter of special concern; and the same might be said of clearness, simplicity, and directness of style. All in all, skillful, effective technical writing is built upon the foundation of skill in writing in general.

Still, general writing ability does not in itself guarantee that your technical writing will be satisfactory. Technical writing is governed by a sense of values that other writing does not usually emphasize so strongly; it often calls for the use of special forms; it requires familiarity with some special techniques.

Dealing as it so often does with the sciences, regular or applied, technical writing sets an unusually high value on objectivity, meticulous accuracy, and restraint. It is directed to the reader's mind and makes little effort to appeal to his emotions. Its purposes are utilitarian, and it is usually intended for readers who already have, to some degree, a special interest in the subject matter. Consequently, though it places a high value on interest, it does not try to be so colorful and entertaining that it runs the risk of becoming flashy and superficial.

Some of the forms in which technical writing appears are the bulletins and pamphlets issued by experiment stations and government bureaus, and the technical or semitechnical papers and articles read at meetings or published in technical periodicals. Many of these forms do not have to be studied as separate types, but there is one form of technical writing—the report—that is particularly important and calls for special study. Though not every technical writer is called upon to produce all of the other forms listed, almost anyone who does any technical writing at all must write reports. Because of this fact and because reports vary so widely on different occasions, they are given special emphasis in this book. Special attention is also given to proposals, which industry and government have come to regard as immensely important.

As you work in any of the forms of technical writing, you will at various times need to give instruction, write a definition, explain a process, describe an object, or analyze a condition or problem. The problems that arise on each of these occasions will receive individual attention in this book.

You will also find it necessary, in technical writing, to know how to use numbered lists, tables, and figures. You will need unusual familiarity with the conventions that govern abbreviations and the use of figures or words for numbers, especially in cases when the conventions of technical writing differ from those that govern writing in general. You will need to acquire skill in the use of topical headings so that the reader can easily grasp the organization of your work and find what he is looking for. And further, since the soundness of technical writing must bear up under critical appraisal, you will need to master the process of documentation.

The treatment of technical writing in this book is divided into three parts. The first is devoted to matters that concern all forms of technical writing. Style is discussed first, because a writer's style can affect the quality of whatever he writes, even a memorandum of two or three sentences. Next, suggestions are made about organization. Succeeding chapters cover mechanical form, special problems, and the use of tables and figures.

The second part of the book is devoted to reports and proposals, which deserve special emphasis for reasons already mentioned.

The third section of the text proper deals with business correspondence. Though letters do not always have to do with technical subjects, they assuredly comprise a substantial portion of the writing demanded of almost anyone in a technical profession, and therefore merit attention.

The three main parts of the book are followed by a Handbook of Fundamentals so that if your general ability in writing is weak, you will have access to the information you are most likely to need in overcoming that weakness.

All in all, the material in this book should provide the help you will need in acquiring a reasonable degree of skill in technical writing. To be sure, neither a book nor a course can rapidly develop your full writing potential. Only long experience and extensive practice can do that. But study of the subject can at least do these things for you:

> 1. It can make you aware of the standards your future employers will expect you to meet—standards of accuracy and compliance with instructions as well as standards of writing;

2. It can familiarize you with practical, tested writing procedures;
3. It can give you a clearer picture of the manner in which writing fits into the total pattern of professional activity;
4. It can reduce the likelihood that at the beginning of your professional work you will form undesirable habits which may harden until they become difficult to break;
5. It can leave you a better writer than you were before; but even more important,
6. It can lay a foundation for self-improvement in the future, so that your skill in technical writing, like your other professional skills, will steadily increase as your career continues.

One final comment is in order. In the present day and age, characterized as it is by rapid change, questions are constantly raised about whether various time-honored courses justify retention in one or another college curriculum. Authorities tell us that in some fields what is learned now will be out of date in ten years. Yet no one has been heard to say that *writing* will cease to be an essential skill; and though language is not static, you may take comfort in the fact that skill in writing acquired today will not become outdated.

Thus the ability to write effectively is well worth the time you will spend in acquiring it. Long after some of the things you are learning have ceased to be useful, you will still be using whatever ability you have acquired in using the English language. It would be an overstatement, perhaps, to say that a career in a technical profession will be impossible if you cannot write effectively. It is no overstatement, however, to say that weakness in writing is a handicap that will weaken your qualifications for many desirable positions, and that skill in writing is an asset that can make your professional advancement faster and easier.

2

EFFECTIVE

STYLE

Many an inexperienced writer gives little thought to style, concerning himself only with getting his ideas down on paper without making errors. If you adopt this attitude—if you think of style as merely an ornament—you are neglecting one of the major opportunities to make your writing more effective. Of all the weaknesses that annoy those who read technical writing, ineffective style is one of the most damaging.

What do we mean by style? It is style that makes the difference between the enthusiasm of a patent medicine advertisement and the restraint of an article in a medical journal, or between the fervor of an attorney pleading a case and the analytical impersonality of a judge in his charge to the jury. In short, style concerns not what a writer says but how he says it.

In technical writing you will not need to develop a distinctive personal style. Rather, you can concentrate on such qualities as clearness, directness, conciseness, and readability. Your style should ordinarily be neither so informal as to sound unprofessional nor so excessively formal as to sound stilted. It should lean toward restraint, for though a reasonable degree of enthusiasm is sometimes an asset, an excess suggests bias. It should not be charged with emotional intensity, for it usually deals with questions that are answered by the mind rather than by the emotions.

Yet even in the limited context of technical writing the subject of style is elusive. A long list of desirable qualities compete for attention. This list includes not only those mentioned—clearness, directness, conciseness, and readability—but also simplicity, precision, and specific, concrete, understandable diction. Some of these qualities overlap. Some of them are the result of others. (For example, readability and clearness are the end results of attention to directness and simplicity.) Also, unfortunately, an excessive effort to attain one quality may cause the loss of other qualities equally desirable. (For example, up to a certain point conciseness improves readability, but carrying conciseness too far will make your writing less readable.)

This chapter will concentrate attention on what you can actually *do* in order to make your style more effective. At a working level, you have two major questions to think about as you try to improve your style: *How can I be sure that my diction (choice of words) will make my style more effective?* and *How can the construction of sentences make my style more effective?*

The two sections that follow are based on these questions, but the subject of conciseness involves both diction and sentence structure to such an extent that it is discussed separately in a third section.

EFFECT OF DICTION ON STYLE

If your diction is to do its part in making your style effective, you will need to bear in mind the following points until you can apply the right principles automatically:

Avoidance of stiff, pompous language
Avoidance of needlessly technical language or jargon
Use of concrete, specific language

Avoidance of Stiff, Pompous Language

One of the main reasons that style in technical writing often becomes ineffective is that writers, perhaps subconsciously, try too hard to be impressive. Consequently they sound stiff and pompous, producing a style that at best is ostentatious and at worst becomes needlessly hard to translate into concrete meaning.

A technical writer's style is pompous when he uses such expressions as *increase the visibility of the incandescent gases* when he might say *make the flame easier to see,* or *has a deleterious effect upon* rather than *harms* or *injures* or *damages.* Some writers cannot resist the temptation to use *optimum* when *best* would convey the idea, or *finalize* when they might use *complete.* A practicing engineer once produced the expression *minimize the expenses to an optimum degree.* (Why not just *hold expenses to a minimum?*) The classic example of a rebuke to such ostentation occurred when President Roosevelt, looking over black-out instructions during World War II, crossed out *terminate the illumination* and substituted *turn out the lights.*

Pompous style also results from the inordinate tendency of many writers to resort to words like *characteristic, property,* and *condition.* The result is such phraseology as "It possesses the quality of malleability" instead of "It is malleable"; or "It lacked the property of hardness" instead of "It was not hard enough" (or "It was too soft"); or again, "It passed into a solid condition" instead of "It solidified."

Avoidance of Needlessly Technical Language

Technical language has come into existence because it is often needed for clear, effective communication. A single technical term will frequently convey an idea that would otherwise call for a long phrase or clause, and there are some subjects that it would be almost impossible to discuss without using technical terminology.

Nevertheless, technical language should not be used when it can be avoided. The purpose of writing is not just to get a message down on paper, but to get it into the mind of the reader; and you may be sure your reader will not always possess your own specialized vocabulary. One of the most common complaints against technical writing is that those who produce it apparently believe that everyone who reads it is a specialist in the subject.

Moreover, the best authorities on writing agree that even when you are writing solely for technically trained readers, it is best to use technical language only when you need it. It is not good style to use *ozostomia* when you might say *bad breath,* or to refer to *imbricate scales* when you might call them *overlapping. Scutellate* could be replaced by *flash* on almost any conceivable occasion. And the writer who wrote *reducing the mean diameter of the spray spectrum* might better have used *making the spray finer*—and saved himself the embarrassment of having his style ridiculed in a technical magazine.

The warning against overly technical language applies not only to words that are formally correct, but also to the technical jargon that specialists in any line of activity use when they talk informally among themselves in a shop or laboratory. Such terms as *woofer* and *tweeter* (common among hi-fi enthusiasts) are examples. For a general audience it would be better to use *bass clef* and *treble clef speakers*. Similarly, *sink street* and *lead sled*, common jargon among sailplane devotees, are less suitable than *area of descending air currents* and *heavy sailplane*. To be sure, such language may lend flavor to a popular article and is entirely natural and unobjectionable in personal conversation or informal memoranda with limited distribution. But it is not suitable when standard English is called for, and you may be sure that the conversational language used by a scientist in the laboratory seldom appears in his technical articles or reports.

Use of Concrete, Specific Language

In any writing, technical or otherwise, concrete and specific language makes a writer's style more vigorous and colorful. In technical writing it is especially desirable because it not only increases liveliness but also gives information more effectively and reduces the likelihood of vagueness or misunderstanding.

Concrete and *specific* are not identical in meaning, as you can see when you consider their respective antonyms, *abstract* and *general*. But in a practical effort to improve your style you can try, at the same time, to find words that are both concrete and specific.

The improvement that can result from such an effort can be seen by comparing the following specimens.

> *Abstract and general:*   We plan to present a proposal covering all important aspects of the problem.
> *Concrete and specific:*   On October 10 we shall present a proposal covering cost, materials, methodology, and personnel.
> *Abstract and general:*   The projected lifespan of the building is adequate for the cost.
> *Concrete and specific:*   The projected 25-year usefulness of the warehouse will allow monthly mortgage payments and still provide a 4 to 7 percent margin of profit.
> *Abstract and general:*   Moderate winds, when accompanied by new snow and near-thawing temperatures, often create avalanche conditions.
> *Concrete and specific:*   Winds between 15 and 30 mph, when accompanied by snow and temperatures between 10° F and 30° F, often create unstable slabs in avalanche-starting zones.

Technical writers must constantly make this kind of choice. They base their decisions on the purpose of the writing and the projected audience. For instance, in the last pair of examples above, the first would be acceptable for some purposes, where a high level of technical information was not necessary, while the second would be better for a highly technical audience. Note the contrast in these additional examples, where the same term can be concrete and specific for one audience and general for another.

Abstract and General	Concrete and Specific
vehicle	fork lift, dolly, truck, automobile
truck	pickup, tractor-trailer, railway truck, stake truck
pickup	½-ton, ¾-ton, longbed, six-passenger
receptacle	jar, beaker, trash can, flask
flask	Erlenmeyer, pocket, volumetric, founding

As we will point out later, use of the concrete, specific diction recommended above is valuable not only for its own sake, but also because it contributes to directness.

Vagueness is a serious defect in technical work. Often, to be sure, vagueness is not just a matter of style. You will sometimes be tempted to write vaguely in a conscious or unconscious effort to conceal your lack of specific information. Yet there will probably be times when you have enough information to avoid vagueness but may neglect to use that information. You may tell us, "*Small* samples are taken at *frequent* intervals," when you know the facts well enough to say, "Samples of *100 cc* are taken at *half-hour* intervals"; or you may write, "A *fairly large* piece of steel is placed in the machine," when you might have written, "A *12-inch* piece of steel." Or you might use, "The pouring of concrete is *seriously* behind schedule" when you might have used *"three weeks* behind schedule."

This last example brings up another consideration. The engineer who supplied it commented, "I want my people to tell me the objective facts; I'll draw my own conclusions about whether three weeks is or is not a serious delay." His attitude should not be taken, however, to mean that *interpretation* of specific facts is never desirable. Sometimes interpretation is really needed. Someone might be told that a certain number of pounds of soil per acre had been lost because of erosion, and be unable to judge whether the amount was alarming. Consequently, a writer might do well to add that this loss amounted to half an inch, and that the loss was serious because the topsoil was only ten inches deep. The significant point is that in offering this interpretation a writer would be providing *additional* objective facts rather than omitting them in favor of vague expressions or unsubstantiated interpretation.

EFFECT OF SENTENCE STRUCTURE ON STYLE

In discussing the effect of sentence structure on style we shall not concern ourselves with the damage done by errors in fundamentals of English. It is true, of course, that such errors often affect style—for example, by reducing clearness—but they are covered in the Handbook. In dealing with the question of how to make sentences not only correct but also effective, we will consider sentence length, simplicity and directness, precision, and the choice between active and passive voice.

Sentence Length

Avoidance of Excessive Length. Though some long sentences are clear because of their simple construction, there are many times when the use of shorter sentences could make some piece of writing clearer and easier to read. The following example shows how confusion was created by an effort to jam too much into a single sentence.

> High operating speed shall be provided by a high-speed, motor-driven switch operator in order to assure the full inherent mechanical and electrical performance of the circuit-switcher such as power operation under 1½-inch ice formation, fault-closing capability, close interphase simultaneity, long life of fault-closing arcing contacts, and avoidance of excessive switching transients which can accompany prolonged or unstable prestrike arcing.

Note how clearness is improved by use of more sentences:

> High operating speed shall be provided by a high-speed, motor-driven switch operator in order to assure full inherent mechanical and electrical performance of the circuit-switcher. The performance features include power operation under 1½-inch ice formation, fault-closing capability, close interphase simultaneity, and long life of fault-closing arcing contacts. The switch operator also assures avoidance of excessive switching transients which can accompany prolonged or unstable prestrike arcing.

Long sentences are not always so involved as the preceding example. Stringy sentences may not be especially hard to read, but breaking them up will result in a more satisfactory style. Note the following, from a government specifications report:

> Upon completion of the work, the Contractor shall operate the pump in the presence of the Government representative to demonstrate its operation; shall establish accurately the performance of the pump, including efficiencies, through measurements of the discharge rate by use of the flow tube and pressure-differential gauge in the system, the pumping level (by use of gauges furnished

with the pump) and horsepower input (by the watt-hour-meter disk-constant method or other equally accurate method); and shall submit a certified report of all such readings, calculations, and measurements.

The content would be easier to grasp if the long sentence were broken into four, as follows:

> Upon completion of the work, the Contractor must test the pump in the presence of the Government representative to establish accurately that it meets all performance specifications. Discharge rate shall be determined by the flow tube and pressure-differential gauge in the system. The pumping level shall be determined from gauges furnished with the pump, and the horsepower shall be determined by the watt-hour-meter disk-constant or equally accurate method. All findings shall be submitted in a certified report.

Avoidance of Primer Style. In the effort to avoid making your sentences too long you should not go to the other extreme and use "primer style." An excessive number of extremely short, simple sentences causes writing to become jerky and monotonous. Moreover, it does not necessarily result in clearness. Each sentence, it is true, may be clear; but the relationship of the ideas contained in the individual sentences is often hard to perceive. The primer style forces the reader to combine for himself the ideas that the writer should have combined for him. Also, it necessitates the repetition of words and thus destroys conciseness, as will be pointed out later.

Note how, in reading the following examples, you are slowed up by being forced to come to a full stop too often.

Primer Style

1. This study will be a continuation of the study carried out on the Glide Ranger District during the summer of 1972. Plots will be established in the Tiller Ranger District for study. They will be prepared and located to correspond to conditions encountered in cleared units. In addition, some plots will be altered to determine the effect of conditions which could be economically imposed but do not occur in nature. One example of alteration would be the application of herbicides. Measurement of plant moisture stress (pms) will be taken in all plots. Plant moisture stress readings will be maintained at specific levels in designated plots by the addition of water. Weather observations will be recorded daily. It is hoped that pms readings and weather observations will show a close correlation.

2. Computerized systems can operate in one of three modes. The operating modes are batch processing, remote batch processing, and time-share processing. Conventional batch processing met all but one

of the design requirements. The exception was accessibility. Batch processing could not provide output to diversified, remote locations in a timely manner. Remote batch processing, as the name implies, could provide batch processing at remote locations. It could do this through the use of teletype terminals as output devices. Unfortunately, the response time for remote batch processing is unpredictable. It can vary from a few minutes to several hours.

More Mature Style

1. The Tiller Ranger District study will be a continuation of the study carried out on the Glide Ranger District during the summer of 1972. Plots will be prepared and located to correspond to conditions encountered in cleared units. In addition, some plots will be altered, as with the application of herbicides, to determine the effect of conditions which could be economically imposed but do not occur naturally. Plant moisture stress (pms) readings will be taken in all plots and maintained at specific levels in designated plots by adding water. Weather observations will be recorded daily to determine whether there is a correlation between weather and pms.
2. Computerized systems can operate in one of three modes: batch processing, remote batch processing, and time-share processing. Except for accessibility, conventional batch processing met all the design requirements; but it could not provide output to diversified, remote locations in a timely manner. Remote batch processing, as the name implies, could provide batch processing at remote locations through teletype terminals. Unfortunately, its response time is unpredictable, varying from a few minutes to several hours.

To sum up the matter of sentence length: Some sentences should be short, some long, some medium. It is very easy, however, to fall into a habit of making all the sentences similar in length and type, and care is necessary to prevent such a pattern from forming.

Simplicity and Directness

Simplicity and directness are closely related—so closely, in fact, that when a passage is simplified it usually becomes more direct as well. The present discussion will not try to draw a line between the two, but since each of them includes certain ideas not included in the other, they will be discussed separately.

Simplicity. Simplicity is important in diction and in sentence structure. It has already been covered in connection with diction, so here we shall be concerned with sentence structure only.

No one would build a complicated mechanism to do a job that a simple mechanism could do equally well, nor analyze a specimen by a complicated method if a simple method would get the same results. By the same token, no one should express an idea by creating a complicated grammatical construction when a simple form would serve the purpose equally well. When you can substitute a word for a phrase or a phrase for a clause or sentence without sacrificing anything of value, you will usually improve your writing by doing so.

The following examples show how sentences that are needlessly complicated can be simplified.

> *Needlessly complicated:*   The next step consisted of looking for methods by which the losses that occurred in the bearings might be eliminated.
>
> *Simplified structure:*   The next step consisted of looking for methods of eliminating the losses in the bearings.
>
> *Needlessly complicated:*   A series of continuous runs made under conditions that were identical often yielded results that were different.
>
> *Simplified structure:*   A series of runs made under identical conditions often yielded different results.

In each case the sentence was simplified by cutting out two clauses— clauses that did not make the relationship of ideas one whit clearer than they could be made by a simpler form.

The effort to write simply should not be limited to avoiding needlessly complicated sentences. A passage may consist of relatively simple sentences yet be too elaborate in its general approach. For example, note the following passage: "The process was performed in two steps. The first of these consisted of drying the sample, and the second consisted of weighing it." A simpler style would use only one sentence: "The process consists of two steps—drying the sample and weighing it."

To be sure, if the list of steps were longer and if some or all of the steps could not be expressed briefly and simply, it might be necessary and desirable to introduce the list more elaborately, but for a short list of brief items, a simple approach is adequate.

It is possible, of course, to carry simplicity too far. The writer who substituted *variable elimination* for *elimination of variables* and *concrete information* for *information about concrete* reduced a phrase to a word in each case but in doing so became ambiguous. Similarly, the writer who used *one way of determining system applicability* for *one way of determining the applicability of the system* simplified the grammatical construction but sacrificed readability. Usually, simplifying the structure of a sentence makes it clearer and more readable; but when it produces the opposite effect, it is not desirable.

These examples achieve their simplicity in a manner that calls for additional comment. In each case the sentence was simplified by converting a noun into an adjective. It is true that the English language contains a great many words that can be used as either; but some technical writers go to fantastic extremes and make adjectives out of nouns that are not used in such a manner by the vast majority of literate people. As a result they develop an annoying mannerism that makes it more difficult for them to communicate with the rest of the world and irritates even many of those who understand them.

Directness. Directness, like simplicity, calls for attention to sentence structure. Consider for example the following indirect sentence: "There are parts of the report in which his statements have been influenced by his feelings." The main statement is, "There are parts." The main *fact,* however, is that a writer's feelings have influenced his statements in a report. To make the style direct we might write, "The report contains statements that have been influenced by his feelings," or "He has made statements in the report that have been influenced by his feelings." Each of these direct versions, you will notice, has a meaningful subject and a meaningful verb.

You might take the following suggestion as a formula for directness: Ask yourself, "What do I want to make a statement about?" and let the answer to this question tell you what your main subject should be. Ask yourself, "What do I want to say about it?" and let your answer tell you what your main verb should be.

Consider the result of applying this formula to some additional sentences:

> It must be expected that there will be complaints from people who live in the neighborhood.
>
> A recent article makes the statement that before 1950 Cuban reserves of manganese had been exploited almost to the point of depletion.

Directness in style would suggest:

> Complaints from people who live in the neighborhood must be expected.
>
> Before 1950, according to a recent article, Cuban reserves of manganese had been exploited almost to the point of depletion.

You have probably thought of other versions that would be just as direct. The significance of your being able to do so becomes apparent when two more sentences are considered.

> The new alloy is harder than the one we have used in the past.
>
> The alloy we have used in the past is softer than the new one.

These two sentences give us exactly the same fact, and they are equally direct. They differ, however, in one respect: the first makes a direct statement about the new alloy and incidentally gives information about the old. The second makes a direct statement about the old alloy and incidentally gives information about the new. As sentences, they are equally direct; *but they are not equally direct about the same thing.* A writer's choice between them should depend on which alloy he is primarily attempting to discuss. In other words, directness is more than just a matter of sentence structure without regard to purpose. It is also a matter of presenting the facts as they bear upon our purpose in writing and the reader's probable reason for reading.

Differences like the one illustrated may seem trivial when we look at only one or two sentences, but many an article or report has extended passages in which the writer seems to have gone off on a tangent because the facts about his real subject are presented only indirectly in sentences that are *direct* statements about other matters. The formula offered above reduces the likelihood that you will obscure your point by this particular kind of indirectness.

Indirectness may result not only from a poorly chosen subject, but also from a poorly chosen verb. By using a weak, general verb a writer may then be forced to work in the verb idea as a noun or adjective. For example, he might write:

> *effect a change in* instead of *change*
> *provide information about* instead of *inform*
> *came to the conclusion that* instead of *concluded*
> *accomplish a reduction* instead of *reduce*
> *place more emphasis on* instead of *emphasize*

This is not to suggest that you must express *every* idea as a direct statement. If you do, you will make it hard to distinguish between the important and the less important. Consider for example: "We are now using the process described in the enclosed leaflet. It is a continuous-flow process. Last year we used the batch process." Every fact is expressed directly, but it would be better to write, "We are now using the continuous-flow process described in the enclosed leaflet rather than the batch process that we used last year." The revised version is more effective because it is a direct statement of the major idea, which stands out distinctly because lesser matters are referred to in an incidental manner only.

Precision

Though precision involves both diction and sentence structure, the subject is all too often approached as if it concerned no more than the use of words with correct understanding of their meanings. Writers are warned, for example, not to confuse *imply* with *infer,* *liable* with *likely,* *fix* with *repair,* or *unique* with *unusual.* Most of the discussions along this line contain valid and useful information—unfortunately combined with many refinements that are based on textbook traditions rather than on first-hand observation of actual usage.

So far as such questions of general vocabulary are concerned, the Glossary of Usage at the end of the book covers this aspect of precision in enough detail to guard against the most likely errors. It includes the textbook favorites because there are still many people who value these traditional distinctions.

Precision in the use of technical terms is not covered in the Glossary because each line of work has its own extensive technical vocabulary. The most that a general discussion can do is to alert you to the need for mastering the meaning of each term in your field as you encounter it, and to urge that if you do much writing in a field where terminology is rapidly changing and developing, you invest in any special publication that lists and defines the terms you must use. Also, don't underestimate the help that you can get from the latest dictionaries—unabridged or even collegiate. And don't underestimate the extent to which you make yourself look immature if you use technical language incorrectly.

But precision is more than just a matter of words; as pointed out above, it is also a matter of sentence structure. Many a writer who knows the meaning of each word that he uses will construct sentences that do not actually say what he intends. Consider for example the statement, "Test plots using various species give an average range of life of 19 to 25 years, with a probable average of 22 years." According to this sentence, it is the *plots* that use the species and give the results. But in point of fact, the plots neither use the species nor give the results. It is the tests, not the plots, that give the results and it is the scientists performing the tests who use the species. One way of expressing the idea precisely—certainly not the only way—would be, "Tests of various species, each performed on a separate plot, indicate that the range of life is 19 to 25 years and that 22 years is the probable average."

Another example of lack of precision is the sentence, "The material should be of a hardness and a toughness that will possess good wearing properties." According to this sentence, the hardness and toughness *possess* wearing properties. In point of fact, they *are* properties. A precise restatement would be, "The material should be hard enough and tough enough that it will wear well."

Objecting to lack of precision in sentence structure is not just quibbling about trivial details. A sentence that lacks precision in structure is like a photograph taken by a camera that is not properly focused. We may be able to see the picture, but the blurred lines are annoying and distracting. It is a writer's job to say exactly what he intends to say rather than just coming close enough to the intended meaning for his reader to figure it out. And the major reason for failure to do this job is writers' failure to check their own work carefully enough to notice what they have said—or in other words, to make sure that the statement made by each verb applies to that verb's grammatical subject.

Choice Between Active and Passive Voice

Many commentators on technical writing devote considerable attention to the frequent superiority of the active voice over the passive. (The difference between them is explained in the Handbook, Part IV.) Following are typical examples of sentences where the passive reduces effectiveness.

1. It is maintained by the contractor that . . .
2. It is believed that . . .
3. Your letter has been received by us and . . .

The first of these sentences has a meaningless subject. The second not only has a meaningless subject but is also evasive about who does the believing—which may be a fact of considerable importance. The third is awkward and unnatural. The three sentences could be improved by change of voice, as follows:

1. The contractor maintains that . . .
2. I believe that . . . (if not *I*, some other word showing *who* believes)
3. We have received your letter and . . .

Yet in spite of the frequent overuse of the passive voice, it is a legitimate and useful form, and sometimes preferable to the active. For example, "The concrete was damaged by the cold weather" is better than "The cold weather damaged the concrete"—assuming that it is the concrete, not the weather, that we are discussing. Likewise, the passive voice in "The foam should be skimmed from the vats once an hour" is entirely satisfactory, since the skimming is the important fact and the question of who performs the action is not significant.

The passive voice is called for in a sentence such as, "The samples are collected by the foremen, examined by the inspector, and sent by air to the main office." To use the active voice we would have to write, "The foremen collect the samples, the inspector examines them, and an airplane carries them to the main office." This sentence, with its three different subjects, obscures the fact that the writer is really trying to tell us about the samples.

It is unfortunate that the terms *active* and *passive* are used to identify the two voices of verbs. Careless thinking makes some writers identify *active voice* with *active verb*. But the facts do not justify such an identification, as we can see when we look at two more sentences:

In the darkened room the mother softly hummed a lullaby to her quietly sleeping baby.
The building was shattered by an explosion.

The verb *hummed* in the first sentence is in the active voice but certainly does not convey a picture of vigorous activity. The verb *shattered* in the second sentence is in the passive voice but conveys just as much a sense of activity as if the sentence ran, "An explosion shattered the building." In brief, the extent to which a verb gives an impression of action depends more upon its actual meaning than upon the writer's choice of voice.

A sound guideline in deciding when to use the active voice and when to use the passive is found in the preceding discussion of directness —*that is, use a subject that shows what your sentence actually concerns, and use a verb that says what you really want to say about that subject.* If you will do this there is little danger that your use of voice will damage your style. Yet the warnings against excessive use of the passive may be justified, for the passive voice actually *is* overworked by many writers, it often *does* weaken their style, and it is often a symptom that calls attention to lack of directness.

CONCISENESS

Lack of conciseness, a defect in the writing of many people whose work is good in other respects, is a major cause of complaints by busy readers. Therefore, you should learn how to trim out every word that can be spared when the occasion justifies your taking the time to do so, and should try to avoid wordiness even when writing a paper that you will have little opportunity to revise.

To be sure, you can carry conciseness too far. You will not help your reader by cutting out so many words that you sacrifice clearness, readability, naturalness, precision, and proper application of emphasis. Nothing said in this discussion should be taken as encouraging such a sacrifice.

Since lack of conciseness is sometimes the result of nothing more specific than a general looseness of style, you can often reduce the length of something that you have written by merely looking it over in a general way. You will do a better job, however, if you know what to look for. Therefore, some of the main causes of wordiness are taken up in the following discussion.

Repetition

Repetition for the sake of clearness or emphasis is sometimes desirable, but unintentional repetition is a major cause of wordiness. Many a writer uses *spring of the year,* for example, when the single word *spring* would be sufficient, or says that costs *rise to a higher level* when he need only say that *costs rise.* Other examples of such repetition are *a period of three months* (repetitious because three months *is* a period) and *the specimen was placed on the balance and weighed* (repetitious because the fact that it was weighed clearly implies that it must have been placed on the balance).

A slightly different kind of repetition is seen in the sentences, "In one respect, the house was not in good condition. Its roof leaked badly." If the writer had been telling us of several defects, a general statement might have been desirable as a topic sentence; but since there was only one defect to mention, it would have been enough just to say that the roof leaked. Words are wasted when we are told the same fact twice, once in a general way and once in a specific way.

Wordy Phrases

The English language abounds in phrases that convey little meaning but are sometimes necessary in order to make grammatical construction clear or to make the flow of words smooth and natural. If your writing is to be concise you should be alert for opportunities to eliminate these phrases. For example, you should avoid *in the event that* when you could substitute *if; due to the fact that* when you could use only *due to* or *because;* and *has proved itself to be* when *has proved* or just *is* would be sufficient. Similarly, *prior to the time that* is a wordy version of *before, at the present time* means only *now,* and *during the years between* could be cut to the single word *between.*

Use of General Rather Than Specific Words

We have already discussed some of the reasons why general words are undesirable; in addition, the damage they cause to conciseness is sometimes ample reason for avoiding them. For example, when a writer begins a sentence by saying, "They were fastened together," he then must add "by means of staples," "with rivets," or something of the sort to tell *how.* If he says, instead, "They were stapled (or welded, or glued, or riveted) together," his verb would give the full information without the assistance of a modifying phrase or clause.

Other examples of wordiness caused by use of needlessly general words would be: *transported by means of trucks, kept out by the construction of fences,* and *made contact with him by telephone.* These could be replaced, respectively, by *trucked, fenced out,* and *telephoned him.*

Needlessly Complicated Structure

Simplicity has been discussed separately because it is desirable for its own sake as well as for conciseness. You probably noticed in the examples that when a sentence was simplified it usually was shortened. The following illustrations will emphasize this effect further.

> *Complicated and wordy:* It is essential that we take these precautions if our crews are to be safe.
>
> *Simple and concise:* These precautions are essential to the safety of our crews.
>
> *Complicated and wordy:* There are four dealers in Kansas City, and any of them can supply us with spare parts.
>
> *Simple and concise:* Any of four dealers in Kansas City can supply spare parts.
>
> *Complicated and wordy:* The condition of the machinery that is now located in the basement is not good enough that renovation would be justified.
>
> *Simple and concise:* The machinery in the basement is not in good enough condition to justify renovating it.

One more fact—often overlooked—should be brought out here. If you simplify your sentences to such an extent that you use primer style, your writing will be wordy even though each sentence may be as short as you can make it. When too many sentences are short and simple, the repetition of certain words and phrases is unavoidable. Thus as you revise for the sake of conciseness, you should watch for opportunities to cut out words and phrases by combining sentences as well as by shortening them—though you will also need to guard against making your sentences too long by such a process. The following specimen shows how combining sentences can lead to conciseness.

Primer Style

You requested that a site investigation be carried out on area 896. The purpose of the investigation was to determine the depth of the water table. This investigation has been conducted by Samuel Scott and myself. We sank several test holes, drilling them to a depth of 20 feet. (49 words)

Mature and Concise Style

As you requested, Samuel Scott and I have investigated area 896 to determine the depth of the water table. We drilled several test holes to a depth of 20 feet. (30 words)

Superconcentrated Style

One final word of caution: When conciseness is carried too far, the result is a superconcentrated style, as in the sentences that follow.

Preventing catalyst selectivity deterioration is difficult.
The machinery repair cost question demands attention.

When a reader encounters sentences like these, he does not grasp their meaning until he stops and figures out the relationship of the words they contain. There simply are not enough functional words (prepositions, conjunctions, and articles) to make the structure clear. A looser style would be better, as for example:

It is difficult to prevent the selectivity of the catalyst from deteriorating.
The cost of repairing the machinery demands attention.

In brief, conciseness is not just a means of saving space. It is also—often primarily—a matter of saving the reader's time. And you will defeat your own purpose if you eliminate so many words that you make your material harder to read and force your reader to spend more time on it than he would have spent on a longer version.

FINAL COMMENT ON STYLE

Two possible headings, *clearness* and *readability*, do not appear in the preceding discussion, but their omission does not mean that those qualities are unimportant. On the contrary, it would be impossible to overstate their importance. Whatever else we may do to improve our style, clearness and readability are the end results we hope to achieve. But neither of these qualities can be attained by applying a separate technique. Our style becomes readable when we choose our words intelligently, write sentences of suitable length, develop such qualities as simplicity and directness and conciseness—in fact do all the things that have been recommended. And clearness, too, is produced by doing all these things and also, of course, by attention to such commonplace matters as reference of pronouns, parallel construction, accurate choice of connectives, punctuation—all the fundamentals of English. Thus clearness and readability have not been overlooked here. Each of them is simply the result of so many different causes that an attempt to discuss it would lead us to the very headings under which the subject of style has been covered.

To exhaust the subject of style would be impossible even in a book-length treatment, but it would be wrong to close without pointing out that in regard to each of the desirable qualities discussed, one warning has always been included: don't try so hard to develop this quality that you sacrifice other qualities equally desirable.

The warning concerns not only the sacrifice of qualities we have discussed but also the sacrifice of other values. Sometimes, for example, if you try to make a sentence or clause as simple and direct as possible, you may prevent it from being parallel to some other sentence or clause that it should resemble. On another occasion, phraseology that looks desirable at first may need to be avoided because it results in unpleasant, accidental rhyme or in excessive repetition of some one sound, word, or phrase. Again, an effort to make a sentence more concise may destroy the intangible grace and rhythm that make writing smooth and pleasant to read.

Our discussion of style should therefore end with one general piece of advice: pay conscientious attention to the principles that have been presented; but when the result of following them bothers you—when you feel that a revision you are considering simply does not seem right even though you do not quite know why—you will probably do well to trust your instinct and express yourself in a way that does not clash with your convictions about what sounds best.

EXERCISES

Exercise 1

Improve the diction of the following sentences as indicated.

A. Make the following sentences less pompous.
   1. Following extubation, the dolphin's condition was determined by gross observations.
   2. Operating the vehicle in reverse, the driver came in contact with the car parked to the rear.
   3. According to the press secretary, all statements issued during that time frame were now termed inoperative.
   4. We shall endeavor to ascertain whether he is pecuniarily motivated in his recommendation.

B. Replace language that is needlessly technical by substituting ordinary language.
   1. In elastic materials, after stress has passed its apogee, hysteresis lag comes into effect.
   2. The low yield per acre was attributed to the necrotic condition of a preponderance of the plants.
   3. Several of the animals suffered from trombidiasis.
   4. As a result of the disease the tree suffered from decortication.
   5. His effort to write impressively resulted in cachinnation when his report reached the editors.

C. Reduce the vagueness of the following sentences. If you must invent facts in order to be specific, do so.
   1. Until it has been in use quite a while, the machine should be operated at only moderate speeds.
   2. The bridge has been declared unsafe for trucks with excessive loaded weight.
   3. The Tussock moth has infested a substantial percentage of the trees in a large area and the result will be the loss of a lot of money.
   4. Sales were down quite a bit for December because of the number of days of heavy snowfall.

D. Decide for yourself what is wrong with diction in the following sentences and make changes necessary for improvement.
   1. In line with your request we have viewed and reviewed the current territorial vacancies with an eye toward minimizing to an optimum extent the cost of the moves necessary to filling the territorial positions.
   2. The existing grade slopes in the right direction at the proper degree so grading work will be minimized.
   3. When heat is added to crystalline H_2O at 32° F, it metamorphoses into liquid H_2O at 32° F.
   4. Excessively long traffic stoppages at gas stations resulted in ireful operators.
   5. We recommend that rotational speeds be kept low to reduce stresses on parts.
   6. Diligence should be taken in reconnoitering the terrain to determine precipitous collapse conditions before crossing any avalanche zone.
   7. Pignoration of valuables is contraindicated in uncertain economic conditions.
   8. As the pump is started, close attention must be paid to the fluid level by the operating engineer.

Exercise 2

Break each of the following long sentences into shorter sentences. If it is necessary to change the order of ideas, feel free to do so.

1. From the Humpback mines, below the Cornucopia mine and closer to the canyon bottom, several important shipments have been made of 10 to 20 percent copper sulphide ore that was transported during the high water period by boat to the nearest shipping point at Mercer, and an interesting showing of this grade of mineral is in evidence at this point that well warrants further development as it contains some high values in both gold and silver.

2. The existing fire alarm system is wired such that if a break occurs in the dc series loop feeding the bell relays, all bells and buzzers sound a general alarm, while if a ground or short circuit occurs in the dc series loop, the bells and buzzers on the downstream end will sound a general alarm, which provides a form of supervision but a poor one since it is impossible to distinguish between a fire alarm and a trouble alarm.

3. The only empirical evidence on the degree of representation given by a districting system compared to an at-large system deals with the problem of representation of minorities in cities, which is a situation not wholly applicable to this city in which one finds different community interests, perhaps, but minority population is relatively small.

4. This company, one of the largest integrated natural gas systems in the world, which supplies energy to 35 utilities serving ten million people, operates a technically advanced pipeline system extending into twelve states and reaching from the Gulf of Mexico to Chicago and Wisconsin.

Exercise 3

Rewrite the following passages so as to eliminate their primer style. The order of ideas may be changed, but each revised version should be smooth, natural, and free from stringiness. Your versions should bring closely related ideas together and leave less closely related ideas in different sentences.

1. The validity of plant moisture stress (pms) measurements has been demonstrated. They are a useful tool in the study of plant response. The applications to forestry are immense. No longer will it be necessary to wait for weeks or months to determine whether a plant is responding to a given set of conditions. It can be readily determined on the spot by a simple measurement. Plant moisture stress measurements can be used in direct application to the problem of forest regeneration. Plant moisture stress readings can be taken in the field to understand and predict the response of recently planted trees in their new environment.

2. Operators of barges find that the minimum American Bureau of Shipping requirements for inland barges are insufficient. Some service conditions are too severe. This is often true when barges are towed in shallow open harbors where the waves are short and steep. Barges towed in these waters by tugs having 1000 or greater horsepower have particular difficulty. These barges must be altered to withstand the stresses. Changes include forward rake, closer frame spacings, and heavier plating.

3. Calcine is fed into each furnace through several pipes. The pipes are made of stainless steel. They are insulated to prevent loss of heat. Some of them carry mixtures of coarse and fine calcine. Others convey dust from the baghouses. The pipes have flexible joints inside the furnaces. This enables the operator to locate the outlet of the pipe in relation to the electrodes. The flow of calcine is controlled by an automatic feeding system. Manual control is also available.

4. In one large plant, over fifty percent of the cattle livers are condemned. Their condemnation results from their being abscessed. The abscesses are caused by *Sphaerophorus Necrophorus*. The existence of these bacteria has been known since the 1890's. In addition to liver abscesses they cause foot rot in cattle and sheep. In humans, they are implicated in a variety of diseases. These include abscesses in both liver and lungs. It seems possible, also, that in humans they are involved in diseases of the gastrointestinal tract and the respiratory tract.

Exercise 4

Improve the following sentences by making their structure simpler, more direct, or both. (As you do so some of them will also become more concise.) If necessary, study a sentence until you clearly understand what the writer was trying to say, and then rewrite it entirely.

1. Other jobs done for us by this contractor have been handled in a manner that met with our satisfaction, and the equipment that he will use has been given prior approval.

2. The experience of the applicants will be taken into consideration when we make a decision in regard to the choice of an appointee.

3. Although the contract contains a specification that the weight of the loads shall not exceed ten tons, the weight of several of them amounted to eleven tons or more.

4. The change-over will involve a cost factor of approximately 3500 dollars.

5. The members of the committee are engaging in procrastination because they apparently entertain a belief that the problem will disappear if ignored.

6. The Eskimo believed that the wolf sprang from a hole that had occurred in the ice to exercise control over the caribou herds so that weak caribou would not reproduce and that the tundra would consequently be overpopulated with poor herds.

7. Street improvements must be accomplished as existing surfaces deteriorate to the point where traffic flow is impeded as a result of erosion of the surface.

8. Organic sludge produced by the treatment process would be burned in power-plant boilers as a fuel component to generate electricity.

Exercise 5

By exercising your imagination you can discover what the writer intended to say in each of the following sentences. Revise each sentence as necessary to achieve precision.

1. The paint recommended by the decorators would consist of white ceilings and yellow walls.
2. The source of gravel for use in the subgrade will be obtained from a pit nine miles away.
3. The plumbing fixtures in the addition to the building were made of inferior quality.
4. The tools we need are a high-speed drill, a power saw, screws of several sizes, a crowbar, and glue.
5. Sample stretches of pavement using loads of different weights have already been tested.
6. The rolling equipment, with the exception of the preceding statement, is highly satisfactory.

Exercise 6

Make each of the following specimens more concise.

1. It is obvious that the tests which were performed for the purpose of determining the resistance to skidding do not permit us to draw a definite conclusion.
2. In spite of the fact that the Spring Valley route is not so long as the others that were considered, it has one serious defect, the possibility that landslides might occur. Consequently, it must be rejected.
3. It will be impossible for us to complete this stretch of highway by the date called for in the schedule. The reason for the delay is that some of the steel for the bridge has not yet been received.
4. The weight of the load was so great as to be in excess of what is permitted by law, and its extreme height caused it to strike the bottom of the underpass.
5. The existing chute walls will confine the spills. The only time when they will not do so will be when the flow rates approach the maximum spillway capacity.
6. The rear axle was raised up by means of jacks until it was high enough to clear the stump.
7. After it had pulled its own load through the mud with ease, the cleated tractor was brought back and fastened to the wheeled tractor's load. It then pulled this load also to solid ground.
8. After four days the amount of dissolved oxygen began to decrease. The probable cause of this development was that the oxygen was utilized by bacteria inside the bladder.
9. We have tried out four cars and have arrived at the conclusion that the Weasel is the best among them. Its margin of superiority, however, is not very wide.

3

ORGANIZATION

If you are to be effective in technical writing you must do more than just express each separate fact and idea clearly. You must also organize those facts and ideas in a manner that will show how they are related to each other and to the paper as a whole—for it is primarily in their relationships that facts and ideas become truly valuable.

Good organization is a result of planning your job of writing in advance. To be sure, when a piece of writing is to be short and simple, planning it may be only a matter of deciding what material to include and how to arrange it—jobs that must be done in any writing process. But when a substantial mass of material is to be presented and the relationship of facts and ideas is complicated, the planning involves more than deciding the sequence of presentation. It also calls for deciding on subordination and coordination and for producing a plan of organization that will make these relationships clear.

According to information theorists, there are three general ways information can be organized: randomly, sequentially, and hierarchically. The order cards fall after being dealt from a shuffled deck is a well-known example of random organization—there is no apparent relation or pattern among the cards. A sequence is any list of items arranged so that each item has the same relationship to each of the other items, such as alphabetical, numerical, or chronological ordering. Hierarchical organization involves more complex multi-level relationships—items have other items subsumed under them. The following schematic, sometimes called a tree structure, illustrates the relationships of items in a hierarchical organization.

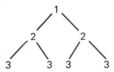

This is basically the relationships of items in an outline, which is another form of tree structure or hierarchical organization. Another example is shown in part below:

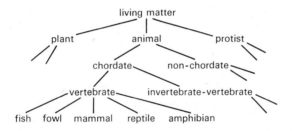

While all three methods of organizing information are useful for certain purposes, such as in computer theory, the random and the sequential methods are unsuitable for organizing information to be conveyed to a reader. One reason is the effort required to retrieve information from them. For instance, if you have a list of 1024 items arranged randomly and your reader wishes to find a single item on the list, he would have to search on the average through about half the items to find the one he wanted. If you classified the items into 32 groups of 32 items, your reader would find the wanted item after looking through an average of about half the classes and half the items within the desired class—a total of about 33 searches instead of about 512.

However, if you organized the items in a hierarchy in groups of four with four levels of classification as shown in abbreviated form below, your reader would have to search 2.5 classes at each of four levels and 2.5 items for an average of 12.5 searches. You will immediately recognize that this example of a hierarchical organization is in the form of an outline.

```
I. Class 1-256
   A. Subclass 1-64
      1. Subclass 1-16
         a. Subclass 1-4
            i. Item 1
            ii. Item 2
            iii. Item 3
            iv. Item 4
         b. Subclass 5-8
            *  *  *
      2. Subclass 17-32
         *  *  *
   B. Subclass 65-128
      *  *  *
II. Class 257-512
    *  *  *
III. Class 513-768
     *  *  *
IV. Class 769-1024
```

Such an organizational scheme requires the addition of 340 classification headings, but studies by learning and memory specialists have shown that adding two times as many appropriate items to a list of random items can improve learning. To be sure, certain mathematical search methods actually make the random and sequential schemes more efficient for retrieving information; but since these methods are not readily available to a reader, their efficiency is irrelevant in this discussion. Also, in writing, other factors are involved that make the hierarchical schemes far more effective.

The first of these factors is the fact that writing is ultimately a teaching process. The writer tries to convey information to the reader that the latter does not have, and the reader tries to learn what the writer has to say. Because the mind seems to retain information best when the relationships among facts are clear, the hierarchical organization, which shows the relationships, seems best for teaching, learning, and remembering purposes. A study in psychology well illustrates its advantage. A group of subjects were given 112 words to memorize. Half the group were given the words arranged randomly, and half were given the words arranged in a tree structure. Both halves were given equal time to learn the words and were tested to see how many words they could recall. After each test, they were allowed to study the words again, and were retested. The table below shows the results.

Test	1	2	3	4	
random	20	38	52	70	words
tree structure	73	106	112	112	recalled

The second factor involves another apparent limitation of the mind. Learning experts have determined that most people can retain in their short-term memory only five to nine items. It is easy to test this assertion. Have a friend or teacher make up a sequence of random numbers, each number with one more digit than the preceding, such as shown below:

```
361
8269
10437
593752
2784926
49738693
610839625
8546071637
```

Have the person read each number aloud to you once, and then attempt to write down the number from memory. The vast majority of people will be unable to reproduce from memory a number with more than nine digits, while the average person will be able to recall only seven digits. The vast majority of people have no difficulty, however, in remembering a number such as 12193824921x732 if it is organized into smaller groups that the mind can comprehend. You will recognize the number when it is properly divided: 1 219 382-4921 x732. The same principle applies in writing. By properly dividing and organizing the information you wish to convey to your reader, you can make the job of learning much easier for him. In addition, since most associations made by the reader require that he retain in his short-term memory what he has just read, it is necessary that you keep the number of points to five or less.

The basis for organization, then, is largely psychological, and the best organizational system for the writer appears to be the hierarchy. The primary hierarchical system for the writer is the outline. Three major points are clear:

1. The tree structure of an outline is best for ease of learning;
2. the tree structure is best for ease of access to the information; and
3. there should seldom be more than five items at any one level for ease of recall, and never more than nine.

Students might take particular note of this for study purposes. An immense amount of information can be made easy to retain by merely organizing it into an outline with no more than five items in any series.

There is one further general point of importance about organization: the organization works best when its nature is clear to the perceiver. This is illustrated by another psychological study: A number of chess masters and novices were presented with chess pieces on a chess board and after a five second glance, they were asked to reproduce the arrangement on another board. When the pieces were arranged randomly, there was no difference in success between the masters and novices. But when the pieces were arranged at the twentieth move of an actual game, the masters achieved a 90 to 100 percent success, while the novices achieved only about 30 percent success. The difference can be accounted for in part by the fact that the masters, within five seconds, perceived the organization, while to the novices, many pieces appeared to be randomly placed. Without perception of the organizational scheme, therefore, the players—masters and novices alike—could not recall the placement of the pieces. The same principle applies in writing and in reading. If the reader understands the relationship among the facts and ideas presented, he will be able to understand what he reads far more easily. The technical writer provides information about his method of organization by use of such devices as outline-type tables of contents, headings and sub-headings, cue words such as *first, second, furthermore,* and so on. The writer's major question for himself should be: Will the reader understand the relationships being discussed?

These are the basic reasons for the need of organization in writing. There remains the problem of *how* to organize. The remainder of the chapter treats that problem, with most of the discussion dealing with the writer's primary organizational device—the outline—and the kinds of difficulties involved in constructing a good one.

Making an outline is an excellent practice not only for the reasons given above but also because the very process of creating a good outline often forces a writer to clarify his thinking and to settle questions of thought relationships that might otherwise go unnoticed. In this chapter, however, the emphasis will be on the outline as a plan that will help you to produce an effective piece of writing rather than as an end in itself. This is no trivial distinction. Many an outline that seems at first glance to be a fine specimen turns out to have flaws when put to use as a writing guide.

Of course an outline may also be used for other purposes. It may be submitted to the person who will eventually receive your completed paper so that he can see whether you are on the right track, or it may serve as a source of headings for inclusion in the completed paper. However, any outline that turned out to be unsatisfactory as a plan to use while writing would also be unsatisfactory for the other uses. Thus, the emphasis placed here on outlining as a method of planning is justified from every point of view.

If an outline is to function effectively, it should meet the following requirements:

1. It should cover the material that you will need to include after you have narrowed down your subject by carefully considering the function of your paper.
2. It should be based not only on your consideration of the subject but also on any knowledge you may have about the readers—their reasons for reading, their attitudes, their familiarity with the subject, and anything else that will affect their reception of your ideas.
3. It should be specially designed to accommodate the specific facts and ideas that you want to present.
4. It should give a sense of continuity—of organic unity—rather than being merely a collection of headings that fall within the scope of the subject. (This fourth requirement is less vital in some kinds of writing —for example an instruction manual intended to be used for reference purposes rather than read as a whole—but even when continuity is not essential, the pattern that you create should be systematic, not haphazard.)

THE PROCESS OF PREPARING AN OUTLINE

It would be foolish to pretend that there is an easy way of organizing a complex mass of facts and ideas, and equally foolish to maintain that such a result may be accomplished by only one method. The present approach to the problem is not offered as either an easy method or the only method. It is simply a systematic procedure that will usually reduce to a minimum the difficulty of getting good results.

One principle should be made clear at the outset: Before you can make an outline that will serve its purpose, you need to decide, in the main, what material you will use in the paper you expect to write. If you disregard this principle you will often find that when you gather your facts and ideas, they will not fit into the organization you have decided upon. It is impossible to be sure of the best organization for your material until you know what that material is.

Another point that will make the process easier is to keep in mind the relationship between the outline and the tree structure. If you write down a brief but full statement of the subject matter of your work, you will have the apex of the tree. Then ask yourself: What are the major components I need to explain if my audience is to understand the subject? The answer will be the second level in the tree. Then concerning each component, ask: What are the elements of this component? For each element, ask: what are the points I need to explain if my audience

Pilot Plant Treatment of Vegetable Wastes

I. Characteristics of raw wastes
   A. Peas
      1. Low volatility
      2. High sodium and chloride
   B. Corn
      1. High volatility
      2. Deficiency of nitrogen
   C. Carrots
      1. Medium volatility
      2. High sodium
      3. Deficiency of nitrogen

II. Continuous operation of treatment system
   A. Acclimatization time and mechanical adjustment
   B. Operation on a continuous basis
      1. Operation with peas
      2. Operation with corn
         a. October corn run
         b. December corn run
      3. Operation with carrots
      4. Accumulation and handling of corn wastes
   C. Hydraulic overflow rates
      1. Overflow after one-day detention
      2. Overflow after six-day detention

III. Nutrient requirements
   A. Results of augmentation of nutrients--October corn run
   B. Results of augmentation of nutrients--December corn run
   C. Limited significance of results obtained

IV. Clarification
   A. Problems created by changes of vegetables processed
   B. Foaming problems

V. Settling characteristics
   A. Poor characteristics compared with solids from substrates of domestic sewage
   B. Possibility of improvement by organic loading

FIGURE 1.   Specimen Outline Showing Correct Physical Form. This specimen illustrates a simple system of indentation and shows the numbers and letters that precede four ranks of points. (For decimal symbols, see Chapter 4.) The specimen is single-spaced except above main points, but double or single spacing throughout would be acceptable.

is to understand this element? When you have finished, you will have an organizational chart of the body of your paper:

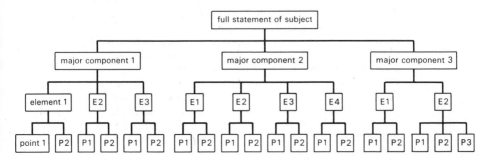

As you can see, no subsection can consist of a single item, for unless you have inadvertently left something out, the single item would logically cover all the ground that its main point covers. Thus, it would not organize your material any further than you had organized it already. The relationship between the tree structure and the outline is by now obvious: each entry in the tree structure is comparable to a point of the same rank in the outline.

It is important, however, that you regard your original plan as only tentative and subject to constant revision. As you proceed you should:

1. Think about your subject and the function your paper is to perform, and gather information about the subject if necessary. As you do so, jot down your facts and ideas in the form of rough notes. You will probably add or eliminate material later, but so far as possible you should make notes that cover the substance of your paper.

2. Study these notes, and group together material that seems to be related. As you do so you can usually settle upon a tentative heading for each group—one that will later form the basis for a point in the outline. If necessary, postpone a decision about where to place doubtful points.

3. Taking into consideration your purpose in writing, your reader, and your raw material (the latter now being grouped under headings that indicate its nature) decide upon the main points of your outline and hence upon the main divisions of your discussion. There should seldom be more than five points, and never more than nine, for the reasons given above. (If you are writing something to be used only for reference rather than to be read as a whole, the desirability of holding down the number of main divisions is not so great.) After you choose your main points, you will probably have to place some of the groups of material as subdivisions of larger groups; in fact, some of the groups were probably made subdivisions when you formed your original groupings. In any event, your main points will not result from

just *cutting up* the subject; they will also result, at least in part, from a process of *building up* main points from the raw material.

4. Bearing in mind not only the logic of the subject but also the attitudes and abilities of your probable readers, arrange your main points in a suitable order.

5. Now apply to the first main point the same treatment that you applied to the entire subject. Thus work out and arrange the highest rank of subpoints under your first main point.

6. Either apply the same treatment to the other main points or else continue working on the first main point until you have organized its contents to the extent that you consider necessary. (At this stage, the order in which you work on one or another part of your outline will probably depend on the order in which your ideas mature and your facts become available.)

The preceding method of outlining is offered as a general guide and should not be regarded as a straitjacket. There is no reason that you should not depart from it when the way in which your ideas develop makes it desirable to do so. For example, you may have in mind a possible list of main points before you begin to jot down notes. You may develop ideas for handling some of your material at lower levels before you are certain what your main points will be. It is advisable, however, to regard such ideas as merely tentative until you see how they fit into the pattern that emerges from the general procedure recommended.

This procedure has the advantage of avoiding two other approaches to the job, both undesirable. One of these is to decide first upon the first main point, develop it in full, then choose and develop a second main point, and so on until the outline is complete. (Some people do this because it seems natural to write out the points of an outline in the order in which they appear on the page, just as they write ordinary text.) The other is to decide upon the main points after considering only the subject as a whole, and then to work out subpoints and sub-subpoints —trusting to luck that the specific facts and ideas to be expressed will fit into the framework thus developed. Although the procedure recommended also consists of forming a general, overall plan and then working out the details, it is based on consideration of the material that must be handled when the plan is put to use.

GIVING SUITABLE RANK TO YOUR POINTS

The question of how to decide on the rank to be given to the various points of an outline calls for further comment. Sometimes the decision about the rank of each point is dictated by facts beyond our control. For example, it might be essential to base a discussion of a college or university on its formal organization. Thus it would be unsuitable to make a point covering a department equal in rank to points covering entire colleges, even if the department taught more courses, handled more students, and spent more money than some of the colleges.

Much of the time, however, a writer is free to use his own judgment. Suppose, for example, that in a certain area the acreage and value of each of two kinds of grain equaled or exceeded the combined value of all the fruits. In this case it might be not only permissible but actually desirable to make the point covering each kind of grain equal in rank to the heading *Fruits*.

With a reasonable amount of ingenuity you can usually give each point the rank that its actual importance justifies without producing an illogical result. To be sure, if topics such as "Grain" and "Wheat" were shown as points of equal rank, the equality would not be defensible because one includes the other. But if wheat happened to be especially important to the purpose of your paper, it would be possible to change "Grain" to "Grains of Lesser Importance," and thus justify equal rank for the topic "Wheat." There would then be nothing wrong with taking up "Barley" and "Oats" as subtopics under "Grains of Lesser Importance." Thus the rank of each topic could be made to reflect accurately the importance of the material that it covered.

In brief, when you are not forced to follow a preexisting pattern over which you have no control, you should create a pattern that gives each point the rank its actual importance justifies, but one that is free from such flaws as faulty subordination and faulty coordination, which will be discussed later in the chapter.

ARRANGEMENT OF POINTS

Sometimes arrangement of material may be based on a recognized principle—the chronological order, the order of location, or the order of increasing difficulty, for example, and sometimes it may be based on reasons peculiar to the particular occasion. But in any event, the arrangement should not be left to chance, but should result from a conscious decision about what is best.

Moreover, unless a piece of writing is to be short, no single system is likely to govern the arrangement of *all* the material in the outline. We may often arrange our main points according to one principle and our subdivisions according to another. Indeed the very nature of our subpoints may make it impossible to arrange them on the basis that governs the main points. When the main points, for example, are based on what is to be done at different times during the year and the subpoints are based on what is to be done at certain places or by certain persons, the chronological arrangement used for the main points could not possibly apply to the subpoints.

In addition to the nature of the material itself, there is another important point to consider. When you make an outline you are not just planning how to write about a subject; you are also planning how to communicate with your readers in a manner that accomplishes the result desired. The best arrangement, therefore, is one that is based not only on the logic of the subject but also on the reader's readiness to understand and accept your facts and ideas. Effective arrangement must often grow out of such considerations as: Who will the reader be? Why is he reading? What attitudes, what prejudices—so far as we know—does he already have on the subject? What questions will come into his mind first?

A good example of such considerations is seen in the widely used pattern that places conclusions and recommendations of a report before the body. According to logic, the evidence should precede the conclusions, but many organizations reverse the order because the reader wants to know at once what is recommended.

CHECKING AN OUTLINE FOR ERRORS

However carefully the first draft of an outline is made, errors in organization are likely to creep in. The first draft should therefore be checked over with the following points in mind:

1. There should never be a single main point. (Do not count *Introduction* or *Conclusion* as main points.) If there is, further examination will show (1) that this single main point covers the entire subject and hence is not a main *division* of the material, (2) that some of the material placed under this main point does not belong there, or (3) that the full subject has not been covered.
2. There should usually be no more than five or six main points. If an outline must be long, the increased length should result from the use of more subpoints and sub-subpoints rather than from the number of main points.
3. There should rarely be a single subpoint. That is, there should rarely be an A without a B or a 1 without a 2. The only justification of a single subpoint is the need of treating two or more closely related points in the same manner.
4. There should be no faulty subordination. That is, no subpoint should be placed under a main point where it does not belong. (It might be equal in rank or might belong under some other main point.)
5. There should be no faulty coordination. That is, no point should be shown as equal to another when it is logically subordinate.

6. There should be no overlapping among the main points or among subpoints of equal rank under the same main point. This flaw may result from faulty coordination or from other causes.

GENERAL SUGGESTIONS

Classification

The term *classification* here refers to the establishing of divisions. For example, an outline on erosion control in the United States might have four main points:

    I. Erosion control in the East
   II. Erosion control in the South
  III. Erosion control in the Midwest
  IV. Erosion control in the West

Each of these points might be subdivided thus:

    A. Erosion control in the past
    B. Erosion control in the present
    C. Erosion control in the future

Some classification of this sort is often necessary and valuable, especially at the upper levels of the outline. Moreover the outline produced by classification is usually clear, orderly, and apparently logical. There is a real danger, however, that because classification is so easy, it will be carried too far. The resulting outline might contain some headings that are unnecessary because there is nothing much to say about them, and other headings that call for the same material in two or three places. Also, such an outline often fails to provide the headings that would be most useful in organizing the specific facts and ideas to be presented.

Another defect resulting from excessive classification is that it leads to subpoints that always bear the same relationship to their main points. In a sense this relationship does always exist; that is, the discussion under a subpoint is always part of the discussion of the main point. But the thought relationship can vary. For example, suppose a main point concerns the decline in the production of metals. Mere classification would probably result in such subpoints as ferrous and nonferrous metals, and sub-subpoints such as lead, silver, and zinc. But perhaps there would be no reason for such division into parts. Perhaps, in a discussion that applied to all the metals, we might wish to present first the evidence that a decline is imminent, second, its causes, and finally its probable consequences. Classification is unlikely to provide headings that cover these latter kinds of material.

One virtue of the procedure recommended earlier in this chapter is that by reducing the likelihood of excessive classification it also reduces the likelihood of our producing an outline that appears logical and orderly but later turns out to be unsuitable for our purposes.

Parallel Treatment of Similar Points

Though the desire for parallelism should not lead you to jam any portion of your paper into an unsuitable organization, it is sound practice to organize similar points in a similar manner so far as the subject matter permits. Such a practice enables your reader to become familiar with the pattern and read more efficiently.

When your outline is to be scrutinized by someone else, parallel phraseology as well as parallel organization becomes desirable. Main points should be phrased alike so far as possible, and subpoints of equal rank falling under the same superior point should also be phrased alike. This treatment is recommended not because arbitrary rules demand it, but because it makes an outline easier to grasp. Parallelism is a principle that applies not just in outlining but everywhere in writing.

Avoidance of Undesirable Extremes

It is often necessary, in preparing an outline, to balance between two extremes, either of which would be undesirable. It is undesirable, for example, to have too many ranks of points; but it is also undesirable to have a long string of subpoints of equal rank—perhaps eight or ten—under the same main point. Such a string might better be divided into two or three parts. Instead of naming in a single list ten advantages of a certain process, for example, you might do well to subdivide them into different *kinds* of advantages and name the specific advantages as points of a lower rank. Whether this would be an improvement, however, would depend on whether this change would create an outline with too many ranks.

Further, it will constantly be necessary to decide whether to make your points long enough so that each is self-sufficient, or whether to keep each point short so it may be grasped at a glance.

Logical Allotment of Space

The space devoted to various points in an outline should be roughly proportionate, if possible, to the amount of space they will occupy in the paper to be written. Thus a reader will not be misled about which points will comprise the bulk of the paper.

Clearness in the Outline

You should take special pains to make your outline a clear enough record of your plan so that a reader can understand it. This should be done even if you yourself are the only probable reader. Many a writer, when he is not able to write from his outline immediately, finds it hard to remember his own intentions later on, and runs into trouble unless his outline is clear enough to refresh his memory.

Each point should be worded so that its meaning is impossible to misconstrue. It should be broad enough to cover all the material you intend to place under it, but not so broad as to cover more material than you intend to present.

Not only the individual points but also the relationship among different points should be clear—especially the thought relationship between a main point and its subpoints. Some outlines fall so far short of meeting this requirement that a main point and its subpoints do not even seem to cover the same material. If such a discrepancy exists in your outline, you need first to decide whether you actually *do* have in mind two different bodies of material, or whether your plan is sound but is not clearly recorded. In the former case, either the main point or the subpoints must be moved—or else discarded. In the latter, the phraseology must be changed so that the soundness of your plan becomes apparent. It is advisable to check this kind of defect by asking yourself two questions about each doubtful portion: (1) If someone first read the main point and then the subpoints, would it be clear that the main point could be discussed under the subpoints used? (2) If someone read the subpoints first and then the main point, would he see that the main point covered the subpoints without going beyond them?

Sentence or Topic Outlines

In many discussions of outlining, the distinction between the sentence and the topic outline is given considerable attention. When no one except the writer will see the outline, this distinction is obviously unimportant. Indeed, it is unlikely to be considered important in any circumstance. In many excellent outlines, sentences are used for some points and topics or other types of phraseology for others. Unless the points of an outline are to be used as topic headings—in which case they should all be topics—there is no reason to avoid variation in form if the principle of parallelism, already discussed, is complied with.

FINAL CHECK OF AN OUTLINE

When you approach the completion of an outline, you can check your work by considering the following questions:

1. Does your outline have a desirable number of main points?
2. Do your main points cover the subject, and will they lead to the best main divisions of material in view of your purpose and your readers?
3. Have you provided for the inclusion of the specific facts and ideas that you want to present?
4. Is the outline free from points that are not really called for?
5. Is your outline free from classification that serves no useful purpose?
6. So far as the subject matter permitted, have you used parallel organization where it would be expected?
7. Have you balanced properly between the desire to avoid long series of equal points and the desire to avoid points of too many ranks?
8. If you have any single subpoints, can you justify them?
9. Have you avoided faulty subordination?
10. Have you avoided faulty coordination?
11. Have you avoided overlapping points?
12. If your outline is to be examined by someone else, will it create an impression of clearness and continuity?
13. If your outline is to be examined by someone else, will it give a reasonably accurate impression about the relative amount of space the various points will occupy?

It should be apparent by now that these are not arbitrary rules dreamed up by schoolmarms or English composition instructors, but grow out of the psychological needs of the reader. The successful writer has always followed these points even if he was not aware of the bases for them.

EXERCISES

Exercise 1

Following are fragments of outlines, each of which contains one or more weaknesses in organization. Point out in one or two sentences whatever weakness you find in each.

(1)

A. Areas in need of repainting
   1. Needs in laboratories

2. Needs in classrooms
3. Needs in offices
4. Satisfactory condition of halls

(2)

A. Means of increasing crop production
   1. Improved fertilization
   2. Development of better species
   3. Development of better pesticides
B. Possibility of increasing acreage
   1. New irrigation projects
   2. Use of land now withheld from production

(3)

A. Permeant fluids used in the tests
   1. Water with 13 ppm dissolved oxygen
   2. Water with 5 to 6 ppm dissolved oxygen
B. Effect of molding-water content
   1. Effect on permeability
C. Effect of permeation on saturation
D. Effect of permeation on dry density

(4)

I. The ridge route for the new stretch of highway
   A. Advantages
      1. Better grade
      2. Ease of coping with snow
   B. Disadvantages
      1. Greater length
      2. Extreme amount of rock work
II. The Lowell Creek route
   A. The problem of landslides
   B. The damage to value of nearby property
   C. The saving of two miles
   D. The better service to nearby communities

(5)

I. Butadiene styrene copolymer paints: resistance to chemicals
   A. Chemicals resisted effectively
      1. Mineral acids during intermittent contacts
      2. Alkalis and strong detergents
   B. Chemicals resisted poorly
      1. Chemical acids during prolonged contact
      2. Organic acids
II. Chlorinated rubber paints
   A. Poor resistance to strong alkalis
   B. Poor resistance to concentrated mineral acids

C. Good resistance to weak mineral acids
D. Good resistance to organic acids

(6)

I. Possible uses for Lambert aluminum buildings
   A. Uses for farm buildings
   B. Uses for industrial buildings
   C. Uses for commercial buildings
II. Means of making appearance attractive
   A. Use of wood or brick for front end
III. Variety of sizes available
   A. The different lengths and widths
   B. The different heights for ends and side walls

(7)

A. Possibility of increased use of coal for power production
   1. Amount available from strip mining
   2. Amount available from underground mining
   3. Amount available in the East
   4. Amount available in the Plains States

(8)

A. Damage caused by 1973 flood
   1. Damage to buildings
   2. Damage to roads and bridges
   3. Damage to farm land
   4. Damage anticipated from future floods

(9)

I. Effects of the prolonged drouth
   A. Damage to crops
   B. Loss of hydroelectric power
   C. Pollution of air by blowing dust
II. Increased danger of forest fires

(10)

I. Deterioration of Lake Wannigan
   A. Pollution from lakeshore homes
   B. Pollution from farm fertilizers
   C. Pollution above the narrows
   D. Pollution below the narrows

Exercise 2

Following is a list of considerations that might influence a decision about whether to change zoning so as to permit a new shopping center adjacent to a residential area. Make an outline based on this information. You may add other considerations if you wish or may subdivide items listed. The arrangement below is intentionally unsystematic.

1. Effect on tax revenue
2. Effect on traffic
3. The question of protection against fire
4. Compatibility with general plans for city development
5. Availability of adequate water supply
6. Type of landscaping for space not built upon
7. Possibility of excessive noise
8. Kind of architecture proposed
9. Effect on downtown business area
10. The question of providing police protection
11. New jobs that would be created
12. The question of sewage disposal
13. Possibility of attracting business that would otherwise go to other towns
14. Other possible uses for land in the area
15. Possibility of excessive dust and litter

Exercise 3

Make an outline for a paper about the feasibility of establishing a salmon hatchery, using as material the unorganized list of facts provided below.

There should not necessarily be a point in your outline for each item in the list. On the other hand, a single point in your outline may cover two or three items in the list; or there may be items in the list that overlap, just as your own notes will occasionally overlap as you jot them down before making an outline for one of your own papers. In short, the items are not points of an outline. They are merely facts that the paper would include. *You* are to make up for yourself the points that an outline would need to contain in order to accommodate the facts.

It is not necessary for your outline to cover every item on the list, but it should include most of them. It may also include any facts that you need to make up for yourself so that your particular treatment will not contain gaps. Remember that there is no single, correct treatment for the material. The purpose of this exercise is merely to give you a chance to make up an outline without having to spend time looking up information, and to provide a basis for class discussion by permitting comparison of the treatments worked out by a number of people.

1. The Bureau of Sports Fisheries initiated an investigation of the feasibility of operating a Federal salmon hatchery on the Makah Indian Reservation.

2. The production of salmon in the Makah Reservation rivers is not capable of maintaining an adequate supply under present ocean fishery demands.

3. In recent years more than 50,000 salmon-angler days were expended annually at Neah Bay in the reservation.

4. The site of the hatchery must have a dependable supply of water.

5. The Sail River is located at the eastern edge of the reservation and runs into the Strait of Juan de Fuca.

6. Low flows of the Sail River, according to the U.S. Geological Survey's water study, would be less than two cubic feet per second.

7. Usable surface-water storage of 425 acre-feet would be necessary to maintain a flow of 3 cfs in Sail River.

8. The Sooes River flow would be 15 cfs one year out of two.

9. The Sooes River has a drainage area of approximately 38 square miles.

10. A preliminary feasibility report suggested two possible site locations: one on Sooes River and one on Sail River.

11. The Makah Reservation is located on the tip of the Olympic Peninsula in Western Washington.

12. The Makah Indians have always relied on fishery resources for their principal livelihood.

13. The hatchery site must be suitable for the construction of the physical facilities for a hatchery development.

14. Most of the Sooes drainage area lies outside the reservation.

15. Test drillings would be required to determine the quantities of ground water that could be developed in the Sooes basin.

16. The Makah Reservation provides an excellent harbor for sport and commercial fishermen.

17. Salmon-fishing fleets from the United States and Canada operate offshore and in the Strait of Juan de Fuca.

18. The Sail River site has geographic advantages.

19. The Sail River drains an area of five and one half square miles, 80 percent of which lies within the reservation.

20. A potential site for ground-water development lies along the Sooes River between the reservation boundary and the proposed hatchery site.

21. The hatchery site should be above the flood plain.

22. The Sooes River drains the western side of the reservation and enters the Pacific Ocean at Mukkaw Bay.

23. The Sail River flows through a narrow valley with a steep gradient.

24. One year out of twenty the Sooes flow would be 9 cfs.

25. Sizable quantities of ground water are unlikely in the Sail River basin.

26. The terrain and the foundation material should be such as to hold down the cost of site preparation.

27. The hatchery must have a dependable water avenue downstream so that brood stock can return.
28. The water supply must not contain toxic elements.
29. The Sail River flow will approach 0.2 cfs only one year in twenty.
30. The water must have optimum quality, volume, and temperature.
31. The stream should be free of obstacles to migration.

Exercise 4

The following table indicates the kind of information that might be contained in a paper on crop conditions. It does not, of course, provide the details that would comprise the substance of the paper. Make three outlines, each with a different basic organization, that might cover the information indicated.

You may not consider all three bases equally good, but that does not matter. The purpose of the assignment is merely to illustrate that the same material may often be handled in different ways. Realization of this fact makes it more likely that when you organize the contents of something that you must write, you will devote some attention to choosing the best possible approach rather than adopting some pattern because it looks easy or is the first one that you think of.

CONDITION OF CROPS IN DODGE COUNTY

| *Kind of Crop* | *Areas* | | | |
	A	B	C	D
Grains	Good	Good	Poor	Poor
Sugar Beets	Good	Poor	Fair	Good
Fruit	Fair	Poor	Fair	Poor
Potatoes	Poor	Fair	Poor	Fair

Exercise 5

Following is a list of general subjects to be used as a basis for outlines, either as they stand or as you may adapt them. Choose one or more of them and make an outline containing from 15 to 30 points. These outlines may be topic, sentence, or a combination, as directed by the instructor. They need not be "thought-content" outlines unless the instructor so requests. Supply a title so that your subject will be more specific than those listed.

1. The advisability—advantages and disadvantages—of seeking summer employment in the field you plan to work in after graduation com-

pared to seeking summer employment in the best-paying area you can find.

2. The steps necessary to perform some analysis in your field of specialization. (Business majors, for example, might analyze the feasibility of starting a pancake house near the campus. Civil Engineering majors might consider the flow of traffic through the campus.)

3. The criteria you will consider when you apply for employment after graduation.

4. The demands of the curriculum in which you are seeking a degree (specific subjects, kinds of subjects included, types of work called for, differences between courses taken in earlier and later years, and so on).

5. Desirability or undesirability of using artificial turf, or choice between different kinds of artificial turf, for a football or baseball field.

4

MECHANICS

Our discussion of mechanics here will be limited to brief instructions on manuscript form, technical style, hyphenation, decimal numbers for headings, equations, and documentation. Punctuation, capitalization, spelling, and other points that concern all writing are covered in the Handbook.

MANUSCRIPT FORM

Much of the time, technical writing does not demand any special manuscript form. In such cases the following general instructions should prove sufficient.

1. Ordinary manuscript should be typed on one side of 8½-by-11-inch white paper of good quality. Double spacing is usually preferable to single, though single spacing with double spacing between paragraphs is standard in letters and often desirable in reports.
2. The margin at the top should be 2 to 2½ inches on the first page and 1 inch on other pages. Other margins should be: left, 1 inch; right, ¾ of an inch to 1 inch; bottom, 1 inch. If a manuscript is to be bound at the side, the left margin should be increased by ½ or ¾ of an inch. If

51

it is to be bound at the top, the top margin should be increased to 2 inches.

3. The beginning of each paragraph should be indented five spaces.

4. Unless a separate title page is used, the title of a paper should be placed on the first page, centered on the first line. It may be entirely in capital letters, or, if the author's name does not follow, may be underlined and written in upper and lower case (capital letters to begin each important word, small letters elsewhere).

5. If the name of an author accompanies the title, it should be centered a double space below the title and should be upper and lower case. There should be three or four blank spaces between it and the text.

6. The page number is ordinarily placed in the upper right corner of each page except the first, from which it is omitted. If a paper is bound (not clipped) at the top, however, the page numbers should be centered in the bottom margin.

7. Long quotations (75 words or more) should be single-spaced except for double spacing between paragraphs. Margins adjoining such quotations should be increased by ½ to ¾ of an inch on each side.

8. Ordinarily, manuscript should be fastened together by means of paper clips or not at all, and should not be folded. Occasionally, however, especially when it is placed in final, permanent form, it should be semipermanently fastened in a cover that opens at the side, or provided with backing paper and stapled together at the top.

9. Manuscript to be submitted for printing should never be fastened permanently. In such manuscript, illustrations should not be attached to the copy. Rather, an identifying number should be written on the figure's back and a note should be inserted in the manuscript to indicate where the figure belongs.

TECHNICAL STYLE

Technical style, in the present context, concerns the form in which numbers are written and the use of abbreviations. In this discussion, suggestions will be offered to help you determine when to use technical style and how to use it correctly.

Use of Figures or Words for Numbers

Numbers in Ordinary Style. Before the use of numbers in technical style can be explained, it is first necessary to recall the conventions that govern their use in ordinary writing. *The basic rule in ordinary writing is: If a number can be expressed in no more than two words, it should be written out. Otherwise, it should be expressed in figures.* (Examples: *four, seventeen, twenty-seven, one hundred, one thousand;*

but *114, 1198, 14,456*.) There are many modifications to this rule, the most important of which follow:

1. Figures are never used at the beginning of a sentence. Such numbers must be written out, or else the sentence must be changed so it does not open with a number.
2. All numbers in a series are written in the same form—preferably in figures if any number in the series is long enough to call for use of figures.
3. There are many special uses in which figures may or should be used regardless of the size of the number. These include degrees of latitude and longitude or of temperature, prices, scores, time of day, dates, and tabular statistics.
4. For extremely large numbers a mixed form is widely used which is extremely easy to write correctly and to read accurately. (Examples: *50 billion, 125 million, 6.4 billion*.)

Numbers in Technical Style. In technical style, numbers are more likely to be written as figures. The main differences between ordinary and technical style are as follows:

1. In technical style, 10 and all numbers above are expressed as figures. Any number below 10 is written out, except as mentioned below.
2. In technical style, a number that precedes a unit of measurement is written as a figure even if it is below 10. (Examples: *6 inches, 4 hours, 8 cubic yards;* but *six hoes, three stories, eight gusset plates, four arches.*) Writing out a number as a word is especially undesirable when it precedes a unit of measurement that is abbreviated.
3. In a passage where numbers are especially frequent, all numbers may be expressed as figures. (Example: *He used a crew of 3 carpenters, 1 plumber, 6 laborers, 1 foreman, and 1 timekeeper.*) This is particularly desirable when statistical information is being presented.
4. When one number appears immediately after another as part of the same phrase, one of the numbers is spelled out. (Examples: *7 six-inch timbers, two 7-man crews.*) It is preferable that the shortest number be the one spelled out; but when two such terms are close together, the same form should be used for both.
5. Numbers that are merely approximations are often written out, regardless of their size, unless the result would be cumbersome. (Examples: *The company has enough timber to operate for twenty years. The building should stand for fifty years.*)
6. Sums of money are expressed in figures. (Examples: *5 dollars* or *$5, $7.95, $0.80,* or perhaps *80 cents.*)
7. Technical style tends to use decimals rather than ordinary fractions because they make it possible to indicate a greater degree of precision. In a decimal fraction with a value less than one, a zero is placed

before the decimal point (0.719). Also, it may at times be desirable to add a zero after the decimal point for the sake of precision (0.6840).

8. Decimal fractions should not be used to express information for which ´ordinary fractions are customary, especially when accuracy would thus be misrepresented. For example, *2.812* inches should not replace *2 13/16 inches*, since this measurement is not accurate to the thousandth of an inch.

The preceding rules do not answer every possible question; if they did, they would be too long and complicated to be useful. Moreover, there are times when two rules may conflict with each other. Hence there are problems you will have to solve by using your own judgment. Most of the time, however, these rules will make it possible for you to choose between figures and words with confidence that the form you decide upon will be acceptable.

Use of Abbreviations in Technical Style

As any reader may observe, abbreviations are used more frequently in technical writing than in writing in general—so frequently, in fact, that their use is clearly the result of a special set of conventions. If you are to decide intelligently when to follow these special conventions, you will need to know the underlying reason for their existence.

Abbreviations are justified not because they make the writer's work easier—though they may have that effect—but because they assist the reader. In technical writing, certain terms related to the field of discussion may be used over and over again. To write them out in full would take more time and space; but more important, it would make what is written harder to read. This is especially true when the terms are phrases rather than single words. It might be necessary, for example, to write *parts per million, feet per second, revolutions per minute,* or *board feet* eight or ten times in a single brief paragraph. Written out in full, these terms might occupy so much space that the remainder of the material would be overwhelmed; yet each of them expresses only a single concept and would actually be clearer and more compact if expressed by a single symbol. Hence abbreviations that shorten such phrases to a few letters have come to be recognized and may be used when they serve a useful purpose.

If terms for which there are abbreviations do not appear often enough to be a problem, however, nothing is gained by abbreviating them. In fact, unless they are numerous, abbreviations are distracting and annoying; they look peculiar, and they hinder rather than assist communication. Therefore it is advisable to refrain from using the technical abbreviations unless the terms appear frequently. When you decide, after considering the points mentioned, to abbreviate in accordance with technical style, you will have sound authority for what you do if you comply with the following rules:

1. Unless it is extremely short, a term denoting a unit of measurement is abbreviated when it follows a figure. Examples are *inch, yard, pound, ounce, gallon, cubic yard, revolution per second, watt, board foot,* and *horsepower.* Unless it follows a figure, however, none of these terms is abbreviated. One would write *63 ft, 2300 rpm, 435 ppm, 125 hp, 50 cc;* but, "It is expressed in *horsepower*" or "The measurement is converted into *cubic centimeters.*"

2. An abbreviation *for a unit of measurement* is always shown as singular. You should use *lb,* not *lbs; bbl,* not *bbls; gal,* not *gals.*

3. A few extremely short terms denoting units of measurement are not abbreviated. Among these are day, mile, and acre. Since usage is not consistent, no exhaustive list can be given. Systematic personal observation is the only way to be sure about the customs in your own intended profession.

4. In many professions there are terms in addition to units of measurement that are used with extreme frequency and consequently are abbreviated when it is reasonable to believe that the readers addressed will grasp the meaning instantly. Some examples are: *a-c* for *alternating-current* used as an adjective, *F* and *C* for *Fahrenheit* and *centigrade, cp* for *chemically pure, el* for *elevation, emf* for *electromotive force.*

5. Even when you are using the technical style of abbreviation, there are many terms that you may abbreviate or not according to the dictates of your own judgment, provided you handle each term consistently.

6. The fact that technical style permits the use of abbreviations does not mean that it is desirable to use arbitrary signs for words. You should write *8 in.,* not *8″; 12 by 15 ft,* not *12′ × 15′; percent,* not *%.*

   There are a few exceptions to this rule. It is correct to use the dollar sign, and in appropriate context the use of ¢ would be permissible. Degrees are another exception. Three forms are widely used to indicate degrees of temperature: *84° C, 84 C,* and *84 deg C.* (The same forms could be used for degrees Fahrenheit.) Degrees, minutes, and seconds of angles or degrees of latitude and longitude may be expressed by signs. It would be correct, for example, to write *21° 55′ 15″.* However, for the sake of easier typing many people avoid the sign for degrees.

7. Capitalization need not be affected by abbreviation. Abbreviations are not ordinarily capitalized unless the terms they stand for are capitalized.

8. In some professions, notably engineering, it is customary to omit the period after many abbreviations. In deciding whether or not to use periods, you should follow the practice of books and magazines in your own field. Even in engineering, however, the period should be used if omitting it could cause confusion, for example in the abbrevi-

ation for *inch*. Also, the period should always be used after abbreviations that do not result from technical style but would be used in writing in general—for examples: *a.m., p.m., c.o.d., B.C., Fig.,* and abbreviations used in footnotes and bibliographies, such as *ibid., op. cit., Vol., p.,* and *ff.*

9. One final rule about abbreviation will settle many questions: When in doubt write the word out.

HYPHENATION OF COMPOUND TERMS

The question of whether some term is or is not a compound, and if so whether it should be hyphenated, arises with unusual frequency in technical writing. Consequently, though the use of hyphens to divide words at the end of lines is covered in the section on punctuation in the Handbook, hyphenation of compounds calls for treatment in this chapter.

Actually, usage is so far from uniform that it is impossible to draw up a definitive set of rules; and even if it were possible, exceptions would be so numerous as to make it advisable to look up some terms in an up-to-date dictionary. When you consult a dictionary, however, a certain amount of caution is necessary. First, you should make sure that the dictionary is not out of date, for some terms that are hyphenated in one edition have come to be treated as solids by the time another edition is published. Second, be sure that you do not mistake a symbol used to indicate the division of a word into syllables for a hyphen. Third, note well what part of speech is shown after the term in the dictionary entry. A term, for example, that is hyphenated when used as a noun might be treated as two words when used as a verb.

Helpful as a dictionary may be, it will not solve all your problems, for as a technical writer you will often create compounds that it does not contain. The discussion that follows will be helpful on such occasions. It is not a set of absolute rules but rather an explanation of tendencies, based on study of authorities. If you apply its principles, you may feel confident that the form you use will be acceptable, even though there are times when some other form would not necessarily be an error.

1. In writing a term formed by placing a prefix in front of a word, you will almost always be safe if you use the solid form. Such a term may be hyphenated, however, to permit internal capitalization (*pre-Cambrian*) or to clarify pronunciation or meaning (*reform, re-form; re-employ; re-anneal; intra-atomic*). *Self* is an exception to the general rule; when used as a prefix it is always set off by a hyphen (*self-sufficient*).

2. A compound adjective is usually hyphenated, especially when it precedes and directly modifies a noun (*time-consuming method; all-*

inclusive statement; 60-horsepower motor). It is somewhat less likely to be hyphenated if used as a predicate adjective (*the man was hard-hearted; the rules were all-inclusive;* but *the job was half completed; the method was up to date).* A chemical term, for example *carbon monoxide* or *calcium chloride,* is not ordinarily hyphenated even when used as a compound adjective. Also, one should not hyphenate such terms as *easily answered* even when they appear in phrases such as *easily answered question.*

3. Compound nouns are more likely to be written as separate words or as solids than to be hyphenated. A few special types that are hyphenated are exemplified by the following terms: *cave-in, motor-generator, weigher-in, I-beam, foot-pound.*

4. Compound verbs vary, but the type shown in the following sentences is hyphenated: *They double-tracked the railroad. They dry-cleaned the canvas.*

5. Usage in hyphenating compound adverbs varies widely, but a hyphen is always used in a compound adverb formed by adding *ly* to a hyphenated compound adjective (*half-heartedly, quick-wittedly).*

6. Try to avoid writing sentences in which you would be forced, if you followed the customs of hyphenation, to create a form that is unclear or illogical. Many problems are best settled not by consulting rules for hyphenation but by revising the sentence so as to get rid of the troublesome term. (Undesirable: *a mercuric chloride-activated compound; a 7 by 16-foot area; the North Dakota-South Dakota boundary.* Improved: *a compound activated by mercuric chloride; an area of 7 by 16 feet; the boundary between North Dakota and South Dakota.)*

7. To carry rule 6·further: Long, unwieldy compounds are poor style, and instead of trying to decide how to hyphenate them it is better to refrain from using them. (Undesirable: *internal-combustion-gasoline-engine cylinders; a piece of 1½-inch-inside-diameter pipe.* Better: *cylinders for internal-combustion gasoline engines; a piece of pipe with an inside diameter of 1½ inches.)*

8. It is important to remember that the same term may be hyphenated in one usage and not in another. (*The blast-furnace crew worked overtime,* but *A new blast furnace is being designed. The blast-off occurred as planned,* but *It will blast off at 1:15 p.m. We plan to hard-surface the highway,* but *The highway has a hard surface.)*

THE DECIMAL SYSTEM OF NUMBERING HEADINGS

You have probably seen the decimal system of numbering headings in instruction manuals, specifications, and perhaps in some of your textbooks. It is useful when frequent reference to some specific section of the material will be necessary. In this system, a number is placed before each heading in the text, and also before each heading in the table of contents if one is used.

The following examples show how this system, by use of Arabic numerals and decimal points, indicates the rank of each point and identifies the superior point under which the lesser points appear.

1 main point
1.1 subpoint under first main point
1.2 second subpoint under first main point
1.2.1 subpoint under 1.2
1.2.2 second subpoint under 1.2
2 second main point

Unlike the numbers used in an ordinary outline, each number in this system is complete and self-sufficient. That is, the number preceding a point includes the numbers of points of higher rank under which it is taken up. Because of this fact, and because its numbers are compact and easy to include in text, the decimal system does its job extremely well. When there is no necessity of frequently referring to some specific section of the text, however, it is undesirable because it gives the text a cluttered appearance and serves no added purpose.

EQUATIONS

When equations appear frequently in technical writing they are set apart from ordinary text, and sometimes they are numbered for easy reference. Following is a typical example.

. . . The formula for heat flow under these conditions is

$$(7) \qquad Q = \frac{k(T_1 - T_2)At}{d}$$

where

Q = heat flow,
k = coefficient of thermal conductivity for the refractory material,
$T_1 - T_2$ = temperature drop from hot face to cold face,
A = area of the wall,
t = time,
d = thickness of the wall.

Note that the alignment is based on the "equal" sign. Note also that punctuation is just as it would be if the material were placed in regular lines of text rather than set apart.

DOCUMENTATION

As a technical writer you will often use information obtained from printed sources. Consequently, you will need to acknowledge your indebtedness, partly as a matter of courtesy and honesty, and partly because you serve your readers' best interests by doing so. This acknowledgment of sources is referred to as documentation.

By acknowledging sources you may perform various functions: You may show the source of quotations or the authority for statements that might arouse skepticism; you may indicate the extent of your study, and thus build confidence in your remarks; or you may show the reader where he can obtain additional information.

The following discussion should assist you in selecting an appropriate system of documentation; it will cover the general conventions and will illustrate forms that it is ordinarily safe to follow. Many fields, however, have their own special conventions; and on any particular occasion you should not hesitate to use whatever system the readers whom you address are probably accustomed to. And if you think your work might be published, you should use the system of documentation you find in the publication where you think it most likely to appear.

Four methods of documentation are common in technical writing: (1) footnotes only; (2) an alphabetical list of references (no footnotes and no citation of this list in the text); (3) a numbered list of references, and citations in the text referring to these numbers; and (4) footnotes plus a bibliography.

In choosing among these methods you should take into consideration both the functions you need to perform and the customs that prevail in your own particular field.

Footnotes

If you do not expect to cite sources very frequently and if your total indebtedness to references is not very great, footnotes alone will often be the most satisfactory system of documentation.

A footnote may be used for various purposes. It may merely explain something in the text that will be clear to most readers but not to all; it may provide additional comments or facts that would destroy continuity if placed in the text. These uses, however, are unrelated to documentation and may occur even in undocumented work. Footnotes used for documentation usually perform one or more of the following functions: (1) identify the source of a quotation; (2) acknowledge indebtedness for facts or ideas; (3) cite authority for some statement that might arouse skepticism; and (4) indicate sources for additional information.

Footnotes should not be used unless they are needed. There is no reason to use a footnote in support of a statement that your readers are unlikely to question or can easily verify without assistance, nor to identify a brief quotation used only because of its striking phraseology.

Footnotes used for documentation should be numbered. Sometimes all the notes applying to each page are placed at the bottom of the page itself. In this event, each page may begin with footnote 1, or there may be a single sequence of numbers for all the footnotes in a paper. Sometimes all the footnotes are placed at the end of the paper, sometimes at the end of each section. In this event a single sequence is used for numbering all the footnotes at any one location. Placing the footnotes on the pages to which they apply makes it easier for the reader to consult them. Placing them all in one location eases the work of the writer, and of the printer if the manuscript is to be printed. Also, it permits the reader to see in one place all the references that have been cited.

The presence of each footnote is indicated by inserting its number in the text at the appropriate place. If the footnote identifies the source of a quotation, the number is placed at the end of the quotation. If the note applies to material that is not directly quoted, the number is placed at the end of the material that the note concerns. (The text must of course be phrased so that it shows where the application of the footnote begins.) Some forms permit a writer to place this number in parentheses on line with the regular text, but more frequently the number is raised half a space above the line, thus: ". . . Using this system, it was possible to produce about 3,000,000 lumens of light."[2] As can be seen, the number identifying the footnote follows all other punctuation and is not preceded by a blank space.

Regardless of their location, footnotes are ordinarily separated from the text by a solid line beginning at the left margin and extending 1½ inches (15 or 18 spaces) toward the center of the page. This line is a double space below the last line of text and a double space above the first footnote. A double space is left between two footnotes, but single spacing is used within the footnote itself.

The form of footnotes varies in minor respects, but most forms are similar in general arrangement and content. A footnote usually answers the following four questions, so far as possible, in the order indicated: (1) Who said it? (2) What was the title of the piece of writing in which he said it? (3) Where or by whom was this writing made available? (4) On what page or pages may the material cited be found? The following examples show widely used forms for footnotes citing various types of material. They are numbered as if they appeared together, but it is unlikely that so many footnotes would be found on any single page.

Forms for Footnotes

For a book with one author:

1 Niels Jensen, *Optical and Photographic Reconnaissance Systems* (New York: John Wiley & Sons, Inc., 1968), p. 144.

For a book with two or more authors:

2 Harold S. Graves and Lyne S. S. Hoffman, *Report Writing*, 4th ed. (Englewood Cliffs, N.J.: Prentice-Hall, Inc., 1965), p. 317.

For a book with no author given:

3 *Preparation of Technical Publications* (Indianapolis: U.S. Naval Avionics Facility, 1966), p. 34.

For an edited book:

4 William K. Tobold and Larry Johnson, eds., *Automotive Encyclopedia*, (South Holland, Ill.: The Goodheart-Willcox Co., Inc., 1970), p. 345.

For an article in a book:

5 Donald W. Head, "Chemical Cycles in the Sea," in *Oceanography*, ed. Richard C. Vetter (New York: Basic Books, Inc., 1973), pp. 344–46.

For a magazine article (signed):

6 R. B. Overend, "The Great Studded Tire Controversy," *Traffic Safety*, December, 1971, p. 18.

For a magazine article (unsigned):

7 "Health Insurance for All," *Time*, February 18, 1974, p. 67.

For an article in an encyclopedia:

8 "Telautograph," *The Encyclopedia Americana* (1969), XXVI, 327.

For an article in a newspaper (unsigned):

9 *The Washington Post*, April 16, 1974, p. 8.

For a numbered report, technical paper, or any similar citation:

10 V. P. Puzinauskas, *Gussasphalt or Pourable Asphaltic Mixtures*, Research Report 70-2, February, 1970, The Asphalt Institute, College Park, Maryland, p. 5.

For a reference to a secondary source:

11 John Diebold, "The World and Doomsday Fads," *New York Times*, February 15, 1973, as cited in M. J. Osborne, "Water Lilies and the Lombard Steersman," *Tappi*, May, 1973, p. 84.

For a thesis or dissertation:

12 Kenneth V. Erie, *John Ruskin as a Prophet of the Conservationist Movement* (Master's thesis, Channing State University, 1972).

For an address:

[13] John R. Stokes, *Problems of Protecting Lakeshore Quality,* unpublished address at a meeting of the Northwestern Conservation League, Yakima, Washington, May 6, 1973.

For a personal interview:

[14] Harvey Forester, Biologist, U.S. Forest Service, in a personal interview in June, 1971.

As mentioned earlier, the form of footnotes varies in detail. Different systems of punctuation are used. Some forms indicate the volume and issue of magazines, and some omit this information. Indentation is sometimes hanging (first line extending to the margin and the other lines indented) rather than in paragraph form. And some highly specialized scientific books and periodicals use abbreviation far more extensively than is customary elsewhere.

In many respects, however, usage is well established and should be complied with. The title of a book, magazine, or other independently printed reference should be underlined (the equivalent of italics in print), but quotation marks should be used for articles in a magazine, separate units of a collective work, and other writings that are not separately printed. Since footnotes are not alphabetized, the name of an author is arranged in normal order and should be shown exactly as it appears in the original. A publisher's name should be used exactly as it appears on the title page of a book. If it is known that a publisher's imprint has changed since publication of the book, it is proper to give the new imprint provided the form used is precisely as sanctioned by the publisher. A title, the first time it is mentioned, should appear exactly as in the original, although it may be shortened in later citations. That is, a footnote referring to a book may be reduced to the name of the author, the title, shortened if necessary, and the page number; and the name of a periodical or organization, if long, may be sharply abbreviated.

Since a footnote need not repeat information already given in the text, if an author's full name appears in the text, it may be omitted from the footnote. If only an author's last name appears in the text, however, the footnote should provide the full name. The question as to whether the name of an author or any other information belongs in the footnote or in the text can be settled by applying the following principle: Whatever is actually part of the message belongs in the text; what is provided merely as documentation should be in the footnote.

Certain abbreviations are widely used in footnotes, the most common being *ibid.* and *op. cit. Ibid.* is the abbreviation for *ibidem,* Latin for "the same." *Op. cit.* is the abbreviation for *opere citato,* Latin for "work cited." (Italics are used because the terms are foreign; in a typed paper they would be underlined.) *Ibid.* without a page number indicates that a reference is exactly the same as the one immediately preceding it. *Ibid.* followed by a page number indicates that the two references are the same except for the page. *Op. cit.*, which is preceded by the name of the author unless he was named in the text, is used when a reference has already been named but other references have intervened. Obviously, *op. cit.* cannot be used if more than one work by the same author has been named. In order to spare the reader the necessity of turning back to preceding pages, it is best to use *ibid.* and *op. cit.* only to refer to a footnote on the same page. *Loc. cit.*, Latin for *loco citato*, place cited, is sometimes used in much the same manner as *ibid.* but is less definite. *Passim*, Latin for here and there, is used to indicate that ideas have been gathered from various places in a reference, perhaps within a stated succession of pages. *Cf.*, Latin for *confer*, compare, is used just as *compare with* might be used.

It seems desirable, however, in the interests of effective communication, to hold the use of a foreign language to a minimum; more and more technical writers are coming to the conviction that *ibid., op. cit.*, and *cf.* are the only Latin abbreviations that are really necessary, and some are even omitting the underline or italics that identify *ibid.* and *op. cit.* as foreign.

Alphabetical List of References

Often, in technical writing, documentation is limited to a list of materials headed *References*. Such a list may include only the sources from which you drew specific information, may be broadened to cover material that contributed to your general background, or may even include references that would merely be helpful to a reader who wants additional information.

Obviously, a list of references not cited in the text conveys only a limited amount of information. It does not show your authority for specific facts but only indicates whether you have examined a satisfactory number of authoritative, up-to-date references before writing. Still, the evidence that you have consulted good authorities makes your statements more convincing, and the use of this system does not prevent you from mentioning, in the text, the sources of specific facts when you think that such information is important. This may be done by merely mentioning the authority in a regular sentence, as for example, "This theory, as Arthur Miller points out. . . ." Another method, widely used in some fields, is to insert the name of the author and the date of the reference in parentheses, as for example, "A recent study of the effect (Miller, 1973) points out . . . ." (When this method is used, the date follows the author's name in the list of references as well.) In any event, whatever its limitations, the use of nothing more than a list of references is an extremely common method of documentation and therefore must be judged adequate to the needs of many occasions.

The form used for the individual entries in such a list of references closely resembles the form used for footnotes. The main difference is that the last name of the author is placed first to facilitate alphabetical arrangement. When more than one title by the same author is listed, a solid line one half inch long is often substituted for the author's name in the second and later entries. Unsigned references are alphabetized by title. Titles are indicated as in footnotes: quotation marks for a title such as that of an article in a magazine, and an underline, the equivalent of italics, for the title of an independent publication. Two or more references by the same author are alphabetized by title.

Details may vary, but the following alphabetical list illustrates a form that is widely used and illustrates the principles that apply to most types of entry.

References

BROWN, JAMES, *Casebook for Technical Writers*. Belmont, Calif.: Wadsworth Publishing Co., Inc., 1961.

———, *Cases in Business Communications*. Belmont, Calif.: Wadsworth Publishing Co., Inc., 1962.

ECKENFELDER, J. WESLEY, JR., *Industrial Water Pollution Control*. New York: McGraw-Hill Book Company, 1966.

Effects of Studded Tires, A Research Progress Report. Minnesota Department of Highways, December, 1970.

FASAL, JOHN, "Forced Decisions for Value," *Product Engineering*, April 12, 1965.

GIBBY, J. C., *Technical Illustration*, 2nd ed. Chicago: American Technical Society, 1962

JONES, E. L., and PHILIP DURHAM, eds., *Readings in Science and Engineering*. New York: Holt, Rinehart and Winston, Inc., 1961.

MANDELL, D. A., G. W. HINMAN, and others, *The Gas-Cooled Breeder Reactor with Gas Turbines and Dry-Cooling Towers*. Washington State University College of Engineering Bulletin 332, Pullman, Washington, 1973.

"Micromanipulation," *Encyclopaedia Britannica* (1970), XV, 376.

MONTARI, F. W., "Legislative and Administrative Measures Needed," *Proceedings of the Fifth Annual Conference of State and Federal Water Officials*. Washington, D.C.: Water Resources Council, 1971.

"Productivity: Key to Our Economic Health," *Changing Times*, February, 1974.

Numbered List of References

To make documentation more explicit without the use of foot-notes a numbered list of references is often used in place of an alpha-betical list. Thus a reference can be cited by merely inserting its number into the text. If this method is used, the list of references is limited to cited materials and the entries are arranged in the order in which they are first cited. The form of individual entries is identical with the form used in an alphabetical list except that authors' names are arranged in the normal order. The following brief list should be a sufficient illus-tration of form.

References

1. J. C. Gibby, *Technical Illustration,* 2nd ed. Chicago: American Technical Society, 1962.
2. John Fasal, "Forced Decisions for Value," *Product Engineering,* April 12, 1965.
3. James Brown, *Casebook for Technical Writers.* Belmont, Calif.: Wadsworth Publishing Co., Inc., 1961.
4. E. L. Jones and Philip Durham, eds., *Readings in Science and Engineering.* New York: Holt, Rinehart and Winston, Inc., 1961.

To insert the number of a reference into the text of a paper, either of the following forms may be used:

> "The temperature loss rarely exceeded 2.9 deg F for 100 lb of ore (7). This information. . . ."
> "The tests performed by L. V. Smith /2/ showed the form of curves to be those reported for austenite steel."

A numbered list of references should not be confused with the use of footnotes at the end of a paper. When footnotes are used, each has a separate number even though it may refer to a reference that has been cited previously. When a numbered list of references is used, however, each reference is listed only once and the same number is used each time the reference is cited. Also, as can be seen from the examples, the page number of the reference from which material was drawn is not indicated when a numbered list of references is cited, for a single listing may be cited more than once and the page number might be different on each occasion.

Like the alphabetical list of references, the list that is numbered and cited by number is widely used—perhaps more widely in technical writing than any other form. Thus it is apparently found satisfactory by a great many writers.

Footnotes Plus Bibliography

The most detailed system of documentation is that in which footnotes are supplemented by a bibliography. This system is used more frequently in scholarly work than in run-of-the-mill technical writing. It is suitable when references to sources are numerous and when a writer wishes not only to give the exact page number of every citation but also to show in one place all the references that he has consulted, regardless of whether he has cited them all specifically.

When footnotes are supplemented by a bibliography, their form is the same as that already illustrated, except that the publisher and place of publication of a book are not included. The form used for the bibliography is the same form that is used when there are no footnotes.

Final Comment on Documentation

It is impossible to discuss the subject of documentation without presenting a great many rules. Omission of any of these rules would have meant failure to answer questions that constantly arise; and the writer who does not know or does not look up the answers to such questions is likely to use forms that impress the reader as strange and awkward. Also, rightly or wrongly, a reader may form the impression that a writer who has not learned how sources are usually cited may be so inexperienced that he does not know how they should be used. And of course the conventional forms are preferable also because a reader will grasp the content faster if the form is familiar.

None of this should make us lose sight, however, of the fact that the purpose of documentation is to give the reader information that he wants or needs. Anyone who does much documentation will occasionally need to present information that does not fit into the conventional forms. When this is your predicament, there is no reason to be disconcerted. Following the conventional forms only so far as they fit the occasion, merely tell the reader the facts in the simplest, most concise manner that will be clear, arranging them, so far as possible, in the order shown near the beginning of the discussion on documentation. One of the comforting facts about knowing the conventional forms is that you can feel free to improvise when necessary, undisturbed by the fear that you are betraying ignorance.

EXERCISES

Exercise 1

Following are sentences involving the use of numbers and abbreviations. Make the changes that technical style would call for, but do not make unnecessary changes.

This is not an exercise in hyphenation and there are no errors in the use of hyphens. Don't introduce such errors when you make changes involving numbers and abbreviations. Ordinarily, hyphenation is not affected by the use of technical style.

1. The supply of timber in the tract will be exhausted in approximately 7 years.
2. The results were best when the dissolved-oxygen content of the water was 2–4 ppm.
3. By arranging the machines as shown in Fig. 14 the space that they occupy can be held to 30′ × 50′.
4. The water supply is stored in 6 50-gal tanks.
5. When the temperature was increased to 500 deg. C, the expansion amounted to .092 inches.
6. The number of units rejected has been cut by ¼ during the last quarter.
7. 22 inches of snow fell during only two days, causing approximately 78 cars to be stranded.
8. The crust built up at a rate of 0.08 of an inch per hour.
9. The leakage had increased to eight gpm.
10. The steel plates are available in thicknesses of ½ in., 1 in., 1.5 in., and 2¼ in.
11. The concentration had reached 8 parts per million.
12. The present flow, 4800 cfs, is alarmingly low.
13. The mixture must settle for at least 12 hrs.
14. In estimating the fbm on the tract, the timber cruiser made a serious error.
15. The tank will hold 360 cu in., which will be adequate.
16. At 7:30 p.m. the lights come on automatically.
17. The fire destroyed seven pumps, 12 batteries, 24 water bags, and 18 sleeping bags.
18. The speed of the motor fell to 1500 r.p.m. before the change was noticed.
19. By noon the river was only 9 in below flood stage.
20. The heat it will generate is to be measured in btu.

Exercise 2

A. Each of the following sentences contains an expression that a writer might feel should be hyphenated or perhaps written as a single word. For the purpose of this exercise they have all been written as separate words. Rewrite in the correct form each expression that needs to be changed. Although the treatment of compounds is far from uniform, you should attempt, in working this exercise, to follow the principles recommended in the discussion in this chapter.

1. The sensor's field proved capability of calibration was satisfactory. (The capability had been proved in the field.)
2. The residue was stored in lead lined boxes.
3. The fish killing pollution must be eliminated.
4. We feel that we should road test the car before we regard the repairs as satisfactory.
5. Arsenic is classified as a semi metal.
6. The reserve supply is stored in 25 gallon tanks.
7. In Europe, two twin wire machines are already producing this stock. (There are twin wires on each machine.)
8. It would be advisable to install self closing doors.
9. They tried to cover up their actions but the attempted cover up was unsuccessful.
10. The company's foreign operations were handicapped by anti American sentiment.
11. Use of wet pressing machinery increases the cost of electricity. (The machinery presses wet material.)
12. It will be necessary to strengthen the front wheel base. (The base of the front wheel.)
13. He was advised to invest in no load mutual funds.
14. The rail road company will double track the line from Fairfield to Laird.
15. The trouble was caused by un co ordinated operations.
16. Everything was broken except the T square.
17. The part is made of nickel silver alloy.
18. One paper, dealing with the economic value of the folder gluer, was especially interesting to us.
19. The off shore drilling equipment was damaged in a storm.
20. The pension fund should not be invested in risk oriented ventures.

B. Improve each italicized expression in the following sentences either by inserting hyphens, changing the construction, or both.

1. *Monocoque and unitized construction heavy gauge* sheet steel was used for the enclosure.
2. The *externally operable three pole* switch established a new standard of security for high voltage transformers.
3. *Adverse soil condition* findings indicate a need for pile foundation.
4. The *scenic waterways systems* analysis was completed satisfactorily.
5. The study concluded by recommending the purchase of *two propeller type 1000 gallon per minute* pumping units.
6. A *bi directional pawl and ratchet manual switching* mechanism provides *trouble free* operation in areas where positive switching is important.

Exercise 3

Tell whether each of the following statements calls for a citation of its source, assuming that it appears in a work that is documented.

1. In spite of the prolonged exposure of hundreds of millions of people to DDT, there is no evidence that it is causing cancer or genetic change.
2. The craftsman with years of experience but little technical comprehension of the processes that he engages in will in a few years be no longer required by large paper mills.
3. There will be a time lag of at least ten years between the inception of a new idea and its large-scale use in the paper industry.
4. Neither solar energy, wind energy, nor tidal energy has the potential for power production except in localized areas.
5. Spraying these infested areas with DDT would do less damage to its ecology than would be done by leaving it unsprayed and thus allowing the Tussock moth to kill off most of the trees.
6. The public cannot count upon fluoride dentifrices to be beneficial unless the Food and Drug Administration requires that they be labeled with an expiration date such as the one on packages containing camera films.
7. When a toothpaste is especially effective in whitening or brightening teeth, the results are gained by the use of harsher abrasives than are used in other toothpastes.
8. The so-called "organic" honey sold in health-food stores is no less likely to contain the residues of pesticides than honey sold at half the price in supermarkets.
9. So far as environment is concerned, those who object to building more power plants are defeating their own ends. For example, thousands of sewage-treatment plants are needed, and a secondary plant serving 90,000 people requires about 500,000 kilowatt-hours of electricity per year.
10. The likelihood that you will catch a cold from the person who has one is smaller if he coughs and sneezes when you are nearby than if he shakes hands with you.
11. "When the Federal Trade Commission considers the advertising of the cosmetic industry, it should remember that not all values are material—that benefits cannot be measured only in graduated beakers, but also reside in feelings and emotions." (Consider primarily whether this should be acknowledged because it is a quotation.)

Exercise 4

Assume that you have written a paper that you are documenting by means of footnotes only and that you must include, on one page or at the end of the paper, the material provided below. Rewrite each item in the proper form.

1. A book entitled Industrial Water Pollution Control published by McGraw-Hill Book Company in New York in 1966. The author is W. Wesley Eckenfelder. The page cited was page 23.
2. An article by R. B. Overend entitled The Great Studded Tire Controversy. It appeared in the magazine Traffic Safety in the issue for December, 1971. The page cited was 35.
3. The same as number 2 except that the material cited was on page 37.
4. The same as number 1 except that the page cited was page 193.
5. An article in a magazine named Tappi in the issue for May, 1973. The article was called Future Trends in the Paper Industry. The passage cited was on page 73. The author's full name, Philip Nethercott, appeared in your text.
6. The reference was to an address, unpublished, by G. H. Andrews, that was delivered in September of 1972 to the Washington State Good Roads Association at a meeting in Spokane.
7. An article in Time magazine. No author was mentioned. The title was Results of a Lifted Embargo. It appeared on page 22 of the March 18, 1974, issue.
8. An article entitled Important American Aircraft Types on page 253 of the Information Please Almanac published in 1963 and edited by Dan Golenpaul.

Exercise 5

A. Assume that you have written a paper that you are documenting by providing a numbered list of references and citing them in the text. The facts about each reference are provided below. Arrange the material in each item in proper form.

1. A book by Ralph K. Iler published by the Cornell University Press, Ithaca, New York in 1955. Its title is The Colloid Chemistry of Silicates.
2. An article entitled Optical Determination of Linear Expansion and Shrinkage of Wood. It appeared in the June 1970 issue of Forest Products Journal.
3. An unsigned article entitled Marketing Perspective that appeared in the magazine Production Engineering for March, 1973.
4. The third edition of a book entitled Manual of Color Aerial Photography, edited by John T. Smith and Abraham Anson. The book was published by the American Society of Photogrammetry, Falls Church, Virginia, in 1968.
5. An experiment station bulletin: Washington State University College of Engineering Research Division Bulletin 330, written by Howard Capp and entitled More Responsive Water Planning is Possible. Its date was March, 1973.
6. An unpublished doctoral dissertation written at the University of Washington in 1971 by Thomas Hukari. Its title: Some Logical Properties of the English Complement System.

7. A book with the title Color as Seen and Photographed, identified as Color Data Book E74, by the Eastman Kodak Company, which published it in Rochester, New York, in 1966. The reference is to the second edition.

8. An article entitled Background to Musical Developments in the 20th Century, written by Stephen Walsh. It was published in a book, Larousse Encyclopedia of Music, edited by Geoffrey Hindley. The date of publication was 1971; the publisher, World Publishing Company of New York.

B. Assume that you have written a paper that you wish to document only by means of an unnumbered list of references. Using the material in A of this exercise, make such a list of references. Provide a suitable heading, arrange the items in proper order, and place the material in each item in the right form.

5

SPECIAL
PROBLEMS

The special problems discussed in this chapter are those related to definitions, technical description, the explanation of processes, instructions, analyses, and technical or semitechnical papers and articles. Except for articles and perhaps instructions, you will find that the kinds of writing discussed do not usually appear as complete, independent papers, but as component parts of reports or other longer pieces. Nevertheless, each of them calls for special skills that may best be acquired by examining it separately.

DEFINITIONS

In writing about technical subjects you will continually find that in order to express your ideas clearly you must use a term that calls for definition. Sometimes a sentence definition will be sufficient, but there are times when an expanded definition is necessary.

The Sentence Definition

A sentence definition is a statement that contains the bare minimum necessary to cover the meaning of a term and thus to reveal the essential nature of the thing that the term stands for. Usually it consists of a single sentence; but if a single sentence would be long, involved, and hard to read, there is no reason that two or three sentences should be considered objectionable, in spite of the name "sentence definition." In addition to the term itself, this type of definition contains two other parts, the genus and the differentia. The genus indicates a classification or group that includes the term, and the differentia discriminates between the term and whatever else the genus includes. These parts are seen in the definition of a microscope as "an optical instrument (genus) consisting of a lens or combination of lenses for making enlarged or magnified images of objects (differentia)."

The genus must be chosen carefully. It must be accurate and should be as narrow as possible so that the differentia will not have to include an excessive amount of information. It would be better to identify asbestos, for example, as a nonmetallic mineral than merely as a substance. The differentia should contain enough information to draw the line between the term being defined and everything else that the genus might include.

A good sentence definition must be based on a clear understanding of the essential nature of the object for which the term stands. The definition of a pearl, for example, would need to emphasize the pearl's origin. Color, beauty, and value might also be mentioned; but unless the definition were based on the fact that a pearl is an abnormal growth within the shell of certain types of mollusk, it would not serve its main purpose.

The following sentence definitions may be studied as additional examples:

A capillary is a thin-walled tube that carries blood between any of the smallest arteries and its corresponding vein.

Isothermal compression is the compression of a gas under such conditions that the heat generated by compression is removed as fast as it is generated, so that the temperature of the gas does not change. The meaning may be seen in that *iso* is a combining form meaning *equal* and that *thermal* refers to heat.

The flash point of oil is the lowest temperature at which the mixture of oil vapor and air above the surface of the oil will flash up if ignited. The fire point is the lowest temperature at which the surface of the oil will catch fire and burn.

The Expanded Definition

Though a well-written sentence definition is logically complete, it often must be expanded if the reader is to realize all its implications, find answers to all the questions that may arise in his mind when he reads it, and see how it does or does not apply in specific cases. Sometimes it must be expanded by means of additional definitions which make clear the meanings of words it contains. Usually, however, it is expanded by one or more of the following methods: use of illustrative examples, comparison and contrast, listing of the component parts, and elimination.

Illustrative examples make a definition more concrete and sometimes, also, clarify its scope. A definition of *parasite* might be expanded by mention of tapeworms, sheep ticks, and other specific parasites each of which lives in or on some host from which it obtains food or shelter. Illustrative examples are helpful in that they make a definition concrete. There is danger, however, that the examples chosen will not be representative and will therefore create a one-sided picture. If all parasites listed were types that live inside the host, a reader might not notice that the term *parasite* includes types that live outside, even though the sentence definition had been broad enough to include them.

Comparison or contrast is a useful method of expanding a sentence definition because it facilitates discrimination between the term defined and other terms with which that term might be confused. The full meaning and exact limitations of *contagious disease*, for example, might not be realized until that term was compared with *infectious disease.*

A list of parts or divisions encompassed by a term will often help a reader to realize the full scope of the term. For example, a definition of *physics* might read, "Physics is the science that deals with those phenomena of matter involving no change of chemical composition." This statement may be complete in itself, but it would be possible to help a reader by pointing out that under this definition physics includes the science of matter and motion, mechanics, heat, light, sound, electricity, and the branches of science devoted to radiation and atomic structure.

Naming the parts that comprise something has a different purpose when done in a definition than when done for the sake of analysis. When we analyze something, it is assumed that the reader knows what the object is but needs to be told what its parts are; but when we name the parts for the sake of definition, the purpose of naming them is to let the reader know what the whole object is. Naming the parts of something for the sake of analysis is comparable to listing the ingredients contained in a cake—it being assumed that the reader knows what a cake is. Naming them for the sake of definition is comparable to listing the counties that comprise an electoral district so that people will know what is meant when the district is mentioned.

Expanding a sentence definition by elimination is a process in which a writer clarifies our understanding of a term by pointing out what it might seem to include but does not. In defining *insanity,* for example, it might be necessary to mention that the condition called insanity does not include feeblemindedness, imbecility, or any other condition of mental deficiency as contrasted with mental derangement. Elimination can never by itself serve as a complete definition, for essentially, definition is a process of telling what something is rather than what it is not. But by specifying some of the meanings that a term does not include, a definition makes it easier to see the *limits* of its meaning.

The foregoing methods of expanding a sentence definition may be used separately or in any combination. Any kind of material may be used, in fact, if it will help the reader to absorb and remember the sentence definition. The following example shows how a sentence definition has been made easier to grasp by being amplified.

> The term *median* signifies the particular value among three or more which has an equal number of other values above it and below it. For example, if individual annual incomes for five people were $6000, $7000, $18,000, $19,000, and $20,000, the median income would be $18,000. The term with which *median* is most likely to be confused is *mean*, which refers to a point midway between the lowest and the highest values in question. In the list of figures above, the *mean* amount would be $13,000. The mean, that is, depends on only two figures and is unaffected by anything else, whereas the median results from taking into consideration not only the highest and the lowest but everything in between. The term *average*, of course, differs from both—referring, as is generally known, to the amount arrived at by finding the sum of all the figures in the list and dividing it by the number of figures.
>
> Though the difference between these three values may often be slight, there are times when it is highly significant. For example, if all but one of twenty people who took a test made scores between 80 and 98, but the single exception scored only 20, the mean score would be only 59, whereas the average and the median would be much higher—the exact amount depending in either case upon the exact distribution of all the scores concerned.

Point of View in a Definition

Many terms vary in meaning according to the context in which they are used. When a lawyer uses the term *insanity* in a courtroom, he refers to legal insanity and intends to convey a meaning that is not usually in the mind of the average person who uses the term. If a psychologist were to define *normal human being*, his definition would differ from the one that would be offered by a physiologist or sociologist. Thus the point of view from which a definition is written may affect its contents.

There is nothing objectionable about a definition written from a special point of view. However, it is important that, unless the special point of view is obvious from the context, the definition itself mention it.

In technical writing, concerned as it is with specialized subject matter, the need of mentioning a special point of view occurs with unusual frequency. In particular, the difference between the scientific and the popular meaning of many terms should not be overlooked. A zoological definition of *insect*, for example, would include butterflies but not spiders; but to a layman the term would probably include spiders but not butterflies.

The popular meanings of words are valid in ordinary circumstances. In technical work, however, precision is expected, and words that have exact scientific meanings should not be used or defined in a loose, casual manner. Consequently, when as a technical writer you address untrained readers, you should provide a scientific definition of any term that the reader might otherwise misunderstand.

The following specimen shows how a writer has recognized the obligation to define a term from a special point of view.

> The term *plant disease*, as used in the science of plant pathology, refers to the plant's being in such condition, regardless of the cause, that it cannot perform its vital functions in a normal manner.

Sometimes a point of view is special to the extent that a definition may become arbitrary. When government authorities, for example, establish grades of livestock and produce, each grade must be defined as exactly as possible. Though there may be a real difference between beef that is *prime* and beef that is merely *good*, the exact point where the distinction between the two grades is made is an arbitrary point. Similarly, to cite a long-established example, the exact amount of alcohol a beverage may contain before it is legally considered intoxicating must be set arbitrarily.

Definitions that draw arbitrary lines are especially common in specifications. Indeed, definitions are so important in specifications that a special section is often devoted to them. A typical example would be the definitions of *coarse aggregate* and *fine aggregate* that might be included in specifications in a contract that involved the pouring of concrete.

Even an individual is sometimes justified in writing a definition that is to some extent arbitrary. If there is disagreement about the meaning of some term, a writer may define it arbitrarily—not with the intention of settling the general controversy but merely to let the reader know which of the possible meanings he has in mind when he uses the term.

Following is an arbitrary definition:

> An Intermediate Care Home is an establishment that provides protection, physical assistance in meeting daily needs, social services, rehabilitation, remotivation, and recreation for no more than forty individuals of moderate to high dependency but does not provide medication that must be administered by a doctor or registered nurse.

The Effect of Purpose on Definitions

At first thought there would seem to be little need for comment on the purpose of a definition. The purpose, it would seem, is merely to make clear the meaning of the term to be defined. But further consideration shows that there is more to say.

Like any other writing, a definition is written because we are trying to communicate with a reader. Consequently, we face the question, "Who is the reader, and why does he need a definition of this term?" Whether a definition is good or bad depends in part upon whether these questions receive enough attention.

For example, you may need to define a term so that readers who are entirely unfamiliar with it, or who at best have a vague idea about it, may grasp its meaning well enough to meet their needs. Or, you may need to define a term so that readers who already understand its general meaning, or who at least have special knowledge in some field where the term is used, may use it as a basis for making fine discriminations.

The basic facts in a definition and the language used should both be affected by the question of which of these two conditions exists. Suppose we defined butanone as "a highly flammable liquid used as a solvent." On many occasions, nothing more would be necessary to accomplish our purposes; but if chemical matters were under discussion and chemists were among the readers, it would probably be desirable also to say that butanone is derived from acetone and to include its chemical formula. The point is that facts of more than one kind can be used to draw the line between what you are defining and all else with which the thing you are defining might be confused. And the question of what facts to include depends on whom you are addressing and on the specific purpose for which your definition will be used.

As for language—the advice in the chapter on style applies to definitions as it applies to other writing. If you want a particular definition to be useful to anyone who is not a specialist in the subject, it is best to avoid the use of words that are more technical than the term you are trying to define. A layman who did not know what is meant by *mumps* would learn very little from one dictionary's definition, "a specific, febrile disorder characterized by a nonsuppurative inflammation of the parotid and sometimes other salivary glands." He would be better served by, "a contagious disease marked by inflammation and painful swelling of the salivary glands below and in front of the ear." Yet if one were giving instructions to medical students, the language as well as the additional facts would make the former definition the better of the two.

Most instructions on how to write definitions warn against the use of highly technical words. This warning should often be heeded, but it is based on the assumption that the purpose of a definition is to give the reader his first introduction to the term defined, and that the reader is unfamiliar with the technical vocabulary of the field involved. Sometimes this is true, but not always. When a definition is written to enable a scientist to pin down the meaning of a term with extreme precision, it may well be expressed in different language than would be used to give a layman a general understanding of the term. There are times when the very reason that the definition is brought into existence is the need for an exactness that ordinary language does not possess.

Still another condition may exist. A definition may be written to establish a special meaning of an ordinary word that science has adopted because no single word in its vocabulary will serve the purpose. An example would be the use of *doping* to mean, "adding minute, carefully controlled amounts of impurities to growing crystals of silicon or germanium." It is sheer nonsense to object to scientific language in a definition that exists in order to tell the special scientific meaning of an ordinary term.

In general, the point to remember about the effect of purpose on a definition is this: in writing a definition, as in writing anything else, always ask yourself, "Who will my readers be, and how can I write so as to meet their needs?" In any kind of writing, it is bad to concentrate on substance alone and disregard the reader and the function of the written material.

General Suggestions

1. Do not base your definition of a term on any word that comes from the same root as the term itself. To define *permeability* as the quality of being *permeable* does no more than tell the reader that permeability is a quality. (This rule will be modified in rules 2 and 3 that follow.)

2. Always phrase a sentence definition to match the grammatical form of the term defined. (This requirement rules out such sloppy phraseology as "Osmosis is when . . ." Even if one were defining a noun that means a time, it would be better to use such phraseology as ". . . is the time when . . . ," thus avoiding the use of an adverbial clause when a noun is demanded by grammar.) Sometimes the need

for grammatical consistency delays the moment when a writer comes to grips with his real problem. For example, the noun *fertility* is derived from the adjective *fertile*, so the first statement one might make in defining *fertility* is that fertility is the quality of being fertile. The real problem of definition would then consist of defining the adjective *fertile*.

3. In defining a term that consists of more than one part, concentrate your attention at the point where attention is needed. For example, in defining *soil physics* your main problem would probably be to show how soil physics differs from physics in general. Or in defining *biochemistry*, the main problem might be to differentiate between biochemistry and other chemistry. Of course, if you felt it necessary, you might first define the basic word (*physics* or *chemistry*) and then go on to the discrimination indicated above.

4. Be sure that a definition includes everything that should be included. If bird were defined as "a warm-blooded animal that flies through the air," the definition would exclude the ostrich, which is a bird even though it does not fly. There is no reason, however, that you should not resort to a device constantly used in dictionaries and insert the word *usually*, so that you can utilize a helpful fact even though it may not apply to every individual thing that the definition includes.

5. Be sure that a definition excludes everything that should be excluded. (The foregoing definition of *bird* did not exclude bats and hence was inadequate.)

6. In writing a sentence definition try to avoid language with which the reader is unlikely to be familiar. The acceptability of whatever language is necessary for the sake of precision has been mentioned; but in a definition, as elsewhere, ordinary language is better than technical language if it will convey your ideas as effectively.

7. In an expanded definition, be sure to point out how the term you are defining differs in meaning from any other term with which it might be confused. For example, if you were defining *toxin* you would probably need to discriminate between *toxin* and the more general term *poison*, and also, perhaps, between *toxin* and the more limited term *venom*. A sentence definition might cover all the facts, but the expanded definition would be written so that all the facts would not only be covered, but would actually reach the consciousness of the reader.

8. Do not forget that you can sometimes make a definition easier to grasp and remember if you mention the root or roots from which the term is derived. For example, *isobar* might be fixed in a reader's mind by a comment on the fact that it derives from *iso*, meaning equal, and *baros*, meaning weight and in this instance referring to the barometer.

TECHNICAL DESCRIPTIONS

To understand how technical description differs from "literary" description, we need only imagine two descriptions of a room that has been ruined by fire. The literary description would enable us to imagine ourselves in the room—impressed by the fierceness of the fire that had wrought the damage, moved by the half-burned remains of personal belongings, startled to see the sky through a hole in the roof, conscious of the smell of wet, charred wood and the feeling of shattered plaster under foot. A technical description, however, would make no effort to create an imaginary experience nor to arouse our emotions. It would merely tell us, in an objective manner, the facts about the condition of floors, wall, ceiling, and contents—perhaps with the intention of enabling us to consider the need for repairs or to judge what might have been the cause of the fire.

A technical description is sometimes complete in itself, but it is more likely to appear as part of a longer paper. Such a description might deal with the damage done by an accident or a flood, or with the construction, facilities, and condition of a building. It might be written to help us judge the possibilities of remodeling or to decide whether safe working conditions prevailed.

A technical description might also be written to tell us about some newly developed machine, some device that had been used in an investigation, or some piece of equipment the purchase of which was contemplated. Our discussion will deal mainly with subjects of this latter type because such subjects will make the greatest demands on your skill as a writer.

The description of an object may concern the type of object in general or one particular example, but in either event, the basic method would be the same. Indeed, it is often best, in describing some device in general, to describe a single characteristic specimen and then mention the respects in which variation from the chosen example is most common.

The following technical description illustrates many of the points that will be made in this section:

The Champion Scrubber-Polisher

The Champion electric floor scrubber-polisher is a machine designed for domestic or light-duty janitorial service. It can scrub a floor by means of two circular brushes driven at high speed by an electric motor while a cleaning fluid is fed to the floor from a dispenser mounted on the handle. It can also be used as a polisher, in which case felt pads are attached and cover the surface of the brushes. It weighs only 18 pounds, is 12 by 6 inches in size at the base, and stands 38 inches high with its handle erect. It consists of five main parts: the brushes, the motor, the hood, the handle, and the dispenser.

The brushes, which turn in opposite directions, are 5 inches in diameter and are positioned side by side, separated by two inches at the inside. They have 1½-inch plastic bristles fastened to stiff plastic plates as wide as the brushes. In the center of each brush is a metal peg, short enough that it clears the floor by an inch, to which the felt pads are attached when the machine is used for polishing.

The motor is a conventional type that runs on ordinary 110- to 120-volt household current, draws 350 watts, and can be operated at two speeds. It is firmly riveted to an aluminum base and is positioned just above the brushes. It receives its current through an electric cord that runs down through the handle, which is hollow, through an opening near the top. The cord is long enough for the machine to be operated 15 feet from an electric outlet and can be wound onto brackets on the under side of the handle.

The motor and the top part of the brushes are covered by a rigid plastic hood that is 8 inches high and clears the floor by 1½ inches. Its horizontal dimensions are 5 by 8 inches where it covers the motor, but it widens to 6 by 12 inches where it covers the top of the brushes. A rubber strip runs around it where its dimensions are largest to protect walls and furniture.

The handle, which is hollow, has an outside diameter of 1 inch and is 34 inches long. At its bottom, two steel strips are riveted to it and extend through slots in the hood down to the base that supports the motor. They are fastened to this base by hinges so that the handle can stand erect for compact storage or can be held at a convenient angle, when the machine is in use, without tilting the brushes and motor. The lower 12-inch section of the handle is made of the same plastic used for the hood, and the rest is stainless steel. A stretch of 8 inches at the top is covered with rubber for the sake of insulation, and there is a slight bend 6 inches from the top so that the user can hold the handle at a natural angle. On the lower side of the handle, 8 inches from the top, is the switch that controls the motor. On the upper side, just above the bend, is a lever that controls the flow of the cleaning liquid.

The cleaning-liquid dispenser is mounted on the upper side of the handle. The reservoir containing the liquid is made of semi-rigid plastic, transparent enough to make the height of the contents easily visible. It holds 1 gallon, is graduated in 1-cup units, and has a capped opening at the top for filling. Its length, which extends along the handle, is 13½ inches. Its other dimensions are 2½ by 6½ at the top and 2½ by 4½ at the bottom. The bottom is supported and held steady by fitting tightly into a base 2 inches deep and made of the same rigid plastic used for the hood. The top is held firm by a cover of the same plastic, fastened to the handle by a ring that can be moved up and down to facilitate removing or attaching the reservoir proper.

Through the length of the reservoir runs a rod that opens and closes a valve and thus controls the flow of the cleaning liquid, which is carried down through a plastic tube to two openings, one near each of the two brushes. The rod itself is controlled by another rod running inside the handle to the lever mentioned above. (The use of two rods is necessary in order to permit removal of the reservoir.) The valve referred to is located in a small protuberance at the bottom of the reservoir that fits against a rubber seal in the base to prevent leakage.

The Champion scrubber-polisher is a light-duty machine, so its main usefulness is for small-area jobs such as in motel rooms, offices, or private homes. It also serves well as an auxiliary machine for use in restricted areas of larger buildings.

Description of the Object as a Whole

Though most of the information in a technical description is likely to concern some single part of the object described, the opening section deals with the object as a whole. In this section you should include any of the following materials that are needed: (1) a definition of the object to be described; (2) an explanation of the general manner in which it performs its functions; (3) a general description of the object; and (4) a list of its major component parts, each being preferably a part that performs some particular function. Logic dictates that the first and fourth items in this list be respectively at the beginning and end of the description, but the order of the second and third might often be reversed.

Your definition of the object, if you decide that one is needed, will be sufficient if it merely explains what the reader might be doubtful about. For example, consider the definition, "A demagnetizer is a device for removing the magnetism from hardened tool-steel parts that have been held on a magnetic chuck and thus have become permanently magnetized." The term *demagnetizer* has an obvious meaning, but the writer considered it necessary to indicate that a demagnetizer is used in a certain kind of work and for the purpose of demagnetizing a certain kind of object. His definition was adequate to the occasion and performed a useful function. On the other hand, it would be pointless to write such a definition as, "An electric brake is a brake that is caused to function by electricity."

Sometimes the reader will know without being told what an object is used for, and sometimes the use of the object will be indicated in the definition. There are times, however, when the use of an object will not be apparent unless you make a point of stating it. Such might be the case, for example, if you were writing about a permeameter, a vane borer, or an autoclave. It will always be part of your job, when you write a technical description, to pass judgment on whether the use of an object is obvious or whether it calls for comment, and to include such information if you feel that the reader needs it.

The explanation in the opening section about how a device works may not need to be extensive, but if your reader is to understand a detailed description, he must not be left entirely in the dark about how the object performs its function. He would need to know that an autoclave sterilizes or cooks by means of superheated steam under pressure; that a centrifuge separates different materials by means of centrifugal force; or that the air brakes of a train function when the air pressure between different cars is lowered or released so that the compressed air stored below each car can enter the cylinders and apply the brakes. Even a minimum of such information can make all that follows far more intelligible.

When a person becomes familiar with some object by inspecting it personally, he notices the object in general before he notices the details. Similarly, it is desirable in a description to give the reader an overall look at the object before presenting him with the information about its parts. He should be told early in the description about the shape and size of the object and also, perhaps, about its finish, its color, and the material of which it consists.

Obviously, however, this information cannot be given about some objects because they are not sufficiently visible. For example, no one could effectively describe the appearance of the hydraulic-braking system of an automobile; and in this instance the appearance is unimportant anyway.

When a reader understands what an object is, what it is used for, how it functions, and perhaps what it looks like, he is ready for a list of its functional parts. This list should not be excessively long. Rather, it should consist of the main functional units, each of which may be broken down into *its* parts, if necessary, when the time comes to discuss it. For example, you might divide an electric washing machine of the nonautomatic type into the tub and gyrator, the wringer, the motor, and the frame. The smaller parts comprising each of these units would not be mentioned until later.

The items in a list of the main parts of an object may be numbered if numbering them seems helpful, and should be arranged in the order in which they are to be discussed. One possible arrangement would be the order in which the parts would be noticed by a person who was looking at the object. A second would be the order in which they perform their respective functions. The latter is usually the best order when the parts function in succession rather than simultaneously. A third possibility might be the order of decreasing importance, for it might be desirable to tell about the important portions of the object at once rather than delaying them while trivial matters are presented.

Description of the Functional Parts

When the main parts or divisions of an object are described, the treatment of each main part is somewhat the same as the treatment of the entire object. That is, it is defined if definition is needed. Its purpose and the general method by which it accomplishes this purpose are made clear. Its general characteristics such as size, shape, and material are indicated. Its position in relation to the object as a whole and to the other main parts of the object is explained. And if it is complicated enough to merit such treatment, it is in turn broken down into its component parts, which are then discussed in the order in which they are listed. This continues until all the functional parts have been described in as much detail as seems necessary.

The Conclusion

Unless the object described is extremely simple, a brief conclusion is desirable, so that the reader's final impression is not limited to the last small detail. No single formula can be prescribed for the conclusion, but its general purpose will be to make the reader visualize the object as a whole, serving its purpose and functioning as it is intended to function. Often, if the object is one that performs some process, the best conclusion is to review one cycle of operation so that the reader can imagine the object in action.

Adaptation to the Reader and Occasion

The reader's knowledge and the demands of the occasion should affect both the contents and the language of a description. The fact that a description of some article is needed does not necessarily mean that all parts of the article must be described in equal detail. For example, electric motors are fairly well known and standardized, and if such a motor is part of a certain mechanism, it should be sufficient to tell the type, the power, and the kind of current it uses.

General Suggestions

1. A technical description, as we have seen, is intended to help the reader to understand an object as well as to picture its appearance. Thus it must include information about parts that may not be visible, and the emphasis is constantly directed to the manner in which the object functions.

2. In actual practice, a technical description is often accompanied by one or more figures. Instructions on the use of figures are given in Chapter 6. It is possible that some assignments in the writing of descriptions, which appear at the end of this chapter, may be postponed by the instructor until Chapter 6 has been studied.

3. It is often possible to make a description clearer by comparing an object with something that the reader is familiar with. For example, by saying that an object had the approximate shape of an electric light globe with the socket end down, you might make shape and position clearer than you could ever make them by direct description. Comparison to the shape of letters is especially likely to be helpful, as seen in the phrases *L-shaped, shaped like an H, somewhat resembling an inverted U,* or *having an S curve.*

4. In telling about size, try to avoid general words such as *large* and *small.* To illustrate their vagueness: a cupel, used in assaying, has been called a *small* cup; likewise a crucible, used in the steel industry, has been called a pot for melting a *small* amount of metal. Yet a cupel would hardly hold a teaspoonful of material, whereas a crucible might hold as much as 200 pounds of steel.

5. The positions of various parts of an object must be indicated with care unless the use of a figure makes the position of each part clear. Words such as *above, below, behind,* and *beside* must be used constantly. Information on position must include not only location but also information on such points as whether a cylinder is horizontal or vertical, or whether a hole in the center of a rod is parallel to the axis or across its diameter. Trivial facts should of course be omitted, but whatever is worth presenting at all is worth presenting clearly.

6. The preceding discussion has dealt with objects that function actively. A technical description that deals with such a subject as the characteristics of a building or the condition of a piece of equipment is simpler but not essentially different. First it covers the subject in general and then breaks the subject into its main divisions, continuing this treatment until it provides all the information necessary.

EXPLANATION OF A PROCESS

In discussing the explanation of a process we shall deal mostly with processes that are performed mainly by human action. Some of the suggestions will also apply to processes that occur in nature and processes performed mainly by machinery, but for reasons that will be obvious during the discussion, not all the techniques of explanation are usable for the two latter types.

First, it is necessary to differentiate between explaining a process and giving instructions for performing it. Instructions are written to enable those who may perform a process to perform it properly. The *explanation* of a process, on the contrary, is addressed to readers who may be unlikely to perform the process themselves but want to understand it so that they can judge its reliability, practicality, or efficiency. The manner in which certain tests were performed might be explained, for example, so the reader can judge whether the results were valid. Or again, a method of doing some job might be explained so the reader can judge whether it would be more efficient or less expensive than some other possible method. In such cases the typical reader is a supervisor or executive rather than the worker who performs the process, and the explanation appears in a report, article, or proposal rather than in an instruction manual. In contrast to a set of instructions, it is not worded in such a manner that it gives commands; it tells what actions are performed so that the reader may understand what is, was, or is to be done rather than so that he may do anything himself.

When you explain a process, you will usually find it desirable to follow a fairly well standardized pattern—one that has become standard not because of arbitrary considerations but because it leads to explanations that are easy to understand. This pattern calls for an introduction, an overall picture including a list of the main steps that make up the process, an explanation of each of the steps listed, and a conclusion if one would be helpful.

The Introduction

The introductions should be limited to information that the reader really needs, and it is unlikely that on any single occasion he will need information on all the points to be mentioned. This qualification should be held in mind as you read the following suggestions. Also, if any of the introductory information applies only to one part of the process, you should feel free to wait until you come to that part before presenting it.

Sometimes the introduction opens with a definition of the process. It indicates why, where, when, and by whom the process is performed. It includes information on the materials, tools, and apparatus needed in performing the process; and if any important apparatus or materials are likely to be unfamiliar to the reader, it identifies them for his benefit. It tells, if necessary, whether those who perform the process need special skills or training. Sometimes, too, it mentions special requirements about the time when the process must be performed, or special conditions that must exist, such as temperature, humidity, freedom from dust, or ventilation. And finally, it sometimes tells about preparations that must be made before the process is performed.

The Overall Picture

The introduction is followed by any comment necessary about the process as a whole—for example, the theory on which it is based—and by a list of the main steps that make up the process. An effort should be made to hold the main steps to five or six, for if they are too numerous it will be extremely hard for the reader to grasp and retain an overall picture. The main difference between the treatment of a long, complicated process and a simple one is that the former is subdivided more extensively, not that it includes more main divisions.

The main divisions of the process should be named in chronological order so far as possible. They should be expressed in parallel form and may be numbered if numbering seems likely to be helpful, as it is almost sure to be if the number exceeds three.

Each main division should be based on the completion of a stage of the work rather than on some arbitrary consideration such as place or time. It would be undesirable, for example, to divide a process into work done in the field and work done in the laboratory. Even if the work done in the field comprised one specific task and the work done in the laboratory comprised another, you would give your reader a better picture of the process by identifying that task rather than merely identifying the place where it was done.

The Explanation of Successive Steps

After listing the main steps, you should next take up these steps one by one and treat each of them somewhat as you treated the process as a whole. For each division, a definition is provided if needed, and the facts about time, conditions, apparatus, personnel, and preparations are made clear. Then, if necessary, the parts of which the step consists should be listed.

When nothing would be gained by further subdivision, you can begin to tell what is really done when the process is performed. In telling what is done, you should emphasize the results that the acts performed are supposed to accomplish rather than the actions themselves. The effect of this emphasis may be seen by comparing the following examples:

> First, the form is filled with mix and then the mix is tamped with a ½-inch rod 25 times, which leaves room for more mix. Again, and finally a third time, the process is repeated, the result being that all samples are compacted to the same degree.
>
> In order to compact all the samples to the same degree, each is tamped in the same manner. First, the form is filled with mix, and then the mix is tamped. . . .

The important fact here was that the samples had to be uniformly compacted. The tamping was performed in a certain manner only because it produced that result. Yet in the first example, the result aimed at seemed added as an afterthought, and the reader was expected to follow a series of actions the purpose of which had not been indicated. The difference between the two methods may seem small when observed in a brief specimen, but in a longer paper it can mean the difference between understanding and confusion. As a reader reads about an action, he should know what that action is intended to accomplish; and when several successive actions are all performed in order to accomplish a single result, that result should be indicated at the beginning of the passage. Otherwise, the reader will soon be lost in a maze of details.

As your explanation progresses from step to step, you should keep the reader aware of his progress. When you take up a new step you should point out that you are doing so, using the phraseology and the numbers, if you used any, that you employed when you originally listed the steps. Special care is necessary when the steps do not follow one another in a regular, chronological order, as when two steps are performed at the same time by different people. By careful use of transitional material, however, you should be able to keep the reader oriented and constantly aware of the process as a whole and of where he stands at any given moment in his reading.

In the explanation of a process, as in any other writing, you will have to decide how much detail is necessary. There is no reason that all parts of a process must be covered in equal detail. Suppose that one step of a process consists of some test which the reader already understands or which is a recognized, standard procedure. It would probably be sufficient merely to say that this test is performed. Yet some other portion of the process might involve unusual procedures that would call for detailed explanation. Thus, in regard to each portion of the process as well as in regard to the process as a whole, it is part of your job as writer to include the material that serves a purpose and to omit what can be omitted without loss.

The Conclusion

There are many occasions when the explanation of a process does not need a conclusion—especially when the process was performed on some specific occasion and the explanation is part of a longer paper. When you feel that a conclusion is desirable, however, you need not hesitate to add one. Your conclusions might summarize the process, perhaps restating the main steps so that the reader's final impression will include the process as a whole rather than only one small part. It might evaluate the process or the results of the process. It might comment on why the process is important or indicate how it fits into some larger process of which it is a part. But unless any of this material fits the occasion, your explanation should probably end with the final remarks about the last action performed.

Explaining a Process Performed by Machinery

When a process is performed mainly by machinery that functions to some extent automatically, explaining it involves telling how the machinery functions. This creates a problem very much like the problem of writing a technical description. The difference is mainly in the facts which are emphasized and about which the details are numerous. The technical description emphasizes the facts about the machine and tells, incidentally, how the machine operates. The explanation of a process emphasizes what the machine does as it operates, and makes the actual description of the machine incidental.

One possibility of handling the job is to tell first what actually is done by the machinery, organizing the information if possible on the basis of stages of the work as we would do if the work were performed by human beings. The next step is to tell what it is necessary to know— no less and no more—about the machinery that does the job—what it is like and how it functions.

Sometimes this treatment can be varied. If what is said about the machinery falls naturally into divisions that are basically the same as the main divisions of the process—that is, if each separate and distinct part of the process is performed by a separate and distinct part of the machinery—it may be advisable to open with an overall picture of the process and then combine the discussion of each part of the process and the necessary information about the machinery that performs that part. Discussion of human action can be limited to information about how many people are involved and what abilities they need.

One might summarize by recommending the following general method. First, use an introduction consisting, so far as it would be appropriate, of the same kind of information you would use in explaining a process performed by human action. Next, give an overall picture of the process by telling its main divisions. Then tell what is needed about each of the divisions, placing emphasis in descending order on *what* is done, *how* it is done, and the kind of machinery that does it.

Some of the processes that this technique would clearly apply to are:

Feeding cattle by automated machinery
The manufacture of paper
Crushing ore
The extraction of minerals by the flotation process
Manufacture of steel by the oxygen process
Harvesting by means of a combine
Cracking petroleum
Extinguishing a fire in an oil well
Killing brush with a machine that applies chemicals
Treating water with various kinds of water-softening equipment

Explanation of a Natural Process

The introductory information for the explanation of a natural process obviously cannot include some kinds of information used to introduce a process performed by human action, but it might call for the following: a definition which may either tell the meaning of the term that identifies the process or clarify the exact nature of the process; a brief statement of the scientific facts, principles, or theories that cause the process to occur; and possibly the conditions under which the process occurs.

This introduction may best be followed by an overall picture of the process indicating the stages in which the process occurs. If possible, the stages should be so presented that there will be no more than five or six main divisions, just as is the case with processes of other types; but such a limitation may be prevented by the facts.

When the general, overall picture of the process has been completed, each stage should be explained, in chronological order. The explanation of a stage would consist merely of telling what happens, and, if possible, why it happens, unless the reason is common knowledge among the readers addressed or has been covered in the introduction. Obviously, the insistence on emphasizing purpose and stressing the point that actions are merely a means to an end does not apply, but we can apply the principle that single incidents are significant mainly because they show how major developments occur. That is, a natural process in which twenty or thirty happenings occur need not be treated as twenty or thirty distinct happenings. Rather, the happenings can be grouped together as they combine to produce the major developments, so that we conceive the process as occurring in a reasonable number of stages.

The explanation of a natural process may or may not call for a conclusion. The suitable kinds of conclusion materials might be a summary of what has preceded, information about the significance or importance of the process, or both. In the last analysis, however, the use or omission of a conclusion should probably depend upon whether the explanation of the process is an independent piece of work which for psychological reasons should not end abruptly, or whether it is a part of some larger piece of writing and leads naturally into whatever follows.

Following is an example of an explanation of a process performed by human action.

Field Care of Deer

A hunter who is successful in killing a deer can look forward to having excellent meat for the table if he cares for his animal properly in the field. Unfortunately, however, substantial amounts of meat are lost because of improper care. To avoid such loss experienced hunters have developed a process of field care consisting of four steps: cutting the deer open, cleaning it, skinning it, and cooling it. They do the job as soon as possible after the kill, needing nothing more than a sharp knife with a strong blade at least four inches long and a large handle, a piece of rope a few feet long, and a sheet or some cheesecloth for wrapping the carcass.

To cut the deer open the hunter first turns it on its back and places large rocks or logs on each side under the shoulders and hips to hold it in position. Then, if the deer is a buck, he cuts out its sex organs, after which he cuts downward between the hind legs all the way to the pelvic bone. Next, he turns the knife over and makes a straight cut up the middle all the way to the breast bone. As he does so, he uses his free hand to hold the skin and flesh back from the entrails. Taking care not to let the knife turn in his hand, he goes on to cut through the breast bone and up the underside of the neck as far as possible. Finally, he cuts the windpipe in two as close to the head as possible.

To clean out the inside of the deer the hunter starts by grasping the windpipe with both hands and pulling hard toward the hind quarters. This causes the insides of the deer to come out all the way down to its midsection. Then he removes the rocks or logs that had held the body on its back, turns it on its side, and cuts all the way to the backbone through the thin layer of meat that holds the entrails to the ribs. Thereafter he turns the deer on the other side and repeats the process. This done, he removes the entrails by merely taking hold of them with both hands and pulling hard.

Before the cleaning can be completed, another cut is necessary, so he turns the deer on its back again, partially raises it by its hind legs, and shoves a log or rock under its rump to spread its hind legs apart. Then, with his knife, he locates the seam in the pelvis where the bones grow together and splits the pelvis apart by pressing his knife down hard or, if necessary, hitting it with a rock. This makes it possible to do any further cleaning that is necessary.

If no tree is handy the deer is next turned with its under side down in a clean place and left for twenty minutes so that the blood in the body cavity can drain out. If a tree is available, however, the hunter hangs the deer to a branch by its head or antlers and can begin skinning it while it is still draining. In any event, he skins it within two hours of the time of the kill, if possible, because the skin comes off more easily if the carcass is still warm.

The first step in skinning is to make a cut extending for the full length of each leg, on the inside. Having done this the hunter next cuts the skin all the way around the neck, starting underneath at the end of the cut made earlier on the deer's under side. Next he grasps the skin with both hands at the back of the neck and by pulling hard pulls the skin loose as far as possible—usually down to the front legs. He then uses his knife when necessary to work the skin off the front legs. When the front legs are skinned he can then pull the skin off from the rest of the carcass, loosening it, if necessary, with his knife.

Cooling the carcass is simple but decidedly important if spoilage is to be prevented. The hunter takes the carcass back to his camp and hangs it to a tree with a sheet or piece of cheese-cloth around it. He hangs it by its hind legs so that the blood from the tiny vessels can drain out by moving in the direction where the heart had been located. After it has cooled for four or five hours the carcass is ready to be cut up and stored, and will provide fine meat for the table.

INSTRUCTIONS

A piece of writing giving instructions has much in common with the explanation of a process, but it is addressed to a different reader and serves a different purpose. The reader is someone who may be expected to perform the process, and the purpose is to enable him to perform it properly rather than just to understand it. In one sense of the word, instructions seem to mean the same thing as orders; but there is a difference between ordering a person to do something and telling him *how* to do something. The two possibilities are not mutually exclusive, and both may be called for on some occasions. In our context, however, giving instructions means giving information rather than giving orders, even though such information may sometimes be expressed in the imperative form.

In technical writing giving instructions is usually a matter of telling how to perform some physical process such as assembling, using, inspecting, or repairing equipment, performing a test, or doing some job in a laboratory. Such instructions are given by manufacturers to dealers who sell and service their products and to consumers who use them; they comprise the bulk of the contents of laboratory manuals; they are produced, in the form of instruction manuals, as a result of enormous labor, by the multimillion-dollar industries that serve the armed forces. But in spite of the variation in complexity, that these extremes represent, it is possible to offer a good many general suggestions that should be helpful.

Introductory Material

The same kind of introductory material used in explaining a process may be called for in giving a reader instructions for performing a process. That is, the term used to identify the process *may* call for definition. The purpose of the process may need to be stated. Other useful information might concern equipment needed, conditions that must exist, skills necessary, and preparations to be made before the job itself is begun. It would be wrong to assume, however, that every item in the list is always to be covered. On the contrary, the introductory information should be limited to what will be useful on the particular occasion.

How Much to Include

When the introduction is complete and the actual instructions begin, one question that always arises is what to include and what to omit. For example, should purposes be explained or should the contents be limited to the actions called for? It is certainly true that instructions are grasped more easily, remembered better, and carried out more intelligently if the reader understands the reason that he is told to do things in a certain manner. But such explanations are a means to an end and should be kept distinct from the instructions proper so that the latter will stand out sharply and be easy to identify. This result can often be accomplished by a simple change in manner, as shown by the following examples.

. . . To charge the expense of a machine against production for a short period would be extremely complicated. Therefore an hourly rental rate period has been set up for each type of machine. It is to be applied as follows:

1. Keep an accurate record of the time a machine is used on each job that has a separate account number.
2. Charge the job account with the amount arrived at by applying the rental rate.
3. Charge the time lost for minor repairs and service to the job on which the machine is working.
4. Charge major repairs or overhaul against operating accounts 700 to 799.

or

1. An accurate record shall be kept of the time the machine has been used on each job that has a separate account number.
2. The job account shall be charged with the amount arrived at by applying the rental rate.
3. Minor repairs and service shall be charged . . . etc.

Use of Numbers and Other Mechanical Devices

Numbering the items is almost always useful in giving instructions —though it is no substitute for careful, systematic arrangement and organization. Sometimes a single sequence of numbers is adequate, but if the subject matter is at all complicated it may call for a system of numbers and letters comparable to the system used in an outline. That is, a major order (A. Assemble the blower unit) might be followed by a numbered list of specific acts involved in the assembling. The basis for treating several actions as a single larger unit should practically always be the fact that they combine to produce a single result, or that they concern a common problem.

In addition to numbers, other mechanical devices are useful in giving instructions more frequently than in most kinds of writing. Figures (in the sense of illustrations) may be almost essential. Other possibilities are extra-wide margins, parentheses, explanatory notes—either footnotes or notes labeled as such and inserted in the text—and variation in the style of lettering (different type faces in printed matter, and the use of capital letters and underlines in typewritten copy).

These devices are useful mainly because they permit the writer to use inserts that clarify the meaning of terms and give warnings, yet still make the instructions proper stand out so that they cannot be missed. It is possible, of course, to overdo these devices, and thus produce a cluttered-looking paper; but judiciously used they make it easier for the reader to focus his attention mainly on the parts that tell the essential action and to recognize the other material for what it is.

Special Elements

In writing instructions it is often necessary to emphasize something that is not part of the instructions proper. For example, in giving instructions on the use of an insecticide it may be necessary to warn against letting it get onto the skin or into the eyes. In giving instructions about using a poisonous substance it may be necessary to tell what antidote is most effective. Such material should be placed under conspicuous headings and inserted wherever it applies. These special headings not only secure for it the attention it deserves, but also make the instructions proper easier to follow.

Less vital than warnings against damage or danger are special instructions that apply in exceptional circumstances. It might be necessary, perhaps, to tell about a special procedure to use in extremely cold weather, or what substitute can be used if material that is normally used is unavailable. Here again the main consideration is to make the information easily available without letting the instructions proper become so cluttered with *ifs* and *when's* that it is needlessly hard to discover what to do under normal circumstances. The exceptional circumstance can sometimes be provided for by the simple expedient of a parenthetical insertion at the appropriate place. Much of the time, however, the best method is to present the instructions that apply under normal circumstances without interruption and then cope with the variations by adding what is necessary, suitably headed or introduced. A suitable introduction might be somewhat as follows: "Though the preceding instructions can ordinarily be followed as they have been given, it may be necessary, under special circumstances, to modify them as follows."

Style and Tone

Because anyone who writes instructions tries to make them simple and easy to follow, and because this involves breaking down whatever is complex into its simplest elements, the sentences used in instructions are likely to be extremely short. And since a person who reads instructions expects them to consist of a series of separate actions, preservation of continuity is less important than in most kinds of writing. This does not mean that gaps in thought are permissible, but merely that connective words and phrases showing thought relationship may be used sparingly, because the relationship among the items is already understood to be that of separate elements in a series.

In keeping your sentences short and simple, you should not ordinarily try to save words by departing from normal language. Unless space is artificially limited, as, for example, on a package, a cookbook style is not desirable. (A characteristic example of the cookbook style is the sentence, "Place 8 ounces water and three pellets in jar and shake until dissolved.") Not enough space is ordinarily saved by the omission of a few words to compensate for the loss of naturalness and readability.

As for tone, sometimes, as in a maintenance manual or a laboratory manual, it calls for little consideration because the situation is entirely impersonal. There are times, however, when a pleasant tone is an asset. If you compare the booklet that an automobile manufacturer prepares for a purchaser with the repair manual prepared for mechanics, you will see that the automobile industry recognizes the desirability of addressing the user of its products in a manner that develops good will. Often, instructions addressed to subordinates also will get better results if they are given pleasantly and courteously with a view to securing willing co-operation. Most of the effort to create a pleasant tone is expended in introductory and explanatory passages rather than in the instructions proper, which can thus be kept clear of everything except the bare minimum necessary to cover the action.

Conclusion

Writing instructions that will be clear to the reader involves, first of all, asking yourself who the reader will be, what he knows already, how much detail he must be given about the action, and how much explanation about reasons he must be given in order to persuade him to do things as they are supposed to be done. You will need to use your imagination in order to foresee the possibilities of errors and misunderstandings, and to provide for every likely contingency. You will need to use good judgment in deciding how simple your overall organization can be. You may usually feel free to sacrifice some of the niceties of style (for example, primer style is often effective) when clarity demands such a sacrifice. And you should often use illustrations and mechanical devices such as headings, notes, and variations in lettering to help your reader see how everything he reads is related to what he is expected to do. If you handle your instructions in this way—and, of course, if you have reasonable skill in writing—you will probably find that your readers can understand your instructions well enough to comply with them.

The following specimen shows how the recommendations for writing instructions work out in practice.

Etching Concrete Substrates

As an alternative to sandblasting, portland cement concrete substrates may be prepared for priming by etching them with muriatic acid to remove the powdery residue on the surface. If such treatment is used, the surface must be in proper condition. Plaster, mortar, loose aggregate, and other such materials must be removed. Grease must be washed off and paint must be removed by any method that will get rid of it. When the surface has been prepared, the etching shall be done as follows:

1. Add one volume of 20° Baume muriatic (hydrochloric) acid to two volumes of water.
2. Pour this solution onto the surface from a plastic watering can, using one gallon of solution for 50 to 100 square feet of surface. (The amount necessary varies with the condition of the surface.) Work the acid into the surface with a stiff-bristle broom to make sure that no part of it remains dry.
3. Brush the entire surface vigorously.
4. Remove the spent acid from the concrete by flushing the surface with water and going over it with a squeegee or by using a commercial vacuum cleaner. (WARNING: If a vacuum cleaner is used, make sure that it is equipped with an acid-resistant tank.)
5. Rewet the entire area with fresh water and brush it vigorously.
6. Remove the excess water as you did before—by use of a squeegee or commercial vacuum cleaner.
7. Examine the surface to see whether the results of the etch are satisfactory. (The surface should have a finely pebbled texture.)

If the surface does not appear pebbled, repeat steps 2 through 6. Otherwise, it is ready for application of the primer.

ANALYSIS

Analysis, as viewed in the context of technical writing, consists of the process of examining something to distinguish its component parts or elements either separately or in their relation to the whole. Situations in which it is necessary to analyze something and present the results in writing are likely to confront anyone in a position where technical writing is called for. It would be impossible to name everything that those who use this book will be called upon to analyze, but the list would include conditions, causes, results, trends, and problems of every kind.

The nature of a written analysis can be made clearer by recalling some of the analyses that all of us encounter in our normal reading. We read analyses of business conditions or the conditions of some particular industry; analyses of the causes of increase or decrease in production; analyses of the manner in which funds have been distributed or the areas in which expenses have increased; analyses of the results of new technology or legislation; analyses of the shift of industry from one region to another; analyses of problems such as the detection or prevention of defects in materials or the design of better products.

For a concrete example, let us see what was involved when an analysis was made of the suggestion that colleges introduce new programs designed to prepare students to do technical writing as a full-time occupation. Some of the questions that arose were as follows:

What existing courses would the program include?

Could these courses accommodate the increased enrollment?

If new courses were needed, would qualified instructors and acceptable textbooks be available?

Would any existing department be prepared to administer the program?

If not, would a new department of Technical Writing offer enough work of its own to justify departmental status?

Would students enter such a program if it existed?

If so, from what existing programs would they be diverted?

If students completed the program, would they enter the vocation it had trained them to enter?

Would the program assist industry if those who entered it were diverted from other fields where there is as acute a shortage of manpower?

This list could be expanded, but it is long enough to show that analysis of the suggestion demanded consideration of the content, staff, administrative organization, students, value, and expense of the program.

In short, analysis showed the numerous points into which an apparently simple suggestion had to be broken down before it was possible to see all that was involved.

In actual practice, of course, many an analysis can follow a pattern established in previous analyses of the same or similar matters. But when there is no established pattern to rely on, one must often use considerable imagination in order to discover what facts might be significant. An analysis of the unemployment situation, for example, would obviously include the total number and percentage of employable persons out of work, but would also need to tell such facts as the geographical distribution of the unemployed; their distribution in towns and cities of different sizes; their distribution as to age, sex, and race; their education and their training in vocational skills; the number in various brackets based on length of time since last employment; the changes since previous figures were released; the effect of the season or of special causes (strikes, for example) on increase or decrease; and the specific industries with which the unemployed might be connected. All these facts and of course many more might affect the validity of any generalizations upon such points as the seriousness, the causes, and the possible means of improving the situation.

Unemployment is of course a vast and complicated subject, and an analysis of it on the scale described above would probably be produced in order to make available all the facts anyone might need for his own special purposes. Most of the time, however, the person who writes an analysis, like the person who writes anything else, should not exhaust the subject but should bear in mind the specific function his product is designed to perform. He should take into consideration the question of who his readers will be and why they need an analysis of the subject. The entire approach to the job of analyzing, the organization, and the amount of detail and emphasis on various aspects of the subject can be determined more intelligently if the reader and purpose as well as the subject receive constant attention.

This brings us to the question of how one goes about the task of writing an analysis. The following method is suggested.

First, concentrate on the question of who will read the finished product and why. Will it be read by those who need information only so as to understand the subject, or will it form a basis for decisions and action? To illustrate the latter possibility—a reader might want an analysis of competing products in order to decide which one to buy; he might want an analysis of expenses so as to decide upon the areas in which to economize; he might want an analysis of the probable side effects of a change in the design of some product in order to decide whether these would offset the benefits; he might want an analysis of the demand for heat and power that would result from some enlargement of operations so as to know whether existing sources would be sufficient.

Next, bearing the function of the analysis firmly in mind, begin to list the points, large and small, on which information will be needed; and as information becomes available, decide upon an appropriate pattern for presenting it. Perhaps the key question in organization is whether to weave together the facts on different aspects of the subject or whether to keep each aspect separate from the others. In an analysis of a record of accidents, for example, facts might be presented on such different matters as the number of accidents in different kinds of jobs, the number in different shifts, and the relative number for workers with different amounts of experience. It would be possible to create a pattern in which these matters comprised respectively, main points, subpoints, and sub-subpoints; but it would also be possible to handle each of them separately.

When an analysis is made for what might be called random use rather than for specific application, the general method suggested for making an outline (see Chapter 3) would be hard to improve upon as a method of starting the job. This consists basically of jotting down rough notes and creating a plan by grouping together related material. When an analysis is made for a specific application—that is, because it is needed for a predetermined purpose—the writer's understanding of what is needed permits him to narrow down his approach and concentrate attention on what is significant in view of the particular occasion. For example, an analysis of the quality of some curriculum in a specific university might be different if made for the general purpose of accreditation than if made by the personnel department of a corporation as a basis for recruitment.

The actual writing of an analysis poses no unusual problems. A brief introduction is usually sufficient to cover, so far as circumstances demand, such points as the purpose for which the analysis has been made, the basis on which the major divisions rest and perhaps the reasons for their choice, the source of facts included and the scope and limitations of whatever investigation was made, and the assumptions, if any, on which the interpretation of the facts is based. To these points may be added individual facts of special importance and general conclusions that are specially significant.

When the bulk of the material in the body of an analysis consists merely of objective facts for their own sake, the main problem in presenting it is the avoidance of monotony. Unless a writer has a natural gift for interesting style, there is not much that can be done about this except to break the prose with informal tables—which would probably be desirable for the sake of clearness even if monotony were not involved.

When the objective facts call for interpretation, the main consideration is to make the relationship of facts to each other entirely clear and to explain the processes of reasoning involved in deciding upon their significance.

An analysis may or may not call for a general conclusion. Sometimes, merely breaking down the total subject into its component parts is all that is necessary. Yet, as mentioned earlier, an analysis indicates parts of something either separately or in their relationship to the whole; and when their relationship to the whole is important—when the analysis is made to provide a basis for decisions and action—a recapitulation of the major facts, a reenforcement of whatever has been said about their significance, and an application of the analysis to the unsettled issues becomes a practical necessity.

Many people who write analyses do not think of them as being a distinct type of writing, but most of us can obtain better results if we recognize what we are doing. And skill in writing an analysis is well worth while because an analysis may not only exist independently but is often an important part of a report, a proposal, an article, or any other kind of technical writing.

Following is an analysis of the amount of electrical power used for various purposes.

> The growing demand for electricity presents an extremely complicated problem—a problem that is constantly increasing because the demand for electric power has been doubling each ten years. The solution will not be found by urging reduction in its use for so-called frivolous and unnecessary purposes such as, for example, the electric toothbrush. This becomes obvious when the actual amount of power used for various purposes is considered.

> In the area served by one power company, for example, 58 percent of the electrical power generated is used not in homes but by such customers as factories, aluminum plants, agriculture, business establishments, and governmental services—street lights included. An excessive cutback in this use would not only deprive the public of needed services but would also wreak havoc on the economy of the region by damaging industry and creating unemployment.

> The rest of the power, 42 percent, goes to residential users. This amount would not be substantially reduced even if all the "frivolous" uses—toothbrushes, can openers, toasters, and the like—were eliminated. The fact is that these "gadgets" account for only about three percent of the power used in residences—about 1¼ per-

cent of the total for all purposes. Practically all the electricity used in residences goes to perform such heavy-duty jobs as heating, cooking, and heating water.

The problem has another aspect that may disturb those who oppose new efforts to build new power plants because most kinds of power production create pollution and environmental problems. The fact is that increased power production is essential to pollution control itself.

It is generally recognized, for example, that thousands of sewage-treatment plants are needed in order to reduce water pollution. A secondary sewage-treatment plant serving a population of 100,000 requires over 500,000 kilowatt-hours of electricity a year. Also, recycling plants to handle paper, plastics, or aluminum cans require large amounts of power—as do plants that compress or cut up old automobile bodies. The same is true of mass-transit arrangements, which are so strongly urged to reduce air pollution and the consumption of gasoline. When the Bay Area Rapid Transit System in the San Francisco region is in full operation, it will require about the amount of power used at present by 43,000 residences.

(The analysis from which this was adapted goes on to discuss methods suggested for increasing power production.)

TECHNICAL ARTICLES

Though a student may think it unlikely that he will ever engage in writing articles, the chances that he will do so—or at least might profit from doing so—are greater than might be imagined. Hundreds of periodicals dealing with technical and scientific subject matter roll steadily from the presses, and the students of today will eventually be the authors who fill their columns.

To be sure, articles will not be written just to keep the periodicals in existence. On the contrary, the articles must be written because in every aspect of science and technology there is a great deal of information to be disseminated, and the periodicals are needed as a place to publish the articles written to disseminate it. To cite a single example: Medical journals exist because without them medical doctors would not have access to the articles that keep them abreast of the latest developments in their profession.

Scope and Importance

The term *technical article* is, of course, an oversimplification. Articles vary in the extent to which they are technical because the publications that print them perform different functions and are intended for different kinds of readers. At one extreme are the highly specialized scientific journals such as those dealing with nuclear physics or higher mathematics. Articles in these journals concern matters that are far beyond the grasp of any readers except those who have done concentrated study in the fields concerned. Not quite in this rarefied atmosphere are other publications whose readers must still be able to comprehend formulas, mathematical calculations, and technical terminology that call for extensive knowledge of the field concerned as well as general grounding in science and technology.

Somewhat farther down the scale of technical and scientific specialization are periodicals containing articles in numerous technical or scientific fields rather than in only one. Also, articles on such subjects appear in many periodicals that are not limited to science and technology in their interests. The widespread public interest in such matters as pollution, ecology, recurring power shortages, and long-range energy problems results in the presence, in these high-quality publications, of a steady stream of articles by authoritative authors, written for intelligent, well-educated readers who have no more reason for reading than a natural interest in the world that they live in. The publications carrying such articles include even the better newspapers; and a newspaper article by such an author as the science editor of a worldwide press service is entitled to respect. Some of these articles in periodicals with general circulation may contain generally accepted facts; some may attempt to give a balanced picture of a controversial subject; some may take one or the other side of a disputed question; but all of them are the work of well-informed authors, some of whom are recognized authorities.

Not entitled to inclusion as technical articles are the purely popular articles that also appear in great numbers. These are a different breed. Whereas the technical article is written by an author with special qualifications whose purpose is to inform, the popular article is likely to be the work of an author whose main occupation is writing and who may have neither an enduring interest nor a claim to expertise in the field where his article lies. His work is read not so much for information as for its entertainment value. It represents an attempt to be highly colorful and may be inclined toward sensationalism. To be sure, it may serve its purpose; but it is less likely than a bona fide technical article to be organized with care and written with enough objectivity and restraint to bear up under the scrutiny of well-informed readers.

The publications mentioned thus far are numerous, and the writers who will eventually fill their pages will include students who are studying this book and others in its field. An even greater number of today's students will contribute, however, to publications that fall into none of the preceding categories. Rather, they will write for periodicals concerned with specific industries or technologies. Some of these, in fact, are published by one or another of the nation's major corporations and deal almost entirely with its own activities.

The number of these industrial periodicals is immense. Every industry, from the automotive to the beverage-dispensing, has one or more. Their importance would be hard to overstate. Though reports are perhaps the main source of information in industry concerning matters that fall in the area of a professional worker's personal responsibility, it is articles in industrial publications that on a broad scale keep him abreast of general developments, perhaps in his own company and certainly in the industry where he is employed. And it is in such publications that those who use this book will probably find an outlet for the articles that they are eventually in a position to write.

Finding Subjects for Articles

Finding subjects to write about will not be a problem for a person holding a professional position rather than preparing to hold one. When a person holds such a position, subjects grow naturally out of the work he is doing. That is, he will write an article because he has developed some information or ideas that are worth sharing, rather than hunting a subject in order that he may write an article. He may describe some newly designed machine or process, for example, or a modification or improvement of one that is already in use. He may explain some new technique or evaluate alternative techniques. He may tell about some method to solve a chronic problem that has long plagued his industry. He may describe the manner in which some project has been carried to completion—the difficulties encountered and how they were overcome. Examples of these and other types of subjects can be found by looking over industrial periodicals in any extensive library.

Often, the subject of an article develops by stages. Work done in the line of research, development, or production will first result in a report. Then a key figure in this work may have an opportunity to read a paper about it at a convention or conference. Finally, he may work over this paper for publication. All in all, the time is likely to come when a person in a professional position will have a ready-made subject for an article if he is alert enough to recognize it.

For a student, the problem of finding a subject is obviously different. Instead of writing an article because he has the subject matter for one, he may need to look for a subject because he is asked to write an article. It should assist you in this quest if you bear in mind that most published articles are of one of four types: (1) the article that concerns a genuinely new subject, (2) the article that presents new information or a new attitude about a subject which has been written about before, (3) the article that concerns a subject made timely by some new trend or event, and (4) the article that is written to instill a certain conviction about a particular question.

To find a subject for an article about either of the first two types a student might start by considering his own school. If it is one where research is carried out, he may be able to learn enough from an instructor to write about one of the research projects. Or again, he might explain some new course, program, or curriculum. He might tell about unusual features of some new building, or the construction of a roof over a stadium, or the means by which a certain kind of artificial turf for an athletic field can be rolled up and stored between seasons.

Looking beyond the school, he may find a subject in the community where it is located, for in almost any community improvement projects are usually being carried out or studied. Occasionally, too, a subject may be found that involves both school and community—as, for example, student participation in a survey of the attitudes of residents in an area toward a land-use plan.

Articles of the third type of subject—the type dealing with a matter brought into attention by some recent event—are also a good possibility. For example, in a year when tornadoes have been especially frequent, even a writer who is not an authority on them can acquire enough information, old and new, to use them as the subject of an article. Most of the information in such articles may not be new, but a minimum of new information may provide an up-to-date flavor. Not to be overlooked are the possibilities of writing articles made timely by the passage of legislation in Washington or by the announcement of a decision by some regulatory board. On a more local level, state legislation or regulations can also provide subject matter. Accordingly, when you are looking for a subject, you can do worse than to give some attention to recent news developments. There is always a chance that they may revive interest in a subject that has been dormant.

The fourth type of article, one that aims to support some particular conviction, may deal either with a new subject or an old one. If your subject is old, however, your idea for an article about it needs to be evaluated with considerable care. Sometimes, to be sure, though your subject may not be new your conviction about it may be new or perhaps unorthodox—in which case you are fortunate. But if your conviction is no longer disputed, the chances that you can write a worthwhile article about it are dim. If your conviction is one that many other people share but that is still controversial, you will need to have something to add to arguments that have been previously advanced. This last statement should perhaps be modified. If the particular periodical that you have

in mind as an outlet for your article has not previously published anything of the sort, yours might meet a need. But even in this event you
should try hard to find something fresh to say rather than just offering a
rehash of what others have said already.

All in all, perhaps the best way to find a subject of the type in
question is to ask yourself, "Do I agree with every attitude that is conventional in my intended profession? Do I have some ideas of my own?"
Many a person has within himself the germ of an article if he would
realize that it lies in the unorthodox remarks he makes in private conversations.

The extent to which a student can make his article technical in
content obviously depends on how far he has progressed in a professional
curriculum. There is no reason, however, that a student who has not yet
taken any highly technical courses should be unable to gain experience
and skill by writing on a subject within his scope. Moreover, he can gain
such skill without necessarily thinking in terms of an article for publication in a periodical with nationwide circulation. Many publications cir-
culate almost entirely within a single state or region, and it is probably
most practical to think of some such periodical as a potential outlet.

Adapting the Article to the Periodical

As the preceding paragraph implies, when you write an article you
should have in mind at least one periodical where it might be published.
This is sound advice even when you write an article as a classroom assignment. After all, when you write an article during your future career
you will certainly have publication in mind, so you should consider now,
so far as possible, the matters that you will have to consider then. And
any experienced writer will advise against writing an article and then
trying to figure out who might publish it. Rather, you should give some
thought to what periodical or periodicals might be interested while the
article is still taking shape in your mind. Having reached a decision, you
should then adapt your product to the one you have singled out.

Both of these steps—choosing a potential outlet and adapting your
article to it—call for carefully examining any target you think is likely.
During your examination you should bear in mind the questions below,
which apply to your study of any publication you consider either as a
student or during your later career.

Will the article that you have in mind concern a subject that the publication is interested in, to judge from the articles you find it to contain?
What kind of readers does the publication seem intended for? How
much knowledge do they apparently possess? Closely related to this is

the question: To what extent should you use technical language (where it might be useful) if your article is to be like those you examine?

How far should you go in the use of formulas, equations, and mathematical calculations?

Should your style be entirely formal? If not, just how informal should you make it? Should you use any jargon?

How prevalent is the use of tables and figures; and if it is frequent, how sophisticated are those used?

How long are the previously published articles? (If you find none outside the range of 500 to 2000 words, for example, and you submit one that runs to 4000 words, the best you can hope for is to have it returned and be invited to cut its length in half and submit it again.)

How long are the paragraphs in most of the articles?

Do the articles seem to be written with restraint or do their authors sometimes give free rein to their enthusiasm?

What kinds of openings seem to be preferred? Have the authors attempted to arouse interest by imaginative openings or do they depend upon the subjects to arouse interest without such assistance?

What kinds of titles are customary? Are they intended to arouse curiosity or are they matter-of-fact announcements of the subjects?

Are the articles documented when there might have been reason for documentation? If so, by what means? (Though this book has discussed documentation, nothing said here should deter you from using whatever system is apparently standard in any publication where you hope to have your work appear.)

The preceding list has been a long one, but each of the items is well worth attention. It is unfortunate when an article fails to be published simply because it needlessly differs in too many respects from the kind of material a periodical customarily uses.

Writing the Article

The process of actually writing an article is not essentially different from the process of writing anything else. Consequently, you should utilize the instructions in the preceding chapters. Since a technical article should be well organized, the suggestions in the chapter on organization should be useful. And though some matters of style may be affected by the adaptation of your article to the prospective outlet, the general recommendations in the chapter on style should be borne in mind—especially after you have written the first draft and are polishing it. The chapter on mechanics will answer many questions that may confront you. (Don't overlook what is said near the opening of the chapter about manuscript form.)

Also, bear in mind the earlier portions of the present chapter. Many an article needs to define some of its terms, explain a process, offer a technical description, or analyze a problem or situation. When these matters were discussed, it was pointed out that the problems in question crop up constantly in such places as reports and articles.

The question of whether to place headings in your article calls for comment. If you do so, they should almost certainly be limited to the main points of your outline. Not all the headings that you see in articles were placed there by the authors. In many publications, headings are inserted by an editor wherever he thinks they are needed to break up long, formidable-looking columns of type. These are not to be confused with headings used systematically to show divisions and subdivisions of books and reports.

Use of tables and figures was mentioned in connection with adapting your article to the periodical that you will send it to. If you decide to use such material, you should know what it will be before you attempt to put your written material into its final form. The use of tables and figures will be covered in more detail in the next chapter. For the moment, it is sufficient to point out that except for the possible use of photographs to arouse interest, the figures, like the tables, in a technical article are an integral part of it and are used because they are the best means of conveying some portion of the information it should contain.

Benefits of Writing Articles

The value of technical articles to industry has already been pointed out, but you are naturally interested, also, in whether you will personally benefit from writing them. Most assuredly they can make a substantial contribution to your professional progress. By writing articles that are published in scientific or technical periodicals you can secure favorable attention from the management of the organization you work for, acquire a reputation among your colleagues, and make a great many people outside your own organization aware of your existence and your professional competence. Consequently, by developing the ability to write technical articles, you acquire a skill that can advance your career.

(An example of a technical article follows.)

SPECIMEN TECHNICAL ARTICLE

One of the primary differences between journal and magazine articles is the necessity of capturing the reader's interest. Unlike a technical report, which people read to gather information necessary to their jobs, an article provides entertainment as well as information. True, professional people read articles to keep up with their fields; nevertheless, the profusion of reading material in every area gives a great latitude to the reader in choosing which articles he will read. The title alone of this article would be enough to attract most readers interested in science, so the authors devote only the first sentence to capturing the reader's interest. In an article on a less exciting subject, or aimed at a more general audience, the authors might have spent far more time at capturing reader attention, or in general explaining why the article should be read.

The second sentence gives a classical definition of the subject, despite the fact that probably no reader of *Science* magazine, in which this article first appeared, would be unfamiliar with the information that the dolphin was not a fish. The sentence serves two purposes, however: it reminds the reader of specific information he may have forgotten, and it firmly delimits the subject of the article. The rest of the paragraph serves as a summary or abstract of the article, and could easily be used without change in a report on the same subject. It gives a brief description of the planned investigation, the problems that hindered it, and the results of the experiments aimed at solving the problem.

The second paragraph provides a history of study in this area, starting with the first known investigations. In a study of an area with a greater number of investigations, the authors might have simply dealt with the more recent or perhaps the most relevant studies in order to save space. In this case, there is enough space to give a brief description of the actual experimental method and its results. Terms such as *intraperitoneally, asphyxia,* and *nasopharyngeal sphincter* are used without definition, since the authors expect that any scientifically sophisticated audience, such as the readers of *Science,* would be familiar with them. For a more general audience, the authors would have defined such terms or attempted to avoid using them.

A transition sentence opens paragraph 3 and moves the reader from general history to consideration of the authors' investigations. Paragraphs 3, 4, 5, and 6 provide a detailed description of the development of equipment and methodology. Exact measurements and the use of trade names of equipment give the kind of specificity required in a scientific report. In a report on the same subject, however, design drawings would probably be included as well, but since the authors do not

ANESTHESIA FOR THE BOTTLENOSE DOLPHIN:
TURSIOPS TRUNCATUS*

1. The great whales, dolphins, and porpoises have recently attracted much interest, particularly because of their apparent intelligence and large, well-developed brains. These animals belong to the order Cetacea and are air-breathing, warm-blooded mammals which spend their entire lives in an aquatic environment. Our investigations have dealt primarily with the neuroanatomy of this animal and, especially, with the construction of a brain atlas from several thousand wholemount brain slides. The application of classical neurophysiological techniques to the massive dolphin brain as planned by this laboratory has been hindered by two major problems: (i) no known safe anesthetics for this animal are available, and (ii) there are no adequate methods for artificially supporting respiration. These difficulties are related in part to the specialized central nervous control of the blowhole mechanisms associated with the complex engagement of the larynx into the nasopharyngeal sphincter and with the unique pattern of respiration seen in these diving mammals. Our experiments have led to a method of anesthesia for the dolphin which has proved safe and reliable.

2. The first attempts at general anesthesia in the dolphin were made by Lilly et al., in 1955. These investigators injected intraperitoneally 30 mg of pentobarbital per kilogram of body weight, the usual primate dose, and found that respiration became uncoordinated and death resulted from asphyxia. When the dose was reduced to 10 mg/kg, air leaked from the mouth, the lungs gradually deflated, and respiratory efforts became ineffectual. In a series of animals receiving various doses of pentobarbital or paraldehyde, leakage of air through the nasopharyngeal sphincter always presaged eventual respiratory failure and death.

3. In October 1963 we began experiments designed to develop a safe method of anesthesia for the dolphin. First we had to perfect suitable endotracheal and respiratory equipment for maintenance of adequate ventilation. Dissections were carried out to study the anatomical relationships of the oral, nasopharyngeal, and respiratory passages. Manual palpation of the laryngeal-nasopharyngeal area was performed in living animals, a specially constructed surgical tank being used in which the animal was firmly secured and kept moist by hosing frequently with sea-water. Four wooden restraints were constructed to fit the contours of the body, and were placed posterior to the blowhole, anterior and posterior to the dorsal fin,

* "Anesthesia for the Bottlenose Dolphin Tursiops Truncatus," E. L. Nagel et al., *Science*, Vol. 146, pp. 1591–93, 18 December 1964. Copyright 1964 by the American Association for the Advancement of Science. Reprinted by permission of the author and publisher. [Paragraphs were not numbered in the article, but numbers have been supplied to facilitate reference.]

expect anyone to attempt to duplicate their work from this article alone, they provide less information.

Despite the generally clear language of the article, the authors sometimes lapse into what could be considered jargon: "gross observations" means simply that they watched the animal. "Extubation" in paragraph four verges on the same fault. Both terms, however, are found frequently in scientific writing.

The discussion of the first set of experiments begins with the seventh paragraph—about halfway through the article. All of the preceding is essentially preliminary information necessary to the reader before he can understand the significance of the actual work. In the following discussion of the experiments, the authors explain not only what they did, but why—an important factor in science writing, all too often omitted. Thus it is clear not only why they attempted a particular method (using the intraperitoneal route) but why they did not use another (intravenous channels). By explaining both, they forestall objections that might rise in the mind of a knowledgeable reader, and provide for all readers confidence in their techniques.

Halfway through the paragraph the authors give the results of the first experiment: the animal died. The rest of the paragraph is devoted to a discussion of the conclusions they drew from this result.

This rather long discussion of an unsuccessful experiment is included in the article for the same reason that the discussion of using intravenous channels is discussed—to indicate that the successful methods were developed for good reasons, and that other methods were either not considered or were considered and rejected for good reasons. This paragraph illustrates the fact that in science, man learns from failures as well as success. Unless the failures are reported, however, others may be forced to repeat them.

and anterior to the flukes. Wet, heavy, foam padding protected the animal's body from the tank and the restraints. The use of a padded, oral speculum (Sands type for horses) allowed the dolphin complete range of jaw motion and provided adequate access to the animal's oral cavity.

4. For the endotracheal catheter we used a clear Tygon tube (2.5 cm outside diameter, 50 cm in length) having a short Murphy bevel and a highly distensible cuff (80 to 100 ml of air for full inflation). A heavy stainless steel coil inside the tube prevented its kinking and collapse. The final design of the respirator was based on a large-animal ventilator, the Bird mark 9X. Preparation for placement of the endotracheal tube was accomplished by insertion of the arm into the animal's oral cavity and through the oropharyngeal sphincter, so that two fingers were hooked around the vertically placed larynx. This structure was then pulled anteroventrally and a finger placed into the glottis for guidance of the endotracheal tube into the larynx. The animals were then maintained on air by means of the respirator for periods of up to 3 hours. Following extubation they swam, vocalized, and fed immediately on return to the home tank. As a precautionary measure, all animals were given three daily administrations intramuscularly of 1,500,000 units of benzathine penicillin G and 10 ml of vitamin B complex, with vitamin C as supportive post-anesthesia care.

5. The respirator was later modified by the addition of an apneustic plateau control unit which permitted variations to be made in the duration of the inflated stage. With this device, the animal was ventilated in a manner mimicking its natural pattern. Thus, the respirator rapidly inflated the animal, maintained it in an inflated state for a preset period of time, and then deflated and inflated the animal in rapid succession.

6. At this time the state of the animal was monitored by: (i) taking the rectal temperature by means of a telethermometer (Yellow Springs model 43) and a thermistor probe; (ii) an electrocardiogram recorded by means of a multichannel recorder (Grass model 7 polygraph); (iii) gross observations of the eye, melon, blowhole, and general body movements; (iv) observing the color of the tongue as a crude, but effective, measure of the degree of oxygen saturation of hemoglobin.

7. We then began experiments with various anesthetic drugs. Since we planned to map the brain in one of the studies, we hoped that barbiturate anesthesia could be utilized to enable adequate comparison with results seen in other animals. Intravenous channels were difficult to obtain in the dolphin, however, other than by means of a percutaneous puncture of the great vessels in the abdominal region. In the first experiment with barbiturate we therefore utilized the intraperitoneal route and short-acting drugs. Thirteen miligrams

In paragraph 8 the authors discuss other anesthetics that might be used, based on the conclusions drawn from the first experiment. Once again, they are careful to explain why they chose one agent rather than another.

In paragraph 9 the preliminary results of the next set of experiments are discussed. These successful results led to a further development of the method, which is described next, and the further results are detailed.

of thiopental per kilogram, in two doses 5 minutes apart, followed in 10 minutes by 5 mg of methohexital per kilogram in two doses 5 minutes apart, produced a loss of all apparent reflex activity for the next 3 hours. Spontaneous movements of the eyes and flukes began 5 hours after drug administration, and voluntary respiration was first noted approximately 2 hours later. The animal was returned to the home tank 9 hours after drug administration. Leakage of air from the mouth proved to be an ominous sign, since death followed in another 1½ hours. However, the animal's lack of buoyancy in air and the increased possibility of developing hypostatic pneumonia presented an increasing risk to the animal with further maintenance on the respirator. This experiment delineated certain major problems concerned with administering anesthesia to the dolphin, namely, that any significant degree of central nervous system depression after anesthesia would probably result in the animal's death. Short of intravenous administration of drugs, further experimentation with rectal, intramuscular, intrapleural, or intraperitoneal injection of hypnotics seemed contraindicated because of uncertain absorption, metabolism, and excretion of these agents in the dolphin.

8. Inhalational anesthetics were tried next. Flammable agents were ruled out because of the explosion hazard in the laboratory. Potent agents were also excluded because of insufficient monitoring to insure safe, controlled administration. Halothane, therefore, was not considered because this agent depresses the cardiovascular system even in light planes of anesthesia. In seeking an agent apparently free of such side effects, nitrous oxide containing varying concentrations of oxygen was then used. Nitrous oxide is a good analgesic and amnesic agent, but a weak anesthetic. It was chosen because of the lack of side effects seen in other species of animals, including man, and because of rapid and complete recovery following its use. The depth of anesthesia is known to be dependent on the species, individual, and the premedication given.

9. In preliminary experiments 50 percent nitrous oxide in oxygen was administered for periods of up to 3 hours, during which time it was well tolerated. This mixture produced a definite degree of analgesia as evidenced by an elevated threshold to noxious stimuli. An uneventful recovery of the animal followed, with resumption of normal breathing pattern, swimming, and feeding activity. To secure deeper planes of anesthesia, the concentration of nitrous oxide in oxygen was then increased to 70 percent, and the following effects were noted in several experiments. Instead of passive acceptance of the respirator, the animal now attempted to breathe around or dislodge the endotracheal tube. These movements seemed uncoordinated and exaggerated during the induction phase of anesthesia. Following this brief excitation phase the animal became largely insensitive to noxious stimuli, such as subcutaneous insertions of needle electrodes or penetration of the blowhole to the nares with the fingers. Such stimuli, in our experience, have always evoked strong evasive reactions in the awake animal.

The last paragraph in the article, paragraph 10, gives the final results of the experiment and discusses directions that future work might take. The authors identify exactly what they have accomplished: they have developed a method that has allowed them, on *four* occasions, with *two* animals, to insert electrode sleeve guides and record brain waves. Since success in such work more than once is adequate to prove the value of the method, they have reported their results at this time. In a later article, they may report what they found from the recordings of the electroencephalographic activity.

Paragraph 10 is a suitable conclusion. The authors had done the work described in the preceding material in order that they might perform certain experiments which would be possible only after anesthetics had been administered. In their last paragraph they show that their methods had been successful by telling about the performance of such experiments. Also, they point out that their methods not only made possible the kind of experiments they themselves wished to perform, but have also opened up the possibility of performing a wide range of other work—major surgery in general. Thus the authors end their article by making sure that their readers realize the full significance of their work.

10. In the last three experiments we extended the monitoring procedures by utilizing the infrared CO_2 gas analyzer (Beckman) in order to measure pCO_2 in the expired air. Two penetrations of blood vessels were performed percutaneously with a 20-cm hypodermic needle. In the last such experiment, 0.25 mg of thiopental per kilogram was injected directly into the abdominal *vena cava*. This small dose of barbiturate did not appear to affect the animal adversely in any perceivable way. Monitoring devices which measure arterial pO_2, pH, and pressure will probably be necessary, however, to test safely such agents as the barbiturates or halothane when given in effective anesthetic doses. Using nitrous oxide anesthesia, we have been able to insert electrode sleeve guides into the skull on four occasions in two animals, and to record successfully cortical electroencephalographic activity. These methods open the possibility of performing major surgery in this species for the first time. However, for some surgical procedures supplementation of nitrous oxide anesthesia with other agents may be necessary.

EXERCISES

Exercise 1

Indicate the genus under which you would place each of the following terms if you were defining it in a sentence definition. Be sure that your decisions are accurate, and try in each case to choose a genus that limits the term as much as possible so that the differentia need not be unnecessarily long and complicated.

(*Note:* In this and other exercises concerned with definition, most of the terms will be familiar to you, so that heavy reliance on a dictionary will not cramp your independent thought. The exercises are not intended to test your knowledge; they are designed to demonstrate the need to avoid accidentally omitting essential facts that you know but do not think about, and the need to express all the essential facts in a precise and orderly manner.)

sand	bacterium
hydroponics	ballistics
stereoscope	fermentation
algae	laser
pizza	radioactivity
grape	slag
heavy water	sun spot
half life	logistics
phlebitis	evolution
ornithology	decibel

Exercise 2

It is often necessary, in writing a definition, to discriminate with special care between the term to be defined or the referent that the term stands for and one or more other terms or referents. For example, if you were defining *tree* you would need to discriminate *tree* and *bush*. If you were defining the following terms, what other terms would call for such special discrimination? (Choose five to answer.)

coma	rayon
hail	fruit
psychiatry	detergent
mass (as used in physics)	ellipse
porpoise	cyclone
tuber	acetylene torch

Exercise 3

Indicate the faults that you find in the following sentence definitions.

(*Note:* The function of a single definition is merely to tell what the term means in one legitimate use rather than to cover all of its possible meanings. For example, a definition of *lock* would not need to cover the meaning of the term as used in *lock of hair, lock in a canal,* and *lock on a door.* The definitions below are faulty not because they fail to include all the different meanings that the term may convey but because they fail to express satisfactorily the single meaning obviously intended on the particular occasion.)

1. An electrical relay is a switch operated by electricity.
2. An anesthetic is a substance used to cause the loss of consciousness and thus prevent pain.
3. An eclipse is when one nonluminous body, such as the earth or moon, passes into the shadow of another.
4. *Laissez-faire* is a term that comes from the French language and is used in connection with economic theory.
5. A mansard roof is a type of roof named after François Mansard, who introduced it in an effort to circumvent Parisian property-tax laws.
6. An optometrist is a doctor who specializes in examining eyes for defects and faults of refraction.
7. Autoclave means to cook or sterilize in an autoclave.
8. A gore is a piece of cloth used in sail- or dress-making.
9. A lens is a transparent piece of glass used in an optical instrument to focus rays of light.
10. Glaucoma is a disease of the eye that impairs vision.

Exercise 4

Write a sentence definition for each of five terms from the following list. None of your definitions needs to cover more than one meaning of the term chosen, and each may be limited, if necessary, to a special point of view.

abscissa	wave (in water)
carburetor	eminent domain
catalyst	typhoon
enzyme	radial tire
erosion	geriatrics
humus	embolism
nymph (biological sense)	empiricism
pyramid	aqualung
propane	witch doctor
depletion allowance	slalom

ASSIGNMENTS

The following assignments cover all the writing problems discussed in the chapter. In almost every case a list of possible subjects is provided. These lists call for comment. *First:* Some of the subjects may impress you as not being especially technical, but there are valid reasons for their inclusion. This book is intended for the use of students in a wide range of fields; and the more technical and specialized a subject is, the smaller would be the percentage of students who could use it. Because it would be impossible to suggest several highly technical subjects in every conceivable field of specialization, a good many subjects have been included that can be handled without specialized knowledge.

It is not necessary to discuss a highly technical subject in order to profit from doing what an assignment calls for. It is entirely practical to learn the technique of handling a technical subject by writing about a subject that is not technical. For example, you could improve your ability to describe any machine by attempting to describe an electric dishwasher, even though you do not need a technical education to learn the facts about one.

Second: In choosing a subject for any assignment you may find it helpful to look over the lists for other assignments also. A certain mechanism, for example, might be listed as the subject for a technical description; but it might also be possible to base an explanation of a process on the way that it functions or to write a set of instructions for using it or repairing it.

Third: As you look over the lists, use your imagination. The lists may be helpful in stimulating you to think of additional subjects for yourself.

Assignment 1

Write one or more expanded definitions of terms listed in Exercise 4, or of other terms if permitted by the instructor. Each definition should be from 100 to 400 words long. (Some of the terms, of course, may not present enough problems to call for an expanded definition.)

Assignment 2

Write a technical description of one of the following objects or of some other relatively simple object with which you are familiar. Aim at a length of 400 to 800 words, adjusting the degree of detail so that the paper falls within these limits. Do not attempt to describe anything so complicated that you cannot do justice to it. (This assignment may or may not be accompanied by one or more figures, as the instructor directs.

If figures are used, it may best be postponed until after the chapter on tables and figures has been studied.)

a barometer

a pressure sprayer to apply insecticide or herbicide

a catalytic heater

an electric razor

an oscillating electric fan

a home movie camera

an electric mixer

a photographic enlarger

a hydraulic door closer

an electric outboard trolling motor

a stapler

a boat launcher

any piece of rock-hound equipment

an oven-toaster

a transit

a door knob-and-latch assembly

a microscope

a back-pack and frame

a compression-type faucet

a snow-blower—or some other small snow-removal machine for home use

a food-waste disposer

Assignment 3

Write an explanation of one or more of the following processes or of some other relatively simple process with which you are familiar. It is suggested that you explain a process that you can cover in 400 to 800 words. If you wish to explain a process that cannot be covered within this length, you may be able to reduce its size by covering only one part of it. List A consists of processes performed mainly by persons, and list B consists of processes performed mainly by machines.

List **A**

1. Using a charcoal broiler
2. Making a photographic enlargement
3. Artificial propagation of fish
4. Sanitary inspection of meat
5. Refinishing furniture
6. Reconditioning a boat
7. Some process performed by rock-hounds
8. Some simple body repair on an automobile
9. Installing some type of floor covering
10. Upholstering a chair or davenport

List B

1. How a vacuum cleaner functions
2. How offset printing functions
3. How a "steam" rug-cleaning machine functions
4. How a heat pump functions

5. How a steam iron functions
6. How an automatic hand gun or semiautomatic shotgun functions
7. How a garbage and waste compactor functions
8. How a chemical-sprayer functions (the type in which the chemical, in a jar, is drawn up and mixed with water from a hose)
9. How a "crawling" type of lawn sprinkler functions
10. How an automobile shock absorber functions

Assignment 4

Write a set of instructions for performing one of the following processes, one of the processes mentioned in list A of Assignment 3, or a process of your own choice that you know how to perform.

1. Panning gold
2. Making a compost pile
3. Antiquing furniture
4. Making a graft on a tree
5. Using a pressure-type weed sprayer
6. Operating a slide projector, motion picture projector, or tape recorder
7. Making yogurt at home
8. Checking some kind of mechanism to find the cause of failure
9. Seeding and transplanting some kind of flower or vegetable
10. Making a casting pattern by use of a sand mold

Assignment 5

Write an analysis of one of the subjects listed below or of a comparable subject about which you can obtain information.

1. Analyze the effects of daylight saving time (on people rather than on use of electricity).
2. Analyze the expenses of attending school.
3. Analyze the factors contributing to traffic congestion at some specific location.
4. Analyze the expense of owning and operating an automobile.
5. Analyze the potential benefits and relative costs of various features of some kind of insurance policy.
6. Analyze the cost of earning a master's degree in relation to the potential benefits.
7. Analyze mobile homes—desirable and undesirable features as places of residence.
8. Analyze the financial aspects of ownership of a mobile home.
9. Analyze the attitudes of students toward collegiate athletics.
10. Analyze the nature and advantages of "no-fault" life insurance.

11. Analyze some periodical—for example, an industrial magazine—in regard to subjects it deals with, length of articles, number of articles, and the like and also space devoted to advertisements.

Assignment 6

Make an outline for a technical or semitechnical article. After the outline has been approved by the instructor, write the article. When you submit the outline, you should name at least one periodical to which the article might be submitted. Your choice of a subject and the nature of the article should result from previously acquired familiarity with the periodical. The length of your article should be approved by the instructor and appropriate for the periodical.

6

TABLES

AND

FIGURES

It is almost impossible to imagine technical writing that does not make frequent use of figures and tables. This is true because some kinds of information—for example, statistical data or the shape, arrangement, and size of objects—are difficult if not impossible to convey by the use of words alone. Hence it is important to know when and how to supplement words by using figures and tables, and how to use words in order to point up the significant features of the figures and tables used.

This chapter will call to your attention the main conventions that govern the use of these nonverbal forms, but there is only one way in which you can really learn to use them to the best advantage. That is, to notice what you see done in the books and articles related to your studies. Not that everything done is desirable; but systematic observation will increase your resourcefulness and will help you to form an intelligent opinion about whether something you are contemplating is done often enough that it will not strike your reader as strange.

TABLES

Of all the nonverbal devices, tables are used most frequently. Usually their function is to enable a reader to compare statistical information more easily. They often take considerable time to compile, and you may often be tempted to take the easy way out and present your statistics in ordinary text, but there are times when it will be almost hopeless for the reader to grasp your information unless he receives it in tabular form.

Though tables usually consist of numerical data, resourceful writers use them to present other kinds of information also. A consumers' magazine uses them, for example, to present the facts about different makes of automobile. It names the different makes of car in the column headings, lists the feature being compared—such as brakes, visibility, steering mechanism—at the left, and indicates the facts about each car in the column under its name. This enables a reader, with equal ease, to check up on all the features for any make of car covered or to compare each make to the others in any particular feature. This same type of table can be used for numberless other types of comparison—the characteristics, for example, of different kinds of apples; the identification of insects that damage crops and the measures for controlling them; the results of using different quantities of fluoride in water to reduce tooth decay.

Because tables can be used for so many purposes, a technical writer should always be on the alert for opportunities to utilize them. Also, he should realize that when the construction of a table seems extremely complicated, he can often serve his reader's best interests as well as his own by dividing the material into two or more simpler tables.

For your assistance in using tables, the following suggestions are offered:

1. At the top of every table there should be a title—preceded by a number if more than one table is used. Numbers of tables are sometimes Roman numerals but more frequently Arabic.

2. Unless a table is merely supplementary and is placed in an appendix, it should be referred to in the text so that the reader will know when to give it his attention. Reference to a table *may* be desirable even if the table is in an appendix.

3. Each table (unless tables are relegated to the appendix) should be placed where it is conveniently accessible at the proper moment. Ideally, a table should be placed shortly after the point where it is first referred to—always on the same page if there is room for it. Under no circumstances should a table be placed very far in advance of the point where it is referred to or discussed.

4. The form of tables varies in detail as necessitated by the material to be presented. Examples will appear later in the discussion.

5. Regardless of other details of form, each column should have a heading that shows accurately the nature of the contents below. When necessary, this heading should designate the units in which quantities below are expressed. If there is not enough room for the essential information in a column heading, part of it may be added to the table in notes, as described in rule 10.

6. A table should indicate all the factors that affect the data it contains. (For example, the size of pipe used would affect data on the performance of a pump.)

7. Standard symbols and abbreviations may be used to save space.

8. Figures in columns are usually aligned under similar digits—ordinarily the right-hand digit.

9. Fractions should be expressed as decimals unless decimals are not customary for data on the subject concerned, or unless they would misrepresent the degree of accuracy achieved.

10. When a note is needed to explain some part of a table, its presence is indicated by a lower case *letter* raised half a space above the line at the point where the note applies. The notes applying to a table are placed at the bottom of the table rather than at the bottom of the page, where they might be confused with ordinary footnotes.

11. A table should not break into the normal margins of the page. If necessary it may be placed so that it extends across the length of the page, with the top of the table toward what would normally be the left side of the page.

12. Do not use needless lines, especially solid lines, between columns of data, between lines, or at the bottom of the table.

13. No table should continue from one page to another unless continuation is unavoidable because the table is more than a page long. When such is the case, *continued* or *cont.* should be used at the bottom of the first page to indicate that the table has not been completed, and at the top of the second page to indicate that part of the table has preceded. Column headings must be repeated on the second page. If totals are to be indicated at the bottoms of columns, the subtotals should be at the bottom of the first page and at the top of the second page. The word *forward* should be used at the left side of the subtotals, to show that they are not final totals.

14. A table from an outside source must be acknowledged, as one would acknowledge any other borrowed material. This may be done by naming the source in parentheses after or under the title, or by use of a footnote. If the footnote is used, its presence is indicated by a number raised half a space above the line. If the number and title of the table are on one line, the footnote number follows the title; if they are on separate lines, the number of the footnote follows the number of the table.

The preceding suggestions are modified when a writer uses a form that strictly speaking is not a table. The term *table* means something more than a list, or two or three lists placed side by side. If you examine the work of experienced writers you will find that they rarely call anything a table unless it not only has headings at the top of its columns but has a left-hand column that also consists of headings, as can be seen in the specimens.

Well-informed people may sometimes call material that does not have left-hand headings "informal tables," but they do not use the

Table 2. Average Daily Emission (in tons) of Atmospheric Pollutants

Source	Hydrocarbons and Hydrocarbon Derivatives		Aerosols	Sulfur Oxides	Nitrogen Oxides
	Hydrocarbons	Hydrocarbon Derivatives			
Automotive[a]	1100	150	35	35	775
Petroleum Marketing	70	0	0	0	0
Petroleum Refining and Production	160	2	5	45	45
Solvents and Coatings	210	150	0	0	0
Totals	1540	302	40	80	820

[a]These figures are now being revised to incorporate later data.

Table II

Average Daily Emission (in tons) of Atmospheric Pollutants

Source	Hydrocarbons and Hydrocarbon Derivatives		Aerosols	Sulfur Oxides	Nitrogen Oxides
	Hydrocarbons	Hydrocarbon Derivatives			
Automotive[a]	1100	150	35	35	775
Petroleum Marketing	70	0	0	0	0
Petroleum Refining and Production	160	2	5	45	45
Solvents and Coatings	210	150	0	0	0
Totals	1540	302	40	80	820

[a]These figures are now being revised to incorporate later data.

FIGURE 2. Specimen Tables. The forms shown, unlike most forms seen in print, can be produced on a typewriter. They may be altered as necessary to accommodate different materials. The numbers at the left are not part of the tables but are added in order to indicate vertical spacing. Additional forms and uses of tables should be observed in publications dealing with your intended profession.

word "table" before the titles nor refer to them as tables in the text. These "informal tables" are not numbered and need not be referred to in the text because, unlike true tables, they are actually part of the text, in substance if not in form. What has been pointed out about this form is not intended to discourage its use, but the writer who treats it and

Table 6. Performance of Self-Polishing Wax on Linoleum

Test	Nukoat	Florglo	Waxeeze	Hygrade
Ease of Application	Excellent	Excellent	Fair	Good
Original Appearance	Good	Good	Excellent	Good
Resistance to Scuffing	Good	Fair	Fair	Poor
Resistance to Scratching	Good	Good	Poor	Poor
Resistance to Water Spotting	Fair	Fair	Poor	Fair
Response to Rebuffing	Fair	Fair	Good	Good

FIGURE 3. Specimen Table Showing Tabular Presentation of Nonstatistical Information.

refers to it as if it were really a table makes his work look both stiff and amateurish.

Forms that may be used for tables are illustrated in Figures 2 and 3.

FIGURES

In the following discussion, all forms of illustrations are referred to as figures. Examination of technical books and magazines will justify this terminology.

To the technical writer, figures are functional rather than ornamental. Though in other kinds of writing they are frequently used merely to arouse interest or improve the appearance of a page, in technical writing they serve primarily as a means of giving information. If they also happen to catch attention or make a page look more attractive, so much the better; but that is not the main reason for their use. On the rare occasions when a technical writer must use figures mainly for the purpose of arousing interest, he will disregard many of the rules for use of figures in technical writing.

Whatever types of figures one may use, the following suggestions should be helpful.

1. *Number and caption.* Every figure should have a caption, which should be preceded by a number (usually Arabic) if more than one

figure is used. It is usually best to use a single sequence of numbers even when figures differ in kind. The number and caption are normally placed underneath a figure but may be placed in a convenient location on the figure itself. (This is likely to be desirable on a graph because of the lettering at the bottom.) A caption is ordinarily phrased as a topic (see page 202 for definition of *topic*); but sentences are sometimes used for figures appearing in magazine articles.

2. *The legend.* In addition to a caption, a figure may also have a legend if it needs one. The term *legend* applies to the additional lines, often in smaller type if printed, that follow the caption and explain the figure or identify its parts. If a legend follows a caption that is phrased as a sentence, the two blend together so that in appearance they constitute a single unit.

3. *Spacing of the caption and legend.* When the caption is underneath the figure, there should be at least a double space between the figure and the caption. However, the caption, and the legend if one is used, should be closer to the figure than to the text that follows, so that they will be identified with the figure, not with the text.

4. *Reference to the figure.* Every figure, unless it is purely supplementary and is placed in an appendix, should be mentioned in the text. (It *may* be referred to even if it is in an appendix.) The reference to the figure is placed in parentheses unless it is part of the sentence where it occurs. Note the following examples:

   As can be seen in Fig. 4, the dimensions . . .

   The arrangement of the equipment (see Fig. 5) is planned
   so as to . . .

   If a figure is related to a long stretch of text, it should not be referred to more frequently than is necessary to keep the reader reminded of its bearing on the discussion.

5. *Placement of the figure in the text.* Any figure in the text should be placed, if possible, almost immediately after the point where it is first mentioned. Certainly it should not come very far ahead of that point, for if it does, the reader may either study it too soon or else, having passed it, may neglect to turn back to it. When possible, you should place a figure on the page where you discuss it.

6. *Partial lines of type.* If a figure is narrow enough to leave room for partial lines of type, the type should be placed on the left and the figure on the right. (This suggestion applies to material typed on only one side of the sheet and therefore regarded as the right-hand page of a book. On a left-hand page, the figure is placed on the left.) However, if two or more figures are placed on one page, the layout may be designed in any manner that gives the page a good appearance and places each figure where the reader can consult it conveniently.

7. *Drawing the figure.* Full instructions for drawing figures are beyond the scope of this text. It should be mentioned, however, that figures should be drawn in India ink, that they should be no larger than is necessary for clearness, and that each figure should be placed within a border unless its own shape creates a natural border. The figure should not extend into the margins of the page.

8. *Self-sufficiency of the figure.* A figure should be self-sufficient so that the reader for whom it is intended may study it without being forced to consult the text. Every part of a figure that cannot be identified from its appearance should be labeled or else identified by use of a key. The key may appear on the figure or may be a legend under the title.

9. *Lettering on the figure.* The lettering on a figure should be placed, when possible, so that it can be read with the bottom of the figure down. Lettering that must run at a right angle to the bottom of the figure should read with the right side of the figure down.

10. *Sidewise figure.* If a figure must be turned sidewise on the page, it should be drawn so that the bottom of the figure comes on what is normally the right side of the page. (This can sometimes cause lettering to appear upside down when the page is in normal position.)

11. *Acknowledgment of source.* If a figure is taken from an outside source, the source must be acknowledged. The acknowledgment may be placed in parentheses after the caption or after the legend if you use a legend from the same source. Acknowledgment may also be made by means of a footnote.

12. *Mounting a photograph.* If a photograph or clipping is used as a figure, it should be mounted neatly, preferably by use of rubber cement, unless copy is being prepared for printing.

13. *Preparation of figures for printing.* If copy is to be printed, figures must not be fastened to the pages of the manuscript. The figures and captions are separately prepared and numbered, and the manuscript is marked to show where the figures should be placed. The figures should be drawn larger than they will be on a printed page—perhaps twice the size. Figures for offset reproduction should be drawn, usually, in the size that they will appear in the reproduction. The full planning of layouts for display printing is beyond the scope of this book.

The preceding instructions indicate that a table or figure should be referred to in the text. This does not imply that a mere mention is the only attention it receives. Often the text will discuss a table or a figure in considerable detail, so that the reader can understand it as easily as possible. The text may direct the reader's attention to the points that are most important, or may explain the table or figure in sufficient detail to make sure that its full significance and implications are not overlooked. One test of your skill as a technical writer—a test often overlooked in this kind of discussion—is your use of the text to help the reader benefit from tables and figures, as well as your use of tables and figures to assist the text.

Thus far, all that has been said about figures applies to figures of every type. Some of the common types call for individual attention. Before they are discussed, however, it should be re-emphasized that no amount of discussion can eliminate the need for constant observation of the figures in technical books and magazines.

Bar Charts

One of the simplest and most useful types of figure is the bar chart, with which every reader is familiar. The type of information it presents might also be presented in a table. In fact you may often first make out a table and then convert it into a bar chart to make differences in quantity instantly and unavoidably *visual*. For example, a reader might see numbers such as 841,654 and 418,543 arranged in columns, along with several other numbers, and yet fail to become aware of the relative quantities until he had given the table close attention. If such numbers were presented as bars, one of which was more than twice the length of the other, relative amounts, being clearly visible, could hardly fail to be perceived.

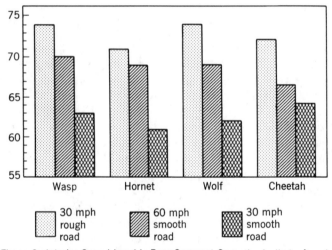

Figure 9. Interior Sound Level in Four Compact Cars (in decibels, A scale). A reading of 50 decibels is comparable to the sound level in a home; 70 decibels is comparable to the sound of a freight train at a distance of 100 feet.

FIGURE 4.   Bar Chart with Vertical Bars.

The bars in a bar chart are sometimes vertical (see Figure 4), sometimes horizontal (see Figure 5). They may be shaded differently in different portions of their length so as to permit comparison of parts as well as of the whole. For example, different bars in the chart might represent the total expenditures of different divisions of some organizations, and each bar might be shaded differently in different sections of its length so as to show what amounts went for specific expenses such as wages, raw materials, power, and overhead.

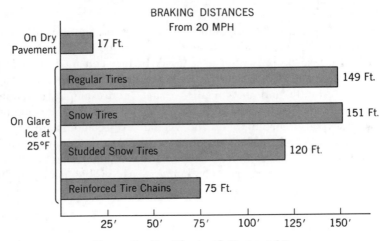

FIGURE 5.   Bar Chart with Horizontal Bars.

There are variations of the bar chart, such as drawing pictures to represent quantities. For example, the increase in the number of passengers carried by an airline might be suggested by pictures of men, or the increase in production of a metal by pictures of ingots. This pictorial treatment is used primarily when an effort is being made to popularize what is written. It is intended to increase interest as well as to provide information. Properly handled, it may on suitable occasions be a useful expedient. Often, unfortunately, it is not properly handled. Instead of drawing three times as many men for one airline as for another in comparing passengers carried, writers sometimes draw one man for each but make one of the two three times as tall—and thus nine times as large in area. Certainly a reader does not easily form a clear, accurate impression of the facts from what he sees. And since a bar chart fails to perform its function unless it uses visual methods to create a clear and accurate impression of the facts, it is best to use ordinary bars or lines, so that no dimension varies except length, or else to show the variation of quantity by varying the number, not the size, of pictures. These methods are the most honest and accurate, and are also the clearest to the reader.

Graphs—Curves

Like the bar chart, the graph or curve is a means of presenting data that might also be presented as a table. In fact it is hardly possible to construct a graph without having constructed a table first. The extra work that it takes to convert a table into a graph is often justified, how-

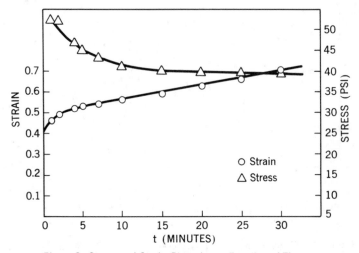

Figure 3. Stress and Strain Plotted as a Function of Time.
Although not shown here, the initial stress is zero.

FIGURE 6.   Curve (Courtesy of General Electric Company).

ever, by the fact that a graph makes it much easier for the reader to grasp the information.

Suppose you wished to show how the price of a commodity varied from month to month over a period of one or two years. To learn the low, the high, and the general trends from a table your reader would find it necessary to study the data at length. Yet all these points would be perceptible at a glance if the table were converted into a graph. If the data concerned three or four commodities, the use of curves would be even more advantageous. It would enable the reader to grasp the situation in a moment, whereas he would have to read and compare dozens of figures to get the desired information from a table.

The graph or curve differs from a bar chart in that the graph always shows changes in two values. It shows how one varies as the other varies—for example, how employment increases or decreases as the year progresses. One of these elements is frequently the passage of time and the other is an amount or quantity. When this is the case, the passage of time is usually shown on the horizontal scale and the variation of quantity on the vertical scale.

A graph always has a horizontal and a vertical scale, the horizontal usually indicating what might be called the independent variable, such as time, and the vertical indicating the dependent variable. Full information about the scales must appear as part of the graph. The specimens shown illustrate the placement of this information. The vertical scale,

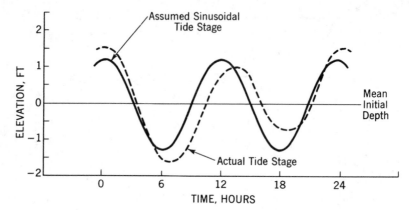

Figure 6. Comparison of Sinusoidal and Actual Tides over Two Complete Tidal Cycles.

FIGURE 7.   Curve (Courtesy of Extension Service, College of Engineering Research Division, Washington State University).

ordinarily placed on the left, may be placed on the right side too if the graph is so large that it is needed there.

Sometimes more than one curve appears on the same graph. In this case, if there is room, each curve should be identified by a short title lettered above it. If room does not permit this lettering, a different kind of line may be used for each curve, and a key identifying the lines can be shown on some unused space on the face of the graph. Other means also—for example a legend under the caption—may be used for identification.

Be sure that when you make a graph you choose the scales, both vertical and horizontal, in such a manner that the visual impression conveyed is justified by the facts. When a line goes up or down at a sharp angle, the reader infers that the rapidity of change is highly significant. If it climbs or falls slowly, he interprets the change as having been gradual. Yet the angle at which the line goes up and down is entirely dependent on the scales. It is therefore a matter of simple honesty for the writer to use scales that create a visual impression in line with the actual importance of what has occurred.

The Pie Diagram

Another widely used device is the "pie" diagram, consisting of a circle divided into segments. It might show, for example, what percentages of a corporation's receipts were allotted to meeting various expenses, to dividends, and to capital gain. More than any other kind of figure it permits comparison, at the same time, of parts both to each other and to the whole.

Like any other drawing the pie diagram should state amounts as well as picture them. These amounts should often be actual quantities as well as percentages.

One difficulty in making a pie diagram is the problem of making the letters identifying the segments all run in the same direction. If there are too many segments or if some of them are too small, this difficulty can become insuperable, making it inadvisable to use the form.

Figure 8 is a typical pie diagram.

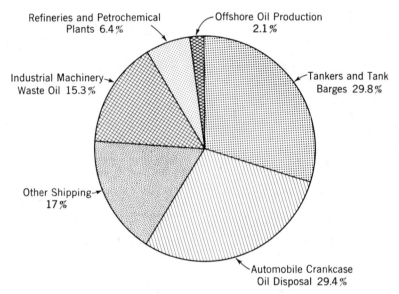

Sources of Oil Polution in the Oceans.
(Source of Information: Society of Naval Architects and Marine Engineers)

FIGURE 8.   Pie Diagram.

Organization Charts and Flow Sheets

Unlike the figures discussed above, the organization chart and the flow sheet are not concerned with statistical information. The very name *organization chart* shows what kind of information such a figure presents —divisions of an organization, usually represented by circles or rectangles so arranged and connected by lines that authority, relationship, and re-

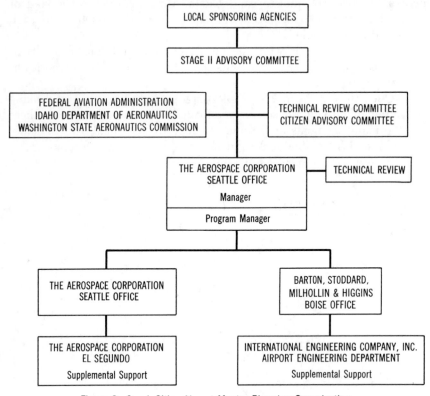

Figure 2. Quad-Cities Airport Master Planning Organization.

FIGURE 9. Organization Chart (Courtesy of Quad-Cities Regional Airport Master Plan Study).

sponsibility are easily seen. Such a chart is far more effective than words in conveying information of that kind. Indeed, many a body has been induced to improve its own organization when an organization chart revealed confused or illogical relationships.

The flow sheet is very similar. Its typical use would be to indicate the method by which some process is carried out. The machines used are sometimes indicated by the conventional circles or rectangles, and sometimes by simplified drawings suggesting their actual appearance. The main difference between the flow sheet and the organization chart is that the flow sheet indicates actual physical movement of materials. The direction of such movement is shown by arrows wherever necessary.

Figures 9 and 10 illustrate an organization chart and a flow sheet.

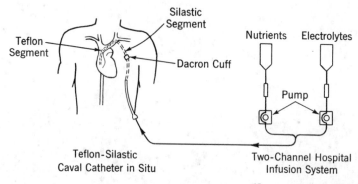

Figure 17. Artificial Lung (See Hill.[15])

FIGURE 10.   Flow Sheet (From an article by J. D. Hill, et al., "Prolonged Extracorporeal Oxygenation for Acute Post-Traumatic Respiratory Failure (Shock-Lung Syndrome)," *New England Medical Journal*. Courtesy of *The Trend in Engineering*, College of Engineering, University of Washington).

Photographs

In technical writing you may need to use photographs for either of two main purposes: to assist verbal description, and to prove the truth of assertions. Photographs would be helpful, for example, in showing characteristic fractures of metal that might result from different causes, or in identifying insects or plant diseases. They might also illustrate the condition of equipment that had not been maintained properly, the breaks that had developed in a pavement or the lining of a canal, or the crowded condition of a factory. Sometimes, as mentioned, the purpose of the photograph might be merely to give information; but often a photograph can be used to prove the truth of an assertion by letting the reader see for himself. For this latter purpose, it has no equal.

The photograph is limited, however, by the fact that it can show only the surface and that it may sometimes unavoidably present both significant and nonsignificant facts of appearance with equal emphasis.

Figure 11 is a photograph used in a typical manner.

Diagrams and Drawings

There are many occasions when diagrams and drawings can help to illustrate the text. For example, electrical wiring plans, schematic diagrams showing how any type of equipment operates, or drawings

Left to right: Pig's Ear (Discena Perlata), Morel (Morchella Elata), Snow Mushroom (Gyromitra Gigas), and False Morel (Gyromitra Esculenta). The false morel, which is nonedible, can be distinguished from the true morel by its false pitting (mere convolutions) and its long stalk in contrast to the real pitting and short stalk of the morel. The pig's ear has neither convolutions, pitting, nor stalk. The snow mushroom has convolutions and a hidden stalk.

FIGURE 11.   Specimen Photograph (Courtesy of Edmund E. Tylutki, Department of Biological Sciences, University of Idaho).

intended to illustrate actual appearance are constantly used by those who write on technical subjects. The shape and relative location of objects as well as the manner in which equipment functions can often be shown better by a diagram than by any other method.

Also, unlike photographs, drawings can picture the interior rather than just the surface of an object. They make it possible to omit what is not significant and to emphasize what is important. As you read the technical writing of others, you should make it a point to observe the manner in which drawings are used; and as you plan your own writing, you should constantly be on the alert for opportunities to make your work more effective by means of drawings.

A characteristic use of a drawing for a suitable purpose is seen in Figure 12.

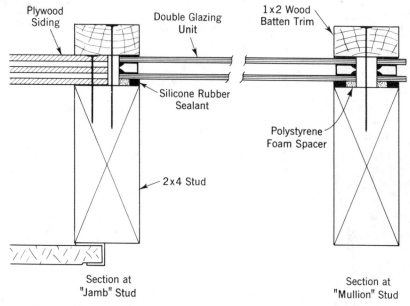

Figure 1. Horizontal Section Through a Frameless Window in Which the Glass Is Adhered Directly to the Building Frame Members.

FIGURE 12.   Diagram (Courtesy of Extension Service, College of Engineering Research Division, Washington State University).

ASSIGNMENTS

Papers in which tables and figures are used can be based on Assignments 2, 3, and 4 at the end of Chapter 5.

PART II

REPORTS, PROPOSALS, AND ORAL PRESENTATION

7

REPORTS—

DEFINITION, IMPORTANCE,

AND QUALITIES

DEFINITION

Anyone who has seen reports from varied sources has noticed that they differ widely in form, length, and many other respects. This gives rise to the question, "What is it that causes reports to be considered a distinct species of writing?" The answer can be found in the following definition: A report is a communication in which the writer (or speaker, if it is an oral report) gives information to some person or organization because it is his responsibility to do so.

As this definition indicates, the common element in all reports, which sets them apart as a distinct species, is the element of responsibility. A report may be made by a committee to its parent organization. It may be made by a consultant to a client. It may be made by the man on the job to his superior. It may be short, simple, and intended for a single reader or it may be long, elaborate, and printed at great expense —for example, the report of a great corporation to its stockholders. But regardless of all these differences, the term *report*, at least in the context of technical writing, should not be applied to a piece of writing unless the writer is responsible for reporting to the reader or readers.

141

IMPORTANCE

In any large organization, reports perform an indispensable service. They enable those in positions of authority to keep track of normal operations, to learn about unexpected developments, and to judge whether progress is satisfactory on new projects. They enable the executive to base his decisions on the advice of specialists. They are the link between research and the practical utilization of its discoveries. They often provide the basis for the writing of proposals and specifications. They sometimes are the only tangible result of long and expensive work. In general, they comprise one of the most important parts of the process of communication that is vital to efficient operation in government and industry.

DIFFERENCE BETWEEN PROFESSIONAL AND SCHOOL REPORTS

Before a person begins to write reports in his professional career, he is almost certain to have written reports in the classroom. Many of the qualities that are desirable in the classroom reports are also desirable in reports written on the job, but there is one important difference. The value of a report written on the job depends mainly on the answer to one question: How well does it meet the needs of those who receive it? Whatever good qualities it may possess are good only because they help it to meet this one test of excellence.

In the classroom, this test does not ordinarily exist. The instructor assigns a report not because he needs or will use the information it contains, but because the student will profit by acquiring that information and perhaps by reporting it. As a result, the student does not form the habit of constantly asking himself, "Does the man I am reporting to need these particular facts?" When at some later date he writes reports on the job, he is likely to include information that is not needed just because it is available and he himself is intensely interested in the subject.

A case in point is the laboratory report. Many instructors insist that a laboratory report contain information on theory and procedure—because one purpose of the report is to show whether the student understands these matters and can explain them. But in industry, it is usually taken for granted that the writer understands the theory on which his work is based and has used an accepted procedure. And since the reader will usually know about these matters already or else will not need to know, he will be just as well satisfied if they are omitted, *unless they involve something new and significant.*

A person who realizes, at the beginning of his career, the difference between what is desirable in reports written on the job and the reports he has written in school has taken an important step toward becoming a good report writer.

QUALITIES TO STRIVE FOR IN REPORTS

Like any other kind of writing, a report must be considered good or bad mainly on its success in performing its function: to affect the readers as it is supposed to affect them. What specific qualities will help it to produce that result?

Some of these qualities have already been discussed because they should be cultivated in other technical writing as well as in reports. These include accuracy, adaptation of the contents and organization to the needs and abilities of the reader, and effective style. But a report makes additional demands. The key to these lies in a single word: *imagination*. As a report writer you will need to visualize what will happen after your report is submitted, and face such questions as: Whom will it reach? How much do the probable readers know about the subject already? How much do they want and need to know? How busy are they? What is the relative importance of the report at the level that the readers occupy? How much time will they give to it?

The following discussion will touch upon several specific qualities. Some of these reinforce each other, but there are times when an excessive effort to achieve one quality will mean the sacrifice of another. Thus in considering the question of how vigorously to cultivate any of them on a specific occasion you will need to foresee the circumstances that will probably prevail when your report is actually used, and decide what the report must be if it is to perform its function under those circumstances.

Self-Sufficiency

A report should be self-sufficient. That is, anyone who is likely to read it should be able to do so without having to rely on his memory or consult his files for extra information.

To make this possible, you may need to include facts that could otherwise be omitted. For example, instead of assuming that you may omit facts you originally received from the person to whom you are reporting, you might well remind yourself that details that were fresh in his mind at an earlier date might be hard for him to recall after the passage of considerable time. Also, it is well to remember that the person you are reporting to may pass the report on to others who lack his own familiarity with the subject. In view of these facts, it is best to make sure that whoever the reader might be and whatever circumstances might exist at the time when he reads, a report should provide what is needed for full understanding.

The material that makes a report self-sufficient comes mainly at the beginning. The opening should make clear the function of the report and the circumstances that called for its preparation so that the question, "What's this all about?" is not lurking in the reader's mind as he plunges into its contents.

Interest

It is natural for a person who concentrates a great deal of attention on anything to become so interested in it that he assumes it is interesting to others. Applied to reports, this is a dangerous assumption. Those who read our reports are busy men. Many matters—probably many reports—compete for their attention. Whether your reports get the attention you hope for will depend in part on whether they make an immediate bid for interest.

Many a report, especially if it does not concern a colorful subject, will have a better chance of arousing interest if it is based on a principle that is a commonplace in sales work: A person often buys something not because he wants what he buys but because he wants the results of possessing it. He buys a suit in the latest style, for example, not because he especially wants the suit but because he wants to be identified with the kind of people who wear suits of that kind. A report writer can attempt to arouse interest in his report by quickly showing that it offers a prospect of producing results that the reader already desires: a better product, for example, or economy of operation or saving of time.

This does not mean that a report should resort to sensationalism in its bid for attention, or that its appeal should be specious. It means only that the interest in reports stems primarily from their practical value, which should therefore be pointed out without delay.

How can this result be accomplished? At least two concrete suggestions can be made. First, organize the body of the report so that the reader discovers as early as possible what benefits may result if your report receives attention. Your skill in organizing in this manner yet at the same time making your plan logical and coherent will be one of the decisive factors. Second, in one or another of the preliminary parts—the letter of transmittal, the covering memorandum, the introduction, or the summary—provide a sentence or two deliberately calculated to arouse interest by foreshadowing the significance of the full report. These sentences should be specific and concrete. Just to say, in effect, "This is interesting and important" is a waste of space.

Whether you can retain interest once you have aroused it depends to some extent on the effectiveness of your writing style. The likelihood that style in itself can *create* interest is slight, for readers read reports for the value of their contents, not for enjoyment. But readability, conciseness, and the color and vigor that can be obtained from concreteness will most assuredly help to retain or increase the interest, and a dull, wordy, obscure style will force interest to engage in a desperate struggle for survival.

Thoroughness

It goes without saying that thoroughness is desirable in a report as elsewhere; but whether this quality is achieved depends in many cases on whether the writer uses his imagination. There are always some obvious questions that a report must answer. But answering the obvious questions is not always enough. For example, an increase or decrease in the number of people taking a training course in two successive years might seemingly point to a conclusion about the success of the course. But difference in weather, the existence of an influenza epidemic during one year, or a change in the makeup of the working force could also affect the number; and the possibility of such outside forces could not safely be overlooked.

There are other factors to be considered for the sake of thoroughness. Assuming that you have demonstrated that the action you recommend would definitely produce the result desired, might there be other and possibly better ways of producing that result? Also, is it possible that the actions you recommend would not only produce the desired result, but would also produce undesired side effects that you have overlooked? For example, you might prove that leaving stubble in the fields rather than burning it would prevent erosion, yet overlook the possibility that leaving it would permit harmful insects to multiply. Or again, a program to destroy predatory animals might succeed in reducing the number of predators but in doing so permit an increase in the population of crop-destroying rodents.

Obviously, if a report is to be thorough, its writing must be preceded by a thorough job of gathering information; but many a writer who has answers to all the essential questions, the less obvious as well as the obvious, fails to include them. He makes the mistake of assuming that because he knows all the necessary questions have been considered, he need not answer them all. He overlooks the fact that the reader will not share his confidence unless he, the reader, is told enough to be sure nothing has been overlooked.

How much detail about secondary questions should a report include? There is no formula that provides an answer to this question, but at least we can realize its existence and use our judgment about whether we have given the reader not only the information he will actually use, but also the extra information that will free him from worry about whether something has been overlooked.

Omission of Unnecessary Material

Like any other desirable quality, thoroughness can be carried too far. We need to ask ourselves not only, "Am I overlooking anything that should be included?" but also, "Am I including anything that could be left out?" If we do not consciously guard against including too much, we are almost certain to err in this direction.

It is not surprising that this should be true. It is natural for a person who devotes a great deal of time and attention to any subject to overestimate the extent to which it will impress others as important. And when we report to those above us, it is easy to forget that something which is a major element in our own area of activity may be a much smaller fraction of the area in which those farther up the line are involved.

The danger of including too much is increased by the fact that, as we gather our material, it is almost impossible to gather exactly what we need and no more. In fact we cannot be positive just what will ultimately be needed and what will not. And since it is painful to see facts and ideas go to waste—facts and ideas that we have worked hard to obtain— it is a temptation to include them just because we have them.

Unfortunately, their inclusion sometimes does more harm than good. A reader is less likely to grasp and remember the facts of real importance if he must separate them from material that he does not really need.

The temptation to include unnecessary material is especially strong when a report leads to negative conclusions and recommendations. To see why this is true, consider the following situation. The question being investigated is whether to harvest a crop of some kind by a different method. The decision rests upon answers to such individual questions as the cost of buying and operating the new machinery, the quality of product it turns out, the skills it demands of its operators, and its reliability. Suppose that the investigation of its reliability shows it to be subject to frequent breakdowns; and since a breakdown at a critical time during harvest would mean a heavy loss because of spoilage, this single defect is enough to bring about a negative decision. If you were reporting on the investigation you would be tempted to tell all you had learned on each aspect of the study. But if you were the reader, you would not wish to waste your time on details about cost of operation, quality of product, and operator skills only to learn later that the machine was not reliable enough to use. You would probably want full facts about the unreliability because those facts would be the basis for a decision, but would want only a minimum of information on the other questions.

Writers of reports or of any other kind of communications will always have to encounter the question, "How much should I say?" And there is no better way to answer this question than by using your imagination—by trying to visualize the moment when those who will read your report pick it up, pressed as they will be for time, and start to read. With this picture in mind, scrutinize the contents of your report and ask yourself constantly, "Would the report perform its function just as well if I left some of this out?"

Freedom From Bias

A person who must investigate a subject and report his findings often starts work with a strong expectation about the probable outcome. When this expectation is based upon sound previous knowledge and experience, it is entirely natural and unobjectionable. When it is based on mere prejudice, however—or even worse, on self-interest—the case is different. There is no surer way to lose a reader's confidence than by letting him gather the impression that you have been swayed by an eagerness to reach a preconceived conclusion.

To be sure, some so-called reports fail to meet the requirement of impartiality. Protagonists of conflicting interests, in preparing reports to use in a controversy, usually present matters from their own point of view. But these reports are not suitable models for a person entering a technical profession. A bona fide technical report is not begun with the purpose of proving something. Rather, its purpose is to present all sides with scrupulous fairness. In the long run, no one in a technical profession will profit by letting his reports be or even appear to be prejudiced. Hard though it may be, you should plan an investigation in such a manner that you will not fail to discover evidence that might run counter to the conclusions you expect and hope to reach, and should present your findings in such a manner as to show that you have been impartial.

Objectivity

The term *objectivity* is sometimes used with practically the same meaning as *freedom from bias*. We will use it here, however, to refer to the quality that is the opposite of *subjectivity*. In urging that reports be objective, we are urging that they be based on concrete facts.

The need for objectivity becomes a matter of special concern whenever a report touches upon human actions or attitudes. For example, reasoning subjectively about what workers would do under certain circumstances might result in a radically different conclusion than would be reached by observing their actual conduct under such circumstances.

Some matters, of course, are basically subjective. The relative beauty of two makes of automobiles cannot be determined by measurement, nor can the question of whether a new process produces better-tasting instant coffee. But even when we deal with questions of taste, objectivity need not be surrendered. It can be attained by getting a substantial number of subjective opinions. Assuming a fair investigation has been made, though we might not be able to say objectively that Brand "A" tastes better than Brand "B," we might be able to make the truly objective statement that 70 percent of the people who tasted both preferred the taste of "A" to the taste of "B."

Restraint

A report should be restrained both in substance and in style. Valuable though enthusiasm may be on some occasions, there is no surer way to lose a reader's confidence than to give him the impression that your enthusiasm has overbalanced your judgment.

Restraint in substance is more than a mere concession to reader psychology. We realize this when we think of the extravagant hopes aroused by the development of drugs that seemed promising but were later found to have undesirable side effects, or the development of insecticides that unexpectedly created problems related to public health.

Restraint in style should lead you to avoid such terms as *astonishing, startling, flagrant,* or *disastrous.* Rather than writing *a tremendous increase* it is advisable to write *a substantial increase,* but even better to say *a 50 percent increase* and let the reader choose his own adjective. Or again, it is better to tell your reader that a flood drove five hundred people from their homes than to refer to it as *calamitous, appalling,* or *catastrophic.*

Restraint is especially important in stating conclusions. This does not mean that you should insert such meaningless phrases as "It is believed that . . ." or "It may be concluded that. . . ." But you should not go beyond the evidence in making claims or predictions, and should recognize the possibility that later developments can always lead to results that no one could have anticipated.

Appropriate Degree of Impersonality and Formality

A personal tone in writing, as contrasted with an impersonal tone, is largely the result of frequent use of the personal pronouns *I* (or perhaps *we*) and *you,* which make it appear that the writer is addressing the reader personally. "I could hear the hum from a distance of 20 feet" is personal. "The hum was audible from a distance of 20 feet" is impersonal.

In some simple situations, especially in letters and sometimes in memorandums, a personal tone—which incidentally conveys the impression of informality too—is natural and desirable, whereas impersonality would seem stiff and aloof. Indeed, there are times when a writer's effort to avoid referring to himself is awkward and even seems evasive. To write "It is believed that . . . ," when your purpose is to express your own belief, seems to be a hedge against accepting responsibility and is vague because it does not answer the question, often important, of who does the believing.

Much of the time, however, impersonality is better. If you write, "I weighed the samples at half-hour intervals," it is apparently your intention to tell us about your own actions. But if you write, "The samples were weighed at half-hour intervals," you are centering attention on what was done to the samples. Of course, if you wished to emphasize that you were the one who did the weighing, the use of "I" might be the best way to do so. Much of the time, however, an impersonal version is more likely to emphasize the facts that are really significant, because a technical writer's purpose usually is to tell about things—the samples, in the case above—rather than about himself.

It is not always necessary, to be sure, that all parts of a report be personal or impersonal to the same degree. There is no reason that on occasion a writer should not use *I, we,* or *you* in the preliminary parts of a report, where he is explaining the circumstances and his own involvement, but stop using them when he begins to present the real substance. He might start, for example, "As you requested, I visited . . . to investigate . . . ," but make no reference to himself in stating the facts that came to light during his investigation.

Of the three pronouns to which attention has been directed, *I* goes farthest toward imparting a personal tone, for reasons that have already been pointed out.

You is also decidedly personal. It conveys the impression that the reader is being directly addressed and thus involves him personally in what is being said. Sometimes, of course, it may appear in a report directly addressed to more than one reader. Obviously, it sets the writer apart from the reader, and it is therefore often suitable in a report written by someone such as an outside consultant. It is less likely to be suitable, however, when the writer and the reader are part of the same organization and the report will go beyond the immediate reader who is addressed as *you.*

The use of both *I* and *you* is affected by another consideration than mere style. The meaning of *I* depends on who is using it, and the meaning of *you* depends on who is being addressed. When it is likely that a report will be incorporated into a larger report by a different writer,

I and *you* should be avoided when doing so would reduce the need for changes.

Sometimes, too, if a report will reach the attention of a good many people besides the person or group immediately addressed, *I* and *you* may be undesirable for another reason. The effect that the two words have on these other readers is the exact opposite of their effect on the person who first received the report. That is, since *you* does not include these others, it is a constant reminder that what they are reading was addressed to someone else, not to themselves.

Indeed, at times a report will be of value, it is hoped, to a good many others besides the original recipients. For example, in an actual case, a report produced by a university research division concerning the waste from a food-processing plant was written in behalf of one company and one community. It was printed, however, and sent by request to a large number of other companies and communities that faced the same problem. A personal, informal style would have been out of place where such a use was foreseen.

We is a term that can have either of two meanings, at least as it might be used in a report. It can refer (1) to the writer and those associated with him, as contrasted with the reader, or (2) to the group or organization to which both the writer and the reader belong. In the first use it is comparable to *I*, but a group such as a committee can use *we* without seeming so personal as an individual seems when he uses *I*. In its second use, *we* is often effective because it subtly suggests that the writer and the reader are both part of the same larger entity—that they are closely associated and have common objectives—presumably the interests of the group or organization as a whole.

In view of all these considerations, what is the best attitude to take in regard to personal and impersonal style? The following recommendations should answer that question.

Do what is natural, and what common sense dictates in view of the situation, the nature of your report, and the use that will be made of it after you write it. If its purpose is to tell about your own actions and hence you cannot avoid *I* without sounding stiff and awkward, use it. If what you write is of primary concern to the immediate reader or readers, use *you* when you need to. On the other hand, when you are telling mainly about *things* and there is no real reason to refer to yourself, the natural and most effective way to do so is in the impersonal manner that you will find in so much of the technical writing actually being turned out in American industry and government. Your style, when you do so, will not be painfully stiff and artificial if you develop reasonable skill as a writer.

As for formality and informality—when a writer uses a personal style throughout a report, that fact in itself conveys an impression of informality; but diction in general is also involved. The following list suggests the difference between formal and informal language. Some of the formal equivalents to informal terms, to be sure, are not the only ones that might have been used, because some of the informal examples are variable or general in meaning.

Informal	Formal
raised the roof	objected strenuously
knocked its performance	criticized its performance
out of kilter	in poor condition
O.K.	satisfactory
green operator	inexperienced operator
is going to	will
won't stand still for	will oppose
not about to	will not
pretty good record	fairly good record
chance it	take the risk
kick (or gripe)	complain

Just how formal should you be in writing a report? Certainly there is no reason to avoid informality if your personal relationship with the reader is informal, if he is likely to be the only reader, and if the matter is one of only temporary concern. But if a report is likely to go beyond the immediate reader or become a matter for future reference, informality is usually inappropriate. It is usually not advisable to refer to a microphone as a *mike*, to a laboratory as a *lab*, or to television as TV. "He lost his temper," is better than "He blew his top." "Not very serious," is better than "not all that serious." "Will not," is better than, "is not about to." Even contractions are usually avoided when the style of a report is not, in general, informal, because they are considered colloquial.

It is neither necessary or desirable, however, that in avoiding informality you should go to the other extreme and use language that is stiff or pompous. The ultimate degree of formality, found for example in legal papers, is never necessary and is seldom even appropriate in technical writing.

Changing informal to acceptably formal style is not difficult. Suppose, for example, that you wish to replace *you* used informally for *one* in the sentence, "You cannot make the trip in less than four hours." If *one* seems to be stiff, it is possible to write, "The trip takes at least four hours," or "It is impossible to make the trip in less than four hours." In brief, there is no reason that formality should be painfully conspicuous. The only necessity, when informality would not be appropriate, is to write with enough dignity to suit the occasion. And if you engage in professional work, you need to learn how to avoid excessive informality without letting your writing become stiff and awkward.

Avoidance of Chronological Order

Many a report annoys its readers because it is written as a narrative with its contents arranged in a chronological order. A typical example was a report actually written in a government bureau concerning three cabins that had been built illegally on government land in a remote area on the shore of a lake formed by a power dam. The report was submitted in late September.

The writer told first of discovering one cabin in June while he was on a horseback trip with another bureau member. The cabin was vacant and contained no evidence of its ownership. In August, on a boat trip, he found that two more cabins had been built nearby. A newspaper with a name and address on it led to a family who were able to identify the three owners. The owner of the first cabin was called upon and claimed that he considered it all right to build his cabin because a friend had told him that the power company had no objections. The bureau representative—the same one who wrote the report—informed the owner that he knew of no arrangement involving the power company, that the matter would be further investigated, but that the cabin would probably have to be removed. The owner said that in such a case he would move the cabin onto piles in the lake (which incidentally would also be illegal and would not be permitted). The owner was told that he would be informed of the results of the investigation before being required to move the cabin, but that removal would almost certainly be necessary. The report closed by saying that the other two owners had not been contacted because a survey was necessary to establish the fact that their cabins were definitely on government land.

Put yourself in the position of the reader—as the writer should have done—and ask yourself what, in that case, you would really be most interested in finding out. If you do so, you will probably conclude that the reader would have been better served by a report along these lines:

1. Identification of the subject and reference to any previous memoranda concerning it.
2. The fact that the ownership of the cabins had finally been established and that one owner had been interviewed.
3. The attitude of the owner and what he had been told.
4. The fact that interviews with the other two owners were being postponed until a survey made sure whether their cabins were on government land.

As it was written, the report consisted of two and one half pages of single-spaced type. The reader was forced to wade through more than half of its length before he came to what really counted—the way matters stood when the report was written.

Charles Dickens once said that the way to write a successful novel was to "Make 'em laugh, make 'em cry, make 'em wait." This may have been good advice to an aspiring novelist, but a report writer should by no means "Make 'em wait" for what they are most interested in learning.

A report that consists of a chronological narrative is easy to write because it poses no problem of organization. All that the writer needs to do is to start at the beginning and continue to the end. Once in a while, when a writer is reporting only to tell of his own activities, such an order may be suitable. Also, there are times when the format of a report takes care of the problem. In a laboratory report, for example, the headings identify each section and enable the reader to read them in whatever order he pleases. A formal or semiformal report has special sections at the beginning to present the heart of its contents.

But on the many occasions when the report is not one of these special types, a writer should resist the temptation to do what is easiest. He should realize that the reader usually wants to begin not where the writer began but where he arrived at the end. He should tell first what the reader will be most interested in knowing first, and only when he has done this should he provide the other information that he considers necessary.

Consideration of Emotional Factors

Reports, like other kinds of technical writing, should appeal to the mind rather than to the emotions. When it is our responsibility merely to present objective facts without making any attempt to say what should be done about them, this principle can be applied without modification. There are times, however, when the emotions cannot be disregarded. Though this point receives little attention in many discussions of technical writing, it is often emphasized within the broad field of industrial communication.

To be sure, an appeal to the emotions cannot be relied upon to do the constructive job of selling our ideas; but there are many occasions when *ignoring* people's emotions can destroy our chance of selling them. Why should this be true? The answer is that many actions recommended in reports affect not only things but also people—their egos, their pride, their status, their attachment to established ways. Suppose, for example, that an investigation shows a new method of doing something to be far more efficient than the method already in use. If the person who writes the report, in his eagerness to prove this, is too devastating in his attack on the old method, he may force people to defend it because the attack seems to reflect discredit upon them personally.

After all, those to whom we report are not always entirely independent in deciding what is to be done. They often need to secure the concurrence of other persons, whose objections could easily delay or seriously endanger the chance of favorable action. If we can present our findings in such a manner that we do not antagonize these other persons, we increase the chance that our recommendations will be followed.

The preceding warning is necessary because it is always a temptation to build the strongest possible case for what we believe in, and it is consequently a temptation to demonstrate that a contrary position is unthinkable. If we yield to this temptation, however, we may needlessly arouse resentment and opposition. And even if the opposition can be overcome, this antagonism may reduce the likelihood that our recommended measures will be successful.

This does not mean that when you write a report, you should be insincere in order to prevent people from disagreeing with you. However, you should give some thought to the question of who will be affected by your report. So far as possible, present your case without hurting these people, angering them, or humiliating them. Don't make anyone look bad when you can avoid doing so. Remember that even if your case is logical, people are less likely to see its logic if they feel hurt or resentful. In brief, don't increase the opposition to your ideas by ignoring the fact that people have feelings.

EXERCISES

Exercise 1

Rewrite each of the following sentences so that it shows the freedom from emotionalism, the objectivity, and the restraint that would be appropriate in a report. If necessary, you may make up objective facts that desirable style would call for.

1. Purchase of this versatile machine will free us from the headache of keeping on hand dozens of sprays, foams, special powders, and liquids that we now use for cleaning.
2. Formula X-98 possesses a fantastic ability to annihilate bacteria, fungi, and yeasts associated with slime.
3. Preserving a stretch of frozen, barren tundra in its present desolate condition is a trivial consideration in comparison with having people suffer from cold because of scarcity of heating oil.
4. Wanton destruction of the delicate balance of natural conditions means nothing to the greedy magnates in comparison with their insatiable desire to line their pockets.
5. Use of the Excello liquid dye system would result in an amazing reduction of clean-up time and would therefore be a splendid decision.
6. This plant was in deplorable condition. It consisted of a ram-

shackle, run-down building in which rickety machinery was operated by crews under lackadaisical supervision.

7. Soundproofing is desperately needed, for jet planes from a nearby airport constantly fly over the building at low altitude wailing like banshees.

8. To reach the barge-loading facilities trucks would be forced to creep along at a snail's pace over a narrow road that is little more than a goat track.

9. Ever since the first man-made plume of smoke rose into the sky, the human race—unconcernedly until the last few years—has been cheerfully polluting the air.

10. Those who object to the production of so many washing machines, refrigerators, and television sets seem blithely unconcerned about snuffing out the aspirations of the poor and underprivileged to have what most of us take for granted.

Exercise 2

Rewrite each of the following specimens so that the language is standard English and the style is impersonal. (There is no intention to imply that every expression you eliminate would be wrong in circumstances where informality was suitable.)

1. Though we had expected to catch him off base, someone had apparently tipped him off so he was able to cover up his irregularities.

2. Early rumors about the proposal caused a lot of static, but when it was explained in detail the objections fizzled out.

3. At first the supervisor tried to give us the runaround, but finally he leveled with us.

4. As a result of the scuttlebutt he came to have a peculiar slant on things.

5. Some of his recommendations were pretty far out, but when anyone objected he lost his cool.

6. In areas where dragging out logs with cats would tear up the ground, it might be practical to lift them out with choppers.

7. When the motor kept conking out it became obvious that the mechanic had fouled up the repair job.

8. When the griping continued he tried to do a snow job on the crew but they didn't buy it.

9. Fuel supplies will be on the skimpy side but we will make out okay if no one presses the panic button.

10. We had a tough time to get the lowdown, for though we urged the supervisor to lay it on the line he clammed up on us.

8

NONFORMAL REPORTS—
THEIR VARIATION
IN FORM AND PURPOSE

Though the qualities discussed in the preceding chapter are always worthy of consideration, the reports in which they are cultivated vary widely in form and purpose. If you work for a large corporation or government bureau, you may find that it has developed certain types to suit its own particular needs, and you should obviously conform to its preferences. Those who use this book, however, will work for a wide range of employers and clients. Consequently there will be no attempt, in this chapter, to fit all reports into a neat set of classifications like a botanist's classification of plants. Rather, though some of the widely recognized types will be discussed, the purpose of the chapter will be:

1. To give you a better understanding of the wide ramifications of reports in form and purpose.
2. To make the nature of the commonest types familiar enough that if you are asked to write a report of a certain type but have available no further guidance, you will know what is probably expected and also realize how much latitude you enjoy.
3. To encourage resourcefulness and self-confidence, so that if you are asked to write a report but given no instructions about what kind, you will know whether some generally recognized type is called for and if so know how to produce it; and if no particular type seems to fit the occasion, you will realize that not all reports are of some

established type and will proceed with assurance to produce what the occasion calls for.

The kinds of report to be covered are: reports made by filling in a form, reports in the form of letters, reports in the form of memoranda, progress and status reports, periodic reports, laboratory reports, and nonformal reports that cannot be considered to be of any special type. You will find that these types are not mutually exclusive, for some of them owe their identity to form and others to purpose.

Formal and semiformal reports are briefly identified, but since they involve a substantial number of special elements, full discussion of them is postponed to a later chapter.

REPORTS MADE BY FILLING IN A BLANK FORM

Many reports are made by merely filling in the spaces on a blank form on which headings are already provided. This can happen when, in a particular organization, some activity becomes so routine that the information about it will always occupy approximately the same amount of space and will always be covered by the same headings.

Frequently the information filled in consists merely of figures. In this case making the report is not a problem of writing at all. Yet sometimes the blanks must be filled in with ordinary prose. On these occasions the printed forms are used to make sure that the organization of the reports will always be the same, to prevent the writer from overlooking any essential routine point, and to discourage any tendency to make the report too long. Such reports are good in that those who read them always know exactly where to find what they are looking for. Their main limitation is their lack of flexibility. Sometimes flexibility is increased, however, by a section for comments.

Writing a report on a blank calls for no skill except the ability to gather the necessary information and to write clearly, accurately, and concisely. The style should be reasonably formal in a report that will become part of some organization's permanent records, but may be informal in a report dealing with a matter of only temporary concern. A report made on a form appears on page later in this chapter.

REPORTS IN THE FORM OF LETTERS

When there is no need for a report to include tables, figures, or any kind of stiff and bulky material, the natural form to use is often a letter. Most of what you need to know about reports that are letters can be learned when letters are discussed later in the book, for the same good practices apply for any letter, regardless of its use. The form for a report is no different from the form of other letters except that the use of a subject line is more likely to be desirable. In language, the letter serving as a report may not call for quite so much concern about tone, negative material, and good will, for the reader is likely to be in your own organization. However, if a letter is used in making a report to someone on the outside—a client rather than an employer—tone and good will are as important as in any other letter.

A letter can be used to report on almost any type of subject. Letters that serve as reports often run longer than the average business letter. Long or short, a report in the form of a letter should be just as carefully organized as any other piece of writing of comparable length and complexity. Such devices as topic sentences for paragraphs should not be neglected. (Topic sentences are discussed in the Handbook under Paragraphs.) Indeed, in a letter report of two or three pages, it may be desirable to use internal headings, as in any other report. Their use is covered in Chapter 9.

Reports in the form of letters can vary widely in formality. When addressing a close associate, you might often be justified in making your style extremely informal. There are occasions, however, when your sense of fitness—your imagination in visualizing how the report will be used and whom it might eventually reach—should make it apparent that more dignity is indicated.

One final caution is needed regarding the use of the letter form for a report. If the subject is likely to be of permanent concern, the letter form is sometimes undesirable because it may cause the report to be filed along with ordinary correspondence rather than as part of the organization's permanent records. When this likelihood exists, it may be advisable to prepare a report in some other form and write a covering letter to accompany it.

REPORTS IN THE FORM OF MEMORANDA

In many ways a memorandum resembles a letter. Like a letter, it is one of the forms used most frequently. It is not usually used for a report that is long or one that calls for the use of other material than ordinary text, and it may range from informal to formal in style. But unlike a letter, a memorandum is strictly for use within the organization—for communication within a department or between departments. Consequently, a memorandum may be unceremonious in style and may sacrifice niceties for conciseness. This does not mean that bluntness and tactlessness are to be condoned. Tact and diplomacy help to get results in dealings within an organization just as they do in outside contacts.

Originally the term "memorandum" implied something of a temporary nature. This is no longer true; yet there are times when a memorandum is written to make immediately available some information that will later be included in a longer, more formal report. Some of the largest corporations in the nation make this use of memoranda.

For use in writing memoranda, printed forms are usually provided. If no form is provided, however, you should provide the necessary information at the top of the first page, using the following form.

```
To:                              Date:

From:

Subject:
```

PROGRESS AND STATUS REPORTS

Progress and status reports, if indeed they should be regarded as separate types rather than a single type called by different names in different places, resemble each other closely enough that it seems advisable to discuss them in the same section.

The term *progress report* has been widely used for so many years that what has been said in the past about the reports it identifies is still appropriate.

In recent years more and more use has been made of the term *status reports.* If the reports so designated are to be regarded as anything except traditional progress reports under a different name, and if the use of the two terms is what one might expect, the key to the difference between the kinds of reports lies in the basic meanings of the two terms. The term *progress* suggests that the purpose of the report is primarily to tell about change—about headway made in a desired direction. *Status,* on the other hand, suggests a report in which the main purpose is to describe existing conditions.

Actually, this means primarily a difference in emphasis. One type emphasizes what has been happening; the other, what condition exists because of those happenings. There is no reason to feel, however, that the two types—if such they are—contain entirely different information. It would be difficult if not impossible to tell about progress toward a desired end without telling what condition had come about as a result of that progress. In fact, it is natural to begin with a comment about that condition. Likewise, it would, to say the least, be unnatural to tell about existing conditions—status, that is—without telling what had happened since some earlier time to bring those conditions into existence.

Occasionally the discrimination between the two terms is made on a different basis. That is, *status report* is used in reference to a report that concerns a specific project that will eventually be completed—placing power lines underground, for example, where they have formerly been on poles. *Progress report* is used to identify reports that tell about progress made in achieving the general objectives of the company, such as improvement in service, that will presumably continue as long as the company remains in existence.

In any event, you should be aware of the existence of both terms and feel at ease in using one, the other, or both as is customary in whatever organization you become connected with. If usage in that organization provides no definite guidance, you should feel free to use whichever term is more appropriate in view of what your report emphasizes. Regardless of your choice of terms, your main concern is to give the reader the information that he needs.

THE PERIODIC REPORT

A periodic report, like a progress report, is distinctive because of its purpose. It is merely a report made as a matter of course at regular intervals, frequently but not solely for the purpose of keeping records. If it deals with a project that will eventually be completed, it is also a progress report, for the two types are not mutually exclusive. In its simplest form it may be merely a filled in blank form—as when a weekly report is made on accidents, production, or the percentage of production failing to pass inspection. It may also be a letter or a memorandum. From these simple forms it may range upward to the elaborate form used in the annual reports of the board of directors to the stockholders of a great corporation—an expensive, illustrated product, printed and distributed by the thousands.

In any given situation the periodic report tends to settle into a well-established form because the points it must cover are usually similar on successive occasions. Consequently, unless you are the first person to fill a position, you will often be able to pattern your periodic reports on those written in the past. Thus the problem of organization, even in a long periodic report, may not be difficult. It pays, however, to be always on the watch for special points that call for unusual treatment and to remember that even when a report follows an established pattern, the difference between good and bad writing affects its quality. And finally, you should not neglect to change the established pattern if you find that it can be improved and are in a position to make changes.

THE LABORATORY REPORT

One of the commonest purposes of reports likely to be written by those who read this discussion is to tell the results of work done in a laboratory. Sometimes this work consists of merely running tests, the results of which may be stated without reference to the reason that the report was requested. On other occasions the report must not only tell the results of laboratory work but must also apply them to some specific problem and even recommend action. In the latter case the laboratory-report form may be discarded in favor of some other form more suitable to the specific occasion.

Most of the time, however, a report on work done in a laboratory may be written under headings that have become fairly well standardized. Though there is some variation in different organizations, the following list covers most of the divisions that a laboratory report is likely to contain.

Title page, or merely a title
Object (often called Purpose)
Theory
Method (or Procedure)
Results
Discussion of Results (often called Comments)
Conclusions
Appendix
Original Data

It is not always necessary to include all these divisions. Much of the time there will be no need to explain the theory; and if the method or procedure has become standardized, the section dealing with it may also be dispensed with. The results, of course, would always be given, though they might be placed on data sheets at the end if they consist of figures. There may or may not be reason to discuss the results; but even when such discussion is needed; it may be placed in the section "Results" rather than in a separate section.

Sometimes a section on conclusions will be unnecessary because the results themselves may be the only conclusions arrived at. A section on conclusions is desirable, however, when consideration of the results has led to convictions that go beyond the results themselves. For example, the results of a test of samples of a type of steel might consist of no more than the facts about its tensile strength and other qualities, but consideration of those results might lead to a conviction as to whether the steel would be suitable for some prospective use.

An appendix need not be used unless it is advisable to add materials omitted from the main report because they are too detailed, too technical, or too bulky. As for "Original Data," much of the time the full data are included under "Results." If the full data seem needlessly detailed, however, the figures under "Results" may be reduced to mere averages and the full data may either be omitted or be placed at the end to permit verification. If placed at the end, they may be included as part or all of an appendix.

The arrangement, as well as the selection of points, in a laboratory report may be varied. The order listed above is logically sound, but sometimes changes may be desirable for the reader's convenience. Conclusions might be placed early in the report, perhaps following the statement of the purpose, for the conclusions are likely to be what the reader wants to know at once. In fact, some of those whom a laboratory report reaches may read the conclusions and nothing else.

To sum up: If given a specific form you should follow it as closely as possible, for the reader will grasp the contents better if they are placed in a form with which he is familiar. If no particular form has been requested, you may use the heads listed above to the extent that you need them. The result will be standard in form so far as any general standard exists.

FORMAL AND SEMIFORMAL REPORTS

For the sake of completeness two more types of reports, formal and semiformal, must be identified here, although they are more fully discussed in Chapter 11.

A formal report is a report that is dignified and impersonal in tone and that presents its material in a pattern resembling the pattern used in books. The extent of this resemblance and the techniques that the pattern calls for will be expanded upon later. It is unfortunate that the term *formal* is used to identify these reports, for its use seems to imply that all other reports are informal. Actually, many reports not classified as formal are free from any flavor of informality. *Nonformal* rather than *informal* is a better term to apply to other types.

The term *semiformal report* has not been used widely enough to have acquired a generally accepted meaning. In this book *semiformal report* means a report that has the tone and general qualities of a formal report but that is simpler in design because it dispenses with some of the parts. Recognition of these reports as a distinct type is useful, because there are numerous occasions when the full apparatus of a formal report is needlessly elaborate but the use of some of its parts and compliance with its general conventions will produce exactly what is needed.

REPORTS THAT FIT INTO NO CLASSIFICATION

There are many varieties of reports that fit into no special classification, especially in their form. They are not made on blanks; they are not letters, memoranda, or laboratory reports; and they do not contain all the parts usually found in formal reports. This often causes uneasiness in inexperienced writers, who often worry too much about form and too little about the reader's needs.

There is no reason, however, to be unduly concerned if you are asked to write a report and none of the recognized types seems to fit the occasion. You should be able to get satisfactory results by handling the job as follows.

First of all, consider the function that your report is supposed to perform. And since it cannot perform its function unless it secures the desired reaction from its readers, think also about who the readers will be—what their interests are and what they already know about the specific subject and the general area in which the subject lies.

Next, turn to the question of what material you will need to include if the report is to perform its function—what points must be covered and what information on each of these points the readers will want or need. As you decide what information to include, always ask yourself how long a report the readers will wish to read.

After you have gathered the necessary information, organize it by using the method recommended in Chapter 3. This method is flexible enough to let you form a plan which organizes your material logically and at the same time makes it easy for a reader to find answers to the questions that are likely to be uppermost in his mind.

Unless the report will be short, decide whether the inclusion of any of the special elements discussed in the next chapter would make it more effective, and use any of them that seems likely to fit the occasion. In particular, use plenty of headings. If you have made a good outline, it will provide headings that cover whatever information you wish to present.

As you start the actual writing, decide whether the occasion calls for a style that is formal or informal—personal or impersonal. Having made your decision, be consistent in using the style that you have chosen as appropriate.

Start your report in such a manner that the reader knows at once that it *is* a report, can tell its specific subject, and can see what function it is supposed to perform.

Mention, near the beginning, any earlier reports or other written material that you will need to refer to or that are specially connected with the report you are writing. This material is sometimes headed

References and listed before the text of a report begins. (Such a list of references is not to be confused, however, with a list of references placed at the end of the report for the sake of documentation.)

Always be sure that the date of your report is indicated at a point where it will be easy to see.

If you are writing the report as an individual, always identify yourself as the writer. One way to do this is to use the form you would use if you were writing an article. For example.

```
A Report on the Breakage of Laboratory Equipment
                           by
         Robert A. Watts, Head of the Testing Bureau
                      July 15, 197-
```

When such a method seems needlessly elaborate, the necessary information can be placed in the first paragraph. If your report consists of a single page, you can identify yourself by telling your position at the opening and signing it at the end. (The phrase *Respectfully submitted* is often placed before the signature under these circumstances.) If the report is more than one page long, however, it is better to mention your position at the beginning and supply your name by signing it at the end. A personally written signature is often desirable, under any circumstances, to indicate that the writer accepts responsibility for the report.

If you are writing in behalf of a committee, it is customary not only to identify the committee at the beginning but also to list the members at the end. Sometimes, in fact, it is desirable to have the members personally sign the original, official copy. The names of members are ordinarily arranged alphabetically except that the name of the chairman comes last and his position as chairman is indicated.

The simplicity of the preceding suggestions demonstrates that writing a report that is not one of the special types is not particularly difficult. All you need to do is observe a few simple conventions as you give the reader the information that he needs in a clear and easily understandable manner. If you do this, the question of whether your report can be classified as to type is likely to be inconsequential.

SPECIMENS OF NONFORMAL REPORTS

The following reports are not offered as models. Moreover, though they include most of the types discussed, they are not intended to encourage you to draw sharp boundary lines between types. To draw such lines would be unrealistic.

The real purpose of including examples is in part to make the business of report writing more concrete. Further, it is probable that among the specimens you will find some devices that you can utilize in your own report writing, even though the reports in which they occur may not serve as patterns to follow in full. Perhaps the most important function of the examples, however, is to show how little regimentation exists in report writing. The more experience you have with reports, the more you will realize that good report writing is a matter of intelligent analysis of the reader's needs, effective organization, and clear, concise style rather than a matter of blind compliance with rigid specifications. Accordingly, you should feel encouraged to use your own judgment when you write reports.

As you study the examples, remember that no report writer ever achieves perfection. Consequently, be on the alert for alternative methods of presentation by which effectiveness might have been increased.

(Illustrative specimens follow.)

SPECIMEN NO. 1*

Report on a Blank Form

The report on the opposite page performs a dual function. As the print at its top shows, it serves as the title page of a longer report, but it is also a self-sufficient report in its own right, to be distributed more widely than the pages it precedes.

In providing the facts necessary for self-sufficiency, which are not repeated on later pages, it makes them available both to those who receive the main report and to those who receive the single page.

It also provides the gist of the full report—giving those who depend on it what is most essential and serving as an abstract for those who receive everything. It is less technical than the full report.

It was followed by twelve additional pages—seven of discussion, one listing references, one explaining nomenclature, and three bearing a table and four figures.

Such a report as this serves the needs of those who only want to keep informed about company activities, those who are concerned with its contents but do not need full details, and those who need the full story. The page on nomenclature has been increasingly used in recent years. Likewise, the inclusion of a list of key words reflects a relatively new development in report writing.

* Courtesy of the General Electric Company.

GENERAL ⊛ ELECTRIC

MAJOR APPLIANCE BUSINESS GROUP

TECHNICAL INFORMATION SERIES

Title Page

AUTHOR C. D. Denson/ D. L. Crady	SUBJECT Polymer Rheology	NO. 72-MAL-23
TITLE Measurements On The Planar Extensional Viscosity Of Bulk Polymers: The Inflation Of A Thin		DATE 11/20/72
		GE CLASS I
		NO. PAGES 12
ORIGINATING COMPONENT	Rectangular Polymer Sheet MAL-Plastics Laboratory	

ABSTRACT

An experimental study of the inflation of a thin rectangular polymer sheet has been conducted to determine whether this technique can be used to measure the planar extensional viscosity of bulk polymers. Viscosities were determined at various extensional strain rates using an undiluted poly (isobutylene) of high molecular weight. Measurements were also made on the complex shear viscosity and these values were found to coincide with a "normalized" planar extensional viscosity when plotted as a function of the square root of the second invariant of the strain rate tensor. Industrial fabrication operations where the planar extensional viscosity is expected to play a role include vacuum forming of large rectangular sheets whose length to width ratio is large and in injection molds having a center gated cavity.

KEY WORDS

Polymers, Plastics Processing, Extensional Flows, Rheology

INFORMATION PREPARED FOR Major Appliance Business Group-Plastics Lab.
(name)   (dept. or lab.)

REPORT APPROVED BY R.S. Hagan   Plastics Lab.
(name)   (dept. or lab.)

Additional Hard Copies Available From Ashley Library

Microfiche Copies Available From - - Technical Information Exchange
P.O. Box 43 Bldg. 5, Schenectady, N.Y., 12301

MA-775 (Rev. 72)

167

SPECIMEN NO. 2*

Report in the Form of a Memorandum

The report on the opposite page is included for two reasons. First, it has an irresistible appeal because of its human interest. Second, it demonstrates the fact that no point in the nation is so remote that it is immune from being the concern of report writers.

This particular specimen was chosen from a substantial number that also tell of field trips to remote Alaskan villages. Its style, like that of the others, is personal and informal—which is entirely appropriate under the circumstances. Though it starts as if it would be a personal narrative, only a very few lines precede the point where the writer begins to present the facts he was supposed to obtain. And actually, the information about the difficulty—insurmountable as it turned out—of reaching the island is highly relevant because of its possible effect on future trips and because it contributes to the reader's understanding of the island and its population of 87 self-reliant human beings.

* Courtesy of the State of Alaska Department of Highways.

State of Alaska

TO:
Donald D. Friend, P.E.   DATE: August 8, 1972
Pre-Construction Engineer
Western District   FILE NO: 92-2519
Department of Highways
                                                SUBJECT: Local Service Road
FROM: Little Diomede Island
Thomas A. Young
Design, Pre-Construction Section
Western District

On August 7, 1972 I attempted a field trip to Little Diomede Island
to determine the extent of road repairs required as per letter of
June 13, 1972 from Davis Menadelook, Village Council Secretary.

Upon arriving in Wales, I was informed that the water was too rough
to take a skin boat to Little Diomede at that time. However, I
learned that the Little Diomede Village Council President, Robert
Soolook, and Davis Menadelook were stranded at Wales. I spoke with
them and acquired the following information:

Little Diomede has a population of 87 and the only vehicle in town
is one snow machine. They have approximately 2400 feet of 10 foot
wide cobblestone type walkway through the village and one gravel
footpath, approximately one half mile long and three feet wide to
their docking area. The entire cobblestone walkway needs repairs
and about half of the footpath requires repairs. All repairs are
to be done by hand. They plan to carry rocks from the beach for the
walkway. The beach is about 200 feet away at the closest point.
They also plan to use hand shovels to repair the footpath utilizing
suitable material located alongside the footpath for almost its
entire length.

They informed me that they only need funding so that the Village
Council can hire approximately 10 to 15 laborers at a rate of
$2.10 per hour. They figure it will take three to four weeks to
complete all repairs. I told them that using 15 people at $2.10 per
hour for eight hours per day for three weeks would amount to $3,780.
However, they did not think it would take more than $1,500 to do the
required work as they anticipate a shortage of laborers and considered
this aspect in their estimate. Robert Soolook and Davis Menadelook
both agreed that $1,500 would be sufficient funds. They are not
aware of any formal Townsite Plan for the village but said that right-
of-way is considered to be no problem.

Existing utilities will not be interfered with during the repair work.

The water was still very rough by late afternoon so I was unable to
reach Little Diomede to visually inspect the proposed repair work.

I suggest that $1,500 is not an adequate amount and that $2,000 would
be more suitable to meet the required expenses of repair work to the
walkway and footpath.

TAY/vbh

169

SPECIMEN NO. 3*

Report in the Form of a Memorandum

Like the preceding report, the one opposite is a single-page memorandum—though a chart and a table accompanied it.

Though it is a memorandum, the heading *Coordination Sheet* indicates its special purpose. The heading also provides for other information not present in most memoranda but needed in such an immense company. There is a place for its number, one (not used) for "Item," one to show the airplane model concerned, one to show distribution beyond those addressed. The heading *Group Index* in place of *From* shows its origin.

After listing references (other pertinent papers), the writer opens by referring to the request that caused the report to be written. Next he tells about the study of the question raised and about its results. Then, he analyzes the results of the study, and he finally states the decision— that no further action is called for.

The discussion is technical but would not be hard to grasp for those to whom it is addressed. Actually, an attentive layman can at least follow the pattern and see that it represents an orderly progression from the problem to the decision.

* Courtesy of The Boeing Company.

TO
A.A. Coe 1A-44
R. B. Allyn   2A-19

C.C.   H. Watson   2A-19   C. E. Pfafman   OL-40   NO. 747-W-2502
        A. H. Webber   OL-32   J. D. Mayer   2A-19
        R. M. Mason   OY-01   ITEM NO.
        M. Loinonen   OY-02
        W. B. Fehring   OY-03   DATE 08-27-69

                                                               MODEL   747

GROUP INDEX   747 Wing

SUBJECT   Walking Beam (W.B.) Support Clearances for 747B

REFERENCE   C/S 747-W-2472, CP-747H-1005A, CP-747H-1092, and CP-747H-1169.
               Layout 747-LO-W-276, Sheets 1 and 2

Reference coordination sheet CP-747H-1169 requested the Wing Group to reconsider
the actuator/hanger clearances as shown on Layout 747-LO-W-276, sheet 1 & 2,
since the results obtained (by Hydraulics Group) from the test rig and airplane
R0001 do not agree with those shown on the aforementioned layout.

With reference to this we make the following comments:

The Wing Group has made a thorough study of the system with reference to adverse
tolerances, pressure in system, strain of links and structural deflection, and
the following are the results of this study:

1. With "0" psi in the system and no structural deflection, the angle
    between the centerline of actuator and centerline of Walking Beam
    Link is 43° 30' (jig position).

2. With 3600 psi pressure in the system, maximum structural flight
    deflection (as supplied by Stress Group) and installation tolerances
    at their "worst" condition, the angle betweeen the centerline of
    actuator and the centerline of the W.B. link is 40° 40'. (See Attachment)

After this study of the entire system, it is our contention that #2 represents
the worst possible condition between the actuator and the W.B. link, and this
angle is 40° 40', not 39° as quoted on coordination sheet CP-747H-1169. Since
this group has provided clearance between the actuator and the walking beam link
down to 38° (centerline to centerline), this clearance is considered to be
adequate and no further work is contemplated.

Regarding the 36° angle as quoted on coordination sheet CP-747H-1169, this is
considered to be an impossible situation, since the location tolerance of the
W.B. is ±.015" (see drawing no. 65B00061 Zone B5) not ±.25 as quoted on
coordination sheet CP-747H-1085A (dated 4-9-69 Thompson and Coe to Webber).

It is also the contention of this group that to perform a dimensional check on
airplane R0001 is not reliable since all jigs and fixtures for installing the
landing gear system were not available and some components were "Hand-located."

No further action is planned by the Wing Group.

L.B. Arnold

L. B. Arnold

           For M. H. Goshemour

SPECIMENS NO. 4 AND NO. 5*

4. Report in the Form of a Memorandum
5. Report in the Form of a Letter

These two reports are taken up together because both of them grow out of the same situation—a traffic study made as the result of a citizen's request for changes in the existing conditions.

The memorandum, No. 4, progresses logically by first listing the studies made as a result of the request and then summing up the facts that these studies revealed. The result, under the heading *Recommendations,* is a recommendation that the requested changes should not be made. The studies showed, however, that making certain other changes would improve the troublesome conditions, and the writer recommended that they be made. These were explained under *Improvements.*

The letter, No. 5, was sent—and is properly regarded as a report—to the citizen who had made the request that led to the studies. Interestingly, Report No. 5 was written by the official who received rather than wrote Report No. 4.

The writer might, of course, have merely sent a copy of the report he himself had received, along with a brief covering letter, but he wisely decided that something less terse and impersonal would be better when sent to someone outside the department.

The letter is excellent. First, it shows that the citizen's request was taken seriously enough to lead to a thorough investigation. Then it presents the facts that the investigation revealed. Next, it tells the decision that the facts led to. Finally, it ends on an upbeat by letting the reader know that his effort has resulted in positive action of another sort and thus has succeeded in bringing about improvements.

* Courtesy of Oakland, California, Dept. of Traffic and Parking Engineering.

CITY OF OAKLAND
Inter-Office Letter

To: _____ Arnold Johnson _ Attention: _____ Date _February 5, 1971_

From: _____ Michael Pickering _____

Subject: _____
Staff Report on Kennedy
Street and 23rd Avenue _____

As the result of requests for additional openings in the median island on Kennedy Street between East 7th Street and 23rd Avenue, the following studies were undertaken:

1. Speed studies on Kennedy Street and 23rd Avenue.

2. Twenty-four hour volumes on 23rd Avenue, 29th Avenue, East 7th Street and Ford Street.

3. Truck and turning movement counts were taken at all intersections, including weaving movements on Ford Street between 29th and 23rd Avenues and at the merge of 23rd with 29th.

4. Accident studies of the entire area -- intersection and non-intersection.

5. Signal need study for 23rd Avenue and East 7th Street.

A summation of the pertinent findings is:

1. Average and 85th percentile speeds for southbound 23rd Avenue are 32 m.p.h. and 37 m.p.h., respectively; for northbound Kennedy are 30 m.p.h. and 39 m.p.h., respectively.

2. The percent of trucks for the entire study area was about 15% for the hours of 11:00 a.m. to 3:00 p.m.

3. An average of two vehicles per hour made U-turns from southbound Kennedy Street through the Rhodes and Jamieson driveway and through the existing median opening to northbound Kennedy Street.

4. Our studies showed little truck activity for the existing driveways on the east side of Kennedy Street with only nine trucks using these four driveways during the six hours we observed.

5. The accident incidences at the intersections of East 7th Street at Kennedy Street and East 7th Street and 23rd Avenue are normal for the intersection volumes at these locations.

6. Forty percent of the accidents at East 7th Street and Kennedy Street are caused by left-turn conflicts.

7. Five percent of the vehicles making left turns from 23rd Avenue to 29th made complete loops via Ford Street back to 23rd. None of these vehicles were trucks.

Recommendations:

1. No median openings be made for existing driveways on Kennedy Street. Usage of these driveways is light. The intersection of East 7th Street and Kennedy Street has demonstrated a left-turn accident pattern with good sight distance. The proposed openings in the median would be located such that sight distances would be reduced for northbound Kennedy Street. In addition, a rear-end accident hazard would be created.

2. Under no circumstances should median openings be provided for the Campanella property on the corner of 23rd Avenue and Kennedy Street. Sight distances from both directions on Kennedy Street will be restricted, especially with any improvements made on the property in the vicinity of the curve on Kennedy Street. The median openings required to service the driveways for this property cannot be located far enough from the curve on Kennedy Street to provide adequate sight distance for vehicles traveling in either direction on Kennedy Street.

Improvements:

1. Left-turn pockets would not appreciably reduce the accidents at Kennedy Street and East 7th Street and may induce more sideswipe type accidents. Lane lines should be installed on Kennedy Street with Stimsonite reflective pavement markers at 48' and 24' centers on the curve. This will help reduce speed and delineate the curve on Kennedy Street.

2. A guide line should be painted to guide vehicles turning left from northbound 23rd Avenue into the proper lane on westbound East 7th Street.

3. New revised advance street name signing should be installed at
   East 7th and Kennedy Streets and at East 7th and 23rd Avenue.
   The existing freeway directional signing at 23rd Avenue is inadequate and misleading. In addition, signing in the vicinity
   of 29th Avenue and Ford Street should be revised.

Work orders for items No. 1 and No. 2 are attached. Item No. 3 involves both State of California and City of Oakland responsibility.
Work orders for the City's portion will be prepared and the State
will be contacted regarding the revision of their freeway guide signs.

Michael F. Pickering
Associate Traffic Engineer

MFP:pc:br

Attachments

February 18, 1971

Mr. Charles L. Campanella
2712 East 7th Street
Oakland, California 94601

Dear Mr. Campanella:

In response to your request for additional openings in the median along
Kennedy Street between East 7th Street and 23rd Avenue, subsequent meetings with Mr. Hurlbut of this department and recommendations suggested
by the Traffic Committee, our staff initiated a comprehensive study of
the region bounded by Kennedy Street, 29th Avenue, 23rd Avenue, Ford
Street and East 7th Street.

The investigation conducted involved speed studies on Kennedy Street,
23rd Avenue and 29th Avenue along with vehicular volume counts on 23rd
Avenue, 29th Avenue, East 7th Street, Ford Street and Kennedy Street.
In addition, truck and turning movement counts were taken at all intersections including weaving movements on Ford Street between 29th and
23rd Avenues, the merge of 23rd Avenue with 29th Avenue and the left
turn movement from 23rd to 29th Avenues. Accident studies for the entire area were analyzed and a signal need study for 23rd Avenue and
East 7th Street was conducted.

It was found that the average speed for southbound 23rd Avenue was 32
m.p.h. and for Kennedy Street the average speed was 30 m.p.h. southbound
and 28 m.p.h. northbound. Truck counts taken over a two-day period
showed that the percentage of trucks for the entire study area was
approximately 15% for the hours of 11:00 a.m. to 3:00 p.m. An average
of two vehicles per hour made U-turns from southbound Kennedy Street
through the Rhodes and Jamieson driveway and through the existing
median opening to northbound Kennedy Street.

Our study showed little truck activity for the existing driveways on
the east side of Kennedy Street with only 9 trucks using this driveway
during the 6 hours we observed. Five percent of the vehicles making left
turns from 23rd to 29th Avenues made complete loops via Ford Street back
to 23rd Avenue. None of these vehicles were trucks. The accident inci-
dence for the intersections of East 7th Street at Kennedy Street and
East 7th Street at 23rd Avenue are normal for the intersection volumes
recorded. Forty percent of the accidents at East 7th and Kennedy Streets
were caused by left turning conflicts.

On February 1, 1971, the Traffic Committee met and the results of our staff study and recommendations were presented. It was the decision of the Traffic Committee that no median openings be made for the existing driveways on Kennedy Street. Usage of these driveways is light, and the intersection of East 7th Street and Kennedy Street has demonstrated a left turn accident pattern even though the existing sight distance is adequate. The proposed openings in the median would be located such that sight distance would be reduced for northbound Kennedy Street and a potential rear end accident hazard would be created. In addition, no median openings will be provided for your property on the corner of 23rd Avenue and Kennedy Street. Sight distances from both directions on Kennedy Street would be restricted especially with any improvements made on the property in the vicinity of the curve on Kennedy Street. The median openings required to serve the driveways for this property could not be located far enough from the curve to provide adequate sight distance for vehicles traveling in either direction on Kennedy Street.

As a result of our study, improvements for better circulation and increased safety will be initiated. Lane lines will be installed on Kennedy Street with reflective pavement markers on the curve. This will help reduce speed and delineate the curve on Kennedy Street. Also, a guide line will be painted to guide vehicles turning left from northbound 23rd Street into the proper lane for westbound East 7th Street. New revised advance street name signs will be installed at East 7th and Kennedy Streets. The existing freeway directional signing at 23rd Avenue and East 7th Street and the signing in the vicinity of 29th Avenue and Ford Street will be revised. Some of this work will be coordinated with the State of California Division of Highways, and it is anticipated that these improvements will be completed within three to four months.

If you have any questions regarding the study conducted by our staff, please contact Mr. Pickering of this office.

Very truly yours,

ARNOLD A. JOHNSON
Traffic Engineer and
Director of Parking

AAJ:MFP:pc

cc: Traffic Committee

177

SPECIMEN NO. 6*

Periodic Report

On the opposite page is the first page of a periodic report written by the Management Development Coordinator of a large Eastern gas company. The pages that followed consisted of similar material.

The function of this report was to draw together monthly reports from various sources—reports on what had been done on one company program—and to present the results to the company officers. Also, since it was sent to all department heads, it served to provide them with ideas and stimulation that might lead to improvements within their own departments.

* Courtesy of Brooklyn Union Gas.

Work Simplification Activity Report   September 17

All Officers & Department Heads

C.A. Kiersted, Management Development Coordinator   Personnel

        This Work Simplification report outlines the improvements
accomplished by departments during August, 1973.

Commercial Department   J. Mullen--Coordinator

        M. Croak has made a change to the procedure for handling the prepa-
ration and mailing of Multiple Dwelling Occupant notices of Discontinuance of
Owner's Service. Under the old procedure an envelope was typed for each of
the occupants living in the building, and a copy of the occupant notice was
enclosed and mailed. Under the new procedure a window type envelope will be
used. The copy of the occupant notice will be folded and inserted into the
envelope so that the name and address portion are visible in the window of
the envelope. This new procedure will eliminate the need to type approximately
48,000 envelopes a year, thus saving some 465 man hours or $1825.

        R. Moore has implemented a change to the procedure for filing and
review of anticipated charge-off memos for periodic locks. Under the old program, these memo cards were filed with all other anticipated charge-off memos
after being reviewed for billing accuracy. Under the new program a file will
be set up in two parts: summer and winter. Memos received during the months
April through September will be put in the summer file and referenced for active accounts twice during the winter period October through March. Memos received October through March will be referenced twice during the summer months
April through September. This will enable the Credit and Collection Section
to recharge unpaid final bills from the previous periodic lock during the time
of the year the customer is most likely to have an active account. It is anticipated that the company will be successful in recharging these accounts 50%
of the time, and as a result will realize a savings of some $18,000 in uncol-
lectable gas bills annually.

Data Processing Operations   W. Smart--Coordinator

        S. Sugarman has made a change to the procedure used for dumping of
programs stored on Disk Packs. Under the original procedure, programs that
were stored on disk packs were dumped onto tape files daily, regardless of
whether or not a change had been made to any of the programs. The new procedure calls for these programs to be dumped onto tape files only by a special
request when a change has been made to one of the programs. A special request "Dump" form has been designed for this purpose. The monetary value of
this program is undeterminable, however there is a savings of one hour a day
of machine time which is now being used to run other programs.

        179

SPECIMEN NO. 7*

Laboratory Report under a Memorandum Heading

Though this report is written on a sheet headed for use as a memorandum, its contents and internal structure are those of a laboratory report. Actually, the laboratory work had been done earlier and a report had been written concerning the results. This new report corrects a previous interpretation and explains what had apparently led to contradictory statements.

* Courtesy of The Potlatch Corporation.

TO: R. E. Zipse   DATE: January 4, 1973

FROM: J. A. Anderson

SUBJECT: Weak Liquor Evaporators

OBJECTIVE:

The purpose of this project was to obtain information on the evaporators relative to their performance during the past three years. This report is being written to clarify and to add to the information given in the previous report, "Weak Liquor Evaporators" by J. A. Anderson, December 14, 1972.

PROCEDURE:

Information was taken from the Evaporator Report and averaged for the months of July and October during 1970, 1971, and 1972.

RESULTS AND DISCUSSION:

In the initial report dated December 14, 1972 it was stated that errors exist within the meters since the pounds of solids being run through the evaporators was shown to be decreasing while the brown stock production was increasing. This statement is in error because it has not taken into account changes in the amount of concentrated liquor added to the weak liquor prior to the weak liquor evaporators. Therefore, there is no reason to doubt the information in Tables I and II of the December 14 report. This would then mean that the evaporators were indeed handling less liquor in 1972 than in 1970.

The reason that this is possible is that the percentage of solids in the liquor coming from the washers was raised during 1972. At the same time the amount of concentrated liquor added to the weak liquor prior to the evaporators was reduced to almost half of what it was in 1970 and 1971.

Table I in this report shows that the amount of solids coming from the washers (prior to the addition of the concentrated liquor) has increased steadily with the increasing brown stock production.

With the addition of #4 recovery furnace, the strain on the concentrated (50%) liquor storage has lessened. Thus, it is now possible to increase the shower water on the brown stock washers without fear of overrunning the 50% liquor storage tank. However, since we have now raised the shower water rate and thus dropped the percentage of solids in the washer filtrate we can expect the volume of liquor going into the evaporators to rise drastically. If, for convenience, it is assumed that the brown stock production in 1973 will be the same as in 1972, that the liquor

181

Weak Liquor Evaporators
J. A. Anderson
January 4, 1973

RESULTS AND DISCUSSION: (continued)

from the washers will average 12.5% solids, and that the liquor going to the evaporators will remain at about 17.4% solids, then the pounds of solids (weak and concentrated) that will be run through the evaporators during October of 1973 will be 114.47 million. This is an increase of 16.7% over October of 1972.

CONCLUSIONS:

(1) The information contained in Tables I and II of the December 14, 1972 report "Weak Liquor Evaporators" is not incorrect as previously reported.

(2) The amount of solids being sent from the washers to the evaporators is approximately 2750 #/ADT. (This figure does not include the concentrated liquor added to the weak liquor.)

(3) The change in washer shower rates will mean that the amount of solids (from the weak and the concentrated liquors) that must pass through the evaporators will increase 16.7% over 1972's amount.

J. A. Anderson

JAA:nm
cc: R. A. Zielinski
    H. Johnson

182

TABLE I

Pound of Solids Per Month Coming from the Washers and Going to the Evaporators (Millions of Pounds)*

Date	#1 Set	#2 Set	#3 Set	#4 Set	Total	#Solids/ADT
July '70	13.29	11.48	13.43	21.56	59.76	3052
Oct. '70	14.37	12.28	10.53	25.99	63.17	2665
July '71	14.04	14.71	15.83	23.56	68.14	2940
Oct. '71	14.92	13.47	14.13	26.76	69.28	2640
July '72	10.95	13.29	17.13	27.73	69.10	2853
Oct. '72	11.19	17.81	16.66	25.83	71.49	2650

*These figures do not include the concentrated liquor added prior to the evaporators. For figures that include the concentrated liquor see Table II of the report "Weak Liquor Evaporators" by J. A. Anderson, December 14, 1972.

SPECIMEN NO. 8*

Report in the Form of a Memorandum

The company where the following report was written had a continuing problem, for which a possible solution had been suggested. This solution had been studied and evaluated. The report tells the results.

The report first recalls the problem to the minds of the readers. Then it lists the company policies that bear on the case, and the "constraints" that affect the practicality of any solution.

Under the heading *Possible Action* it explains what would happen if the proposed solution were attempted, basing its analysis on a hypothetical set of specific, actual facts. It puts the facts in a table.

The decision is negative because the facts as shown in the table would make the solution too costly. This conclusion is stated under the heading *Comments* and is reinforced by two other reasons that make the writer consider the solution impractical.

* Courtesy of National Steel and Shipbuilding Company.

NATIONAL STEEL AND SHIPBUILDING COMPANY

INTER-DEPARTMENT MEMO

File NO: 32/GEN/AKR:vj/72-100

Date.......................

To: .... Dept. ....

Subject: .... Dept.............

From: .... Job No...........

Problem – The area available for steel storage is inadequate. Can this situation be relieved by reducing the number of plate sizes?

Present Policies and Constraints

    1) The breakdown of ship structure into assemblies is governed in part by late lengths. The straking of shell, decks and bulkheads is carefully worked out to control the number of widths.

    2) Plates for built-up shapes are selected for exact lengths and widths, both flanges and webs, so that they can be stripped without producing any excess pieces or large pieces of scrap (because there is no efficient means of taking them out of the flow to the beam-welder and returning them to stock or scrap).

    3) Plates are ordered to the thickness and to the grade called for in the design of the ship.

    4) Before ordering a plate the list of existing sizes is scanned and one of them is selected if the additional scrap is, say, 5%.

    5) Multiple contracts for ships are a basic cause of so many plate sizes; assembly lengths, bulkhead spacing, hull depth and beam, thickness and grade all contribute to proliferation. If all plates were of the same length and width we would still have nearly 100 stacks due to thickness and grade requirements.

    6) The large number of built-up shape sizes, in conjunction with the constraint explained in (2) above, leads to an inordinate number of plate sizes.

    7) The objective of ordering steel for a given sequence so that the plates will be consumed by that sequence reduces handling and storing of excess pieces, but does lead to a larger number of plate sizes.

IDM
To: Distribution
February 8, 1973
(Page 2)

8)   The Wheelabrator imposes an absolute maximum of 124" for
plate widths. The transfer system will not handle plates exceeding
38'-6" long, and the plate rolls, 35' long.

Possible Action

Limiting plate lengths and widths to specific values in fairly
large increments would certainly generate more scrap and should
reduce the number of sizes. In order to evaluate this, we picked
at random our existing requirements for a number of 10.2#, 25.5#, and
40.8# plate sizes and determined what we would have ordered if plate
widths were in one foot increments and lengths in two foot increments.
Here are the results:

	10.2#	25.5#	40.8#
Number of sizes studied	23	50	49
Weight of the sample, tons	117	2,140	3,172
Number if standardized	17	37	37
Number of sizes eliminated	6	13	12
Cost of increased scrap	$2,450	$40,200	$67,800
Percent increase in scrap	12%	10.4%	12%
Each size (stack) eliminated costs, in terms of added scrap:	$\frac{2,450}{6}$=$408	$\frac{40,200}{13}$=$3,090	$\frac{67,800}{12}$=$5,650

Comment-- Although the cost of adopting such standards appears
prohibitive, there are two more factors against the standards:

1)   If the steel for the follow-on ships was brought under the
standards there would be a large increase in storage area required
because the new plates could not be stored on top of any of the
smaller plates they replace.

IDM
To: Distribution
February 8, 1973
(page 3)

2) The AOR-7 uses no higher strength steel of the grade used in large areas of the three other contracts. It does have a bilge strake of HY-80 and we have proposed EH 36 for the sheer and stringer strakes to eliminate riveting.

A. K. R.

A. K. Romberg

To: J. Anthony
cc:
J. V. Banks
J. B. Letherbury
J. McQuaide
E. Schneider
K. Evans
J. Krohn
B. Bouchell
R. Schaneman

SPECIMEN NO. 9*

A Progress (or Status) Report

The report that begins on the opposite page lacks the material normally needed as an opening because, as indicated by its first page being numbered 20, it was incorporated into a longer report covering other departments as well as Environmental Engineering.

In itself, this report cannot be considered as any special form, but the full report that included it (not available for examination) was probably a full-scale formal report.

As regards the longer report—there is nothing unusual about different parts of a long report being written by different people. This possibility accounts for the emphasis in this book on compliance with instructions about form and on the ability to handle an impersonal style. If each of several writers whose work appears in the same report writes in his own personal manner, the report will become a hodgepodge of different styles.

* Courtesy of Douglas United Nuclear Industries, Inc.

ENVIRONMENTAL ENGINEERING

1972 Environmental Release Report

The 1972 Environmental Release Report[12] was documented and issued to the AEC-RL on March 13. During Calendar Year 1972, 4,000,000 pounds of waste materials including 1 gram of radionuclides were discharged to the environs in liquid effluent. This is a reduction of 50 percent from 1971 wastes released in liquid effluent streams. During the same period, 2,000,000 pounds of wastes, including a small fraction of 1 gram of radionuclide, were discharged to the atmosphere in gaseous effluent streams.

Transuranium Waste Procedures

On March 1, 1973 AEC-RL requested that changes be made in procedures for handling transuranium wastes. Evaluation[13] of DUN operations showed that no transuranium wastes are now or will be generated in the near future. Should any program that could generate such wastes be planned in the future, the new procedures would be developed and submitted to AEC-RL.

Radiation Control Signs

The revision to RC-2, "Radiation Source Control" of DUN-M-1, Radiation Control Manual requires that a sign reading "CAUTION HIGH DOSE RATES" be affixed to sources within a Radiation Zone that are a factor of two higher than ambient dose rates. The purpose of the requirement is to assist radiation zone workers to control their exposure by avoiding locations so posted.

One hundred 5 X 9-inch aluminum signs with the above warning have been fabricated and issued for use. Each sign has an alnico magnet fastened to the reverse side so that it can be easily attached to valves or piping as required. Two holes have also been provided in each sign so that it can be suspended where the magnets are ineffective.

N Reactor Liquid Effluent Cleanup

Treatment of N Reactor liquid effluents is currently accomplished by soil. The effluents are routed to a rock-filled crib for percolation through the soil before reaching the Columbia River. Efforts are being made to find a technically more acceptable method for liquid effluent treatment.

The major difficulty in providing an acceptable ion-exchange system for N Reactor liquid effluents is the large volume of such wastes. The volume of N Reactor liquid effluents requiring treatment is more than 100 times greater than that required for other water-cooled reactors (i.e., BWRs and PWRs). And the cost for a complete waste treatment facility for BWRs and PWRs ranged from 5 to 10 million dollars.

---

[12] DUN-8133, "Douglas United Nuclear, Inc., Reactor and Fuel Production Facilities 1972 Environmental Release Report," TE Dabrowski, March 1, 1973. Letter, NR Miller to PG Holsted, "1972 Environmental Releases" dated 3/12/73.

[13] Letter, NR Miller to TA Nemzek, "Additional Guidance, Immediate Action Directive No. 0511-21, Policy Statement Regarding Solid Waste Burial (Ref. Ltr. 4/21/70 DG Williams to Contractors)," dated 3/23/73

-20-

189

ASSIGNMENTS

A Note on Selecting Report Subjects: There is so much variation in the conditions in different schools, and in the courses, training, and opportunities to gather information, that it will probably be necessary to take many liberties with the suggested subjects listed. Each subject is likely to need adaptation and sharper definition.

In adapting a subject you may need to broaden or narrow its scope. You may be forced to assume that you are writing in a different season of the year, or that some recent project in your school or community has not yet been undertaken. Or if a subject proves too large to cover in full, you may need to assume that you are conducting only a preliminary survey intended to show whether a subject should be investigated further or abandoned.

In defining a subject you will probably be assuming that you are writing the report for an actual employer or client. Thus you will need to decide whom your report is being written for and why he needs it— that is, to set up a hypothetical set of circumstances. In actual practice, of course, the information you would include would depend on the needs of the occasion. In the classroom, the hypothetical occasion must be shaped to enable you to utilize whatever information it is possible to obtain.

Many of the subjects suggested could plausibly be used for more than one type of report. There is therefore no reason that in writing any assignment you should not feel free to use a subject from the list suggested for some other assignment, broadening it or narrowing it as the occasion demands.

You may prefer to write about subjects not included in the lists provided. To develop a subject of your own, you should start out by canvassing your own resources. Review, mentally, all that you have learned on any job you have held; all that you know about any technical line of activity some friend or relative might pursue; all that you have learned in any practical courses you have taken. Then try to create a hypothetical situation in which you can utilize that knowledge.

Regardless of the source of your subject, you should guard against making your reports too general, especially if some of your material must be drawn from printed references. In actual practice, a report may use general information but usually applies it to a specific case. For example, general information about killing weeds by use of chemicals is so easily obtainable in print that no one would ever be asked to write a report to present it. But the weed conditions on a specific tract of land could not be found in print, so it is quite plausible that one might be asked to write a report on the conditions on that tract and on the measures necessary to control the weeds there. Or again, anyone could find general information about fluorescent lighting systems, but it is quite possible that a report might be requested on the advisability of using fluorescent lighting in some specific building which was to be modernized or remodeled. In this report, the general, printed information would be applied to the specific facts about the building.

All this should be taken into consideration as you choose your subjects, for the wisdom of your choice will strongly affect the merit of any report you write—an unfortunate condition that will not exist when you are on the job, but cannot be avoided in the classroom. It is possible, of course, to write a poor report on a good subject; but it is impossible to write a good report on a poor subject, for a poorly chosen, unrealistic subject simply does not give you an opportunity to exercise many of the skills that make a report good.

Assignment 1

Write a report in the form of a letter using a subject from the list below or, with the instructor's permission, a subject of your own choice. Subjects 1 and 2 in Assignment 2 might also be suitable if properly adapted. The specimen letters in Chapter 14 should be sufficient guidance as to letter form.

Since this report is a letter, you would need to assume not only that circumstances make it natural for you to report to the reader, but also, if you and the reader are supposedly employed by the same organization, that circumstances make it natural to write a letter rather than a memorandum. For example, you might assume that you are writing the report while on a trip to a different town; or you might, in using some of the subjects, assume that you are reporting to an out-of-town property owner.

It is suggested that this report be 300 to 600 words long.

1. Report on the condition of one or more campsites or other tourist accommodations that you have inspected—perhaps by request. (To make it plausible that this report would be a letter, assume that you are an employee of some state agency and work at—or are visiting—the area of the campsite, but that the agency headquarters is located elsewhere.)
2. Report on the condition of a nursing home, retirement home, or some comparable facility that must meet standards set by the state. (An assumption such as that for Assignment 1 will be necessary.)
3. Assume that you manage a small motel owned by a man who lives some distance away, and that the owner is interested in buying another motel in a town near yours. He has asked you to look it over to determine its condition—and also, if you wish to expand the scope—to determine other relevant facts such as the desirability of the location. Write a report giving him the information that he needs.

4. Assume that you operate a bus station for a company that operates a small bus line, and that the company will soon need to find a new location for its station in a nearby town. The company has heard of one possibility and asks you to look it over and report on its merits. (If you prefer, you may assume that the company has not heard of any location and wants you to look for one.) Write the report that the situation calls for.

5. The company that made or sold some piece of equipment has received a complaint about its performance. You are asked, while on a trip to the town where the complaint originated, to find out what is wrong and submit a report. Maybe the equipment was faulty, but possibly it was wrongly installed or has not been properly maintained. Write the report called for.

Assignment 2

Write a report in the form of a memorandum, using a subject from the list below or, with your instructor's permission, a subject of your own choice. Since your report should be made to a reader to whom you might logically report, you will need to assume that you and the reader are connected with the same organization.

It is suggested that this report should be from 300 to 600 words long.

1. A local group of enthusiasts has, on a volunteer basis, opened a recycling center and operated it for a year or more. This group has now asked that the municipal government take over the operation. You have been asked, as a municipal employee, to look the center over and report on your findings. Write a report about what you have learned during your inspection.

2. Businessmen in a downtown area have complained to the municipal government about the downtown alleys, saying they are filled with litter, clogged by trucks parked needlessly long, and so on. You have been asked, as a municipal employee, to inspect the condition of the alleys at different times of day and on different days of the week to see whether improvement is really needed. Write the report called for.

3. Residents of a mobile home court near the edge of town have complained that in a tavern in the area, the music is so loud late at night as to make sleep impossible. Also, they maintain that the parking lot is the source of loud noise and the scene of various disturbances. As a municipal employee you have been asked to check into the questions of whether conditions are as bad as they say. Write a report on your findings.

4. Your school has decided to install room air conditioners in several offices that become uncomfortably hot in the summer. You have been asked to obtain facts about various makes and models and recommend one or more. Write a report on your findings.

5. Pick one or more buildings on your campus that need extensive

renovation or upkeep, and write a report telling the general amount and type of work needed and the best order for doing it.

Assignment 3

Write a progress report of a length specified by your instructor. It may be in any form you desire unless you receive instructions to the contrary.

Any project that calls for a final report, unless it is small enough to be completed in a very short time, may also call for progress reports. Consequently, the subject on which your progress report is based may be any of the subjects provided for other report assignments. Another possibility is that if you have been assigned a long report in a Technical Writing class or a project that will last for several weeks in some other class, you might write a report telling your instructor about your progress.

Assignment 4

Assume that you are an officer or committee chairman of some organization, and that it is your responsibility to write quarterly, semiannual, or annual reports. Write a report of this nature.

The organization may be connected with your school—a social, professional, or honorary club or fraternity, for example—or a ski club, flying club, religious group, political group, or any other group that is active on the campus. As an alternative you could write a report of similar nature for a school enterprise such as a newspaper, magazine, or yearbook, or for some nonschool organization.

Assignment 5

If your course of study has included a laboratory course to supply you with subject matter, write a laboratory report on some test or experiment you have performed. The subject should call for enough writing to make the report a writing assignment. It may include tables, figures, or filled-in blanks, but should not be limited to such material.

9

SPECIAL ELEMENTS

IN REPORTS

To produce some kinds of writing a writer may need only the ability to select his material, organize it, and express himself effectively. To produce good reports, however, these general skills are not enough. A report writer must also be able to plan the layout of a report—that is, to decide which of certain special elements will be useful in each report and to handle them so that they form an effective framework for his material.

To be sure, some reports are so short and simple that no special elements are needed; and laboratory reports are so standardized that their layout is often in accordance with the pattern already described. Nevertheless, certain special elements are needed so frequently that they call for considerable attention. They are most conspicuous in formal reports, but most writers are not called upon to write full-scale formal reports until long after they have had occasion to use these elements in reports that are less elaborate. Therefore, we will discuss them in connection with reports in general, and supplement the discussion as necessary in the chapter on formal reports.

The elements in question are the covering memorandum or letter, the title page, the table of contents, the headings, the summary, the introduction, the conclusions and recommendations, and the appendix. They have been listed in the order in which they are likely to reach the reader's attention, but they should be used only to the extent that they are needed. The body of the report is not listed because it is not a special element and performs a general rather than a special function; but it would normally comprise a substantial portion of the report and might either precede or follow the conclusions and recommendations.

THE COVERING MEMORANDUM

When a report that is not extremely short and simple is submitted, it is often accompanied by a covering memorandum (or a covering letter if it is sent through the mails). Such a memorandum is especially useful when a report will be of lasting rather than only immediate interest or when it will go beyond the person who first receives it.

A covering memorandum gives the writer an opportunity to make any extra remarks he feels the occasion calls for. These remarks may add up to no more than an identification of the report, a reference to the subject that the report deals with, and a statement of the reason this subject is reported on. Remarks of this limited nature grow out of the same impulse that probably causes you, in a personal contact, to say, "Here is that book (or list, or drawing, or estimate) that you asked for." In making such remarks in a covering memorandum, however, you usually do more than just gratify your natural impulse to say something. You also make it easier for the person who first receives your report to grasp or recall certain facts that concern it.

Sometimes a covering memorandum does a good deal more than this. Since the first recipient of your report is likely to be the person to whom you are immediately responsible, you may wish to tell him things that perhaps should go no further. For example, you may want to explain why you have emphasized some points and played down others. You may want to refer to obstacles you encountered or events that occurred while you were preparing the report. You may want to mention additional information that can be provided if it is needed. In general, you may want to give him facts that it would not be desirable or appropriate to place in the report itself; and the covering memorandum gives you an opportunity to do so.

A typical covering memorandum follows.

Date: May 3, 197–

To: Creston County Board of Commissioners
From: Sydney Wicks
Subject: Rose Lake Recreation Area

Enclosed is my report on what work would need to be done to bring the Rose Lake area back into condition for recreational use, as it was used before it was allowed to deteriorate. This report was requested during your meeting on March 23.

I have taken into consideration the need of removing the silt that has almost filled the lake, improving the water supply, providing picnic tables and cooking facilities, and developing off-the-road parking space. I have also included a rough estimate of the cost of each of these improvements and the approximate time it would take to complete the work.

There is one complication to carrying out this project that I believe I should bring to your attention though it does not fall within the scope of the report and its inclusion would be inappropriate. Some of the nearby property owners maintain that when the area was formerly in use, it was used very late at night by noisy, rowdy groups whose parties amounted to a real disturbance. They insist, too, that their property was repeatedly subjected to vandalism.

I believe that when you discuss the project at a meeting, you should anticipate that several of these property owners will attend and raise strong objections.

Assuming that the project is not held up by this complication, I believe that the information I have provided will enable you to decide whether you wish to pursue the matter further.

THE TITLE PAGE

The title page normally indicates (1) the subject of the report, (2) the person or body to whom the report is made, (3) the person or body making the report, and (4) the date of the report. Also, in large organizations, it may bear the number assigned to the report and information on the project that the report concerns.

The title itself is more than just a label. Though it should not be so long as to be cumbersome, it should give the clearest possible information about the exact subject covered. *Selection of a Fin-Tip Antenna for a KC-135 Airplane,* for example, would be a better title than just *Selection of a Fin-Tip Antenna.*

If a report is a matter of mere routine—for example the periodic report of normal operations of a division of some company—there would be no need for the title page to indicate to whom or by whom the report is made. Identifying it as a monthly or quarterly report makes these facts clear to anyone likely to be concerned.

The material on the title page should be pleasingly arranged, with the optical emphasis on the upper half of the page. Figure 13 is a typical title page for a report in which there was no need to use a specific company form.

COLLEGE OF ENGINEERING RESEARCH DIVISION

Bulletin 324

SIMULATION OF WATER QUALITY ENHANCEMENT

IN A POLLUTED LAKE

A Case Study of Vancouver Lake, Washington

by

Cheng-Nan Lin

Surinder K. Bhagat

John F. Orsborn

Research reported in this publication
was supported by
STATE OF WASHINGTON WATER RESEARCH CENTER
at Washington State University

Published by the
Engineering Extension Service
January 1972

WASHINGTON STATE UNIVERSITY   PULLMAN, WASHINGTON

FIGURE 13.   Well-Designed Title Page for a Report.

197

THE TABLE OF CONTENTS

The table of contents of a report is basically the same as that of a book in that it lists all the headings used and indicates the page on which each heading appears. Since indentation makes the rank of each heading apparent, the table of contents functions as an outline of the report as well as showing where various points are taken up. Occasionally each point is preceded by a number or a letter as it would be in an outline, but most of the time these symbols are discarded or are limited to the Roman numerals identifying main divisions. When the decimal system of numbering headings is used, the numbers appear in the table of contents as well.

Like the table of contents of a book, the table of contents of a report has headings that indicate the presence of appendixes, calculations, detailed data, or any other type of material that follows the report proper. These headings are likely to include the titles of individual tables and figures when such nonverbal devices are all placed after the body of the report. When the tables and figures are scattered through the body, however, their titles appear in a separate list.

Our discussion here has been brief, for two reasons. First, most of the work for a table of contents—deciding on the substance and phraseology of the headings—is done when the outline is made. Second, some of the questions that occur in connection with a table of contents are covered later, in the discussion of formal reports, because that is where they are most likely to arise.

It is important to realize, however, that a table of contents is a valuable part of any report that is more than a few pages long. It lets the reader see at a glance what ground is covered, shows him how the material is organized, and makes it easy for him to turn quickly to whatever he is looking for.

Following is a typical table of contents for a report.

CONTENTS

FIGURES AND TABLES

USE OF HEADINGS

The Value of Headings

One of the useful skills you can acquire as a report writer is skill in the use of internal headings. Such headings, regardless of whether you provide a table of contents, make it easier for your reader to grasp the nature of your material, the general plan governing its organization, and the relationship among its various parts. They help him to find what he is looking for when he wants to find out about some one portion of your subject. They can catch his attention and arouse his interest when he first glances over your work, and thus tempt him into the heart of the report before he has even taken time to read the preliminary sections. They reduce his reluctance to start reading by making the pages of text appear less formidable. And finally, the very fact that you are using them can improve your writing by reducing the temptation to wander away from the point you are supposed to be discussing.

To provide all these benefits, however, headings must be used systematically and must cover everything in the report—a vastly different matter than casually inserting them as they are inserted in newspaper articles or most articles in magazines. The headings used are drawn, of course, from the outline on which the report is based, and their value depends on whether you have done a good job of organizing your material. Headings cannot bring order out of confusion; they can only show that order exists.

Form for Headings of Different Ranks

Since the headings are intended to show how your material is organized, the form of each heading should indicate its rank. Headings of the same rank should be identical in form. Headings of different rank should differ in form. Headings of higher rank should look more important than those of lower rank. And finally, a form used for topic headings should not be used for any other purpose, such as the title of a list or a set of rules.

No single set of forms has become standard, but the following treatment is widely used and would be hard to improve upon.

A *heading of highest rank* (used before special parts such as introduction, conclusions, etc., and for the main divisions of the body) is centered, underlined, and capitalized throughout. It is placed three or four spaces below whatever precedes and a double space above whatever follows. It is not followed by any punctuation.

A *heading of second rank* begins at the left margin. It is underlined and capital letters are used only to begin important words ("upper and lower case capitalization"). It is placed a double or triple space below whatever precedes and a double space above whatever follows. It is not followed by any punctuation.

A *heading of third rank* is identical in form to a heading of the second rank, except that it does not appear on a line by itself and it is followed by a period. It is indented because it comes at the beginning of a paragraph.

Underlines are strongly recommended for typewritten work because they make the headings stand out more distinctly. Figure 14 illustrates the form described.

One feature that commends this system is that it does not include a centered upper and lower case heading. Consequently, that form can be used for the titles of lists and rules without being confused with topic headings.

VARIABLE-SPEED SABER SAWS

Two variable-speed saws, the Zenith X-400 and the Monument 65 A were found to be superior to the others examined. Neither of them caused much effort in feeding. Each has a full-length handle above the motor housing. Each has a slide type of on-off switch that also controls the speed and can be operated easily by the hand holding the handle. Each has a rip-and-circle guide adjustable from a minimum of 1 in. to a maximum of 6 in. Each can be adjusted easily to make bevel cuts up to 45 degrees. The differences between them are mainly as follows.

Zenith X-400

Superior Features. The Zenith X-400 is relatively quiet in comparison with most saws examined and gives less than average trouble with vibration. It handles the sawdust without letting it either accumulate on the board and obscure the guide line or blow up into the operator's face. It is provided with a shoe insert to reduce splintering.

Undesirable Features. Though the on-off switch is conveniently located, the device for locking it into position for a selected speed is a little awkward to use. The shape of the shoe does not provide good support when a cut is started or when a thin strip is cut off the edge of a board. It is somewhat heavier than average. Inspection and replacement of the brushes in the motor requires considerable disassembly and reassembly. (Such inspection and possible replacement is important because running the saw with brushes shorter than 1/4 in. can damage the motor.)

Monument 65 A

Superior Features. The Monument 65 A is a relatively quiet saw that is easier than average to use. It is light enough that it can be used to cut vertical surfaces without causing more than a minimum of fatigue. It has an auxiliary handle on the left side of the housing. The design of the shoe makes it easier than average for the operator to start his cut and also provides support when he cuts a thin strip off the side of the board. It can accommodate an oversize blade that permits cutting lumber up to 6 in. thick.

Undesirable Features. The lightness of the saw results in higher than average vibration. The sawdust has a tendency to collect on the board so that the guide line is sometimes hard to see, and also sometimes blows up into the operator's face. One of the least desirable features is the fact that when the saw must be disassembled, improper positioning of the wire during reassembly causes a danger of electric shock.

4

FIGURE 14.   Specimen Page from a Report Showing Headings of Three Ranks.

Phraseology of Headings

With rare exceptions, headings that are used systematically to show the organization of a report or any other kind of writing are worded as topics. That is, each heading is a noun (or nouns) plus any modifiers needed. It is sometimes assumed that any form that is not a sentence is a topic, but this is not the case.

Though it is doubtful that most writers analyze the grammatical structure of their headings, the fact remains that their headings almost never turn out to be prepositional phrases, adjectives, infinitive phrases, or sentences. Occasionally, in a brief piece of semipopular work, headings are phrased as questions or declarative sentences; but for the most part, discriminating writers use bona fide topics because they instinctively feel that other forms do not sound right. And most readers feel a sense of strangeness when they encounter headings that are not topics, even though they may not analyze the reason for their feelings.

The following examples should further clarify the difference between topics and nontopics.

Suitable and Unsuitable Phraseology for Headings

(1)

Projected Use of Water   (suitable)
    Industrial   (an adjective and hence unsuitable)
    Domestic   (an adjective and hence unsuitable)
Projected Use of Water   (suitable)
    Industrial Use[1]   (suitable, for *Use* is a noun)
    Domestic Use   (suitable, for *Use* is a noun)

(2)

Undesirability of Kilzol Insecticide   (suitable)
    For Use near Beehives   (a prepositional phrase and hence unsuitable)
    For Use near Streams and Ponds   (a prepositional phrase and hence unsuitable)

Undesirable Uses of Kilzol Insecticide   (suitable)
    Undesirability near Beehives   (suitable because *Undesirability* is a noun)
    Undesirability near Streams and Ponds   (suitable for same reason)

---

[1] If subpoints are likely to be far removed from covering main points, adding a few words might be desirable in some cases for the sake of self-sufficiency. For example, in (1), "Projected Industrial Use of Water."

(3)
Reasons for Recycling (suitable)
 To Dispose of Solid Wastes (an infinitive phrase and hence
           unsuitable)
 To Salvage Valuable Materials (unsuitable for same reason)
Reasons for Recycling (suitable)
 Disposal of Solid Wastes (suitable because *Disposal* is a
          noun)
 Salvage of Valuable Materials (suitable because *Salvage* is
          a noun)

(4)
Difficulties with Kiley Trucks (suitable)
 Repairs Expensive (unsuitable; an elliptical sentence—no
       verb)[2]
 Have Insufficient Power (unsuitable; an elliptical sentence
       —no subject)[2]
Difficulties with Kiley Trucks (suitable)
 Expense of Repairs (suitable because *Expense* is a noun)
 Lack of Power (suitable because *Lack* is a noun)

If you doubt your own ability to phrase your headings suitably by applying grammatical principles, you can probably get good results by asking yourself as you look at each heading, "Does this sound like a name? Would it seem natural to say, 'The title of this section of my report is . . .'?" If the answer is no, a change in phraseology is probably indicated.

Another point to remember about the phraseology of headings is that each heading should be as explicit as possible. When a subpoint appears in the text of a report, its main point may be on an earlier page. Consequently, a heading should be as nearly self-sufficient as you can make it without letting its length become cumbersome. For example, a reader seeing *Power Lines* in an outline as a point under *Removal of Obstacles* would know at once exactly what material it covered. But if he encountered *Power Lines* as a heading on page 12 of a report when *Removal of Obstacles* was on page 6, he would have only a vague idea of what to expect. On the other hand, if he encountered the heading *Removal of Power Lines,* he would instantly know the exact nature of the material it covered. What might appear to be needless repetition in an outline can be highly desirable when the points of that outline appear in the report itself.

A report writer should never forget that, except when they are ex-

---

[2] Complete sentences would also be unsuitable but are less likely to be used where most heads are topics.

tremely short, reports are not always read from beginning to end as one reads a story or an editorial. On the contrary, they are repeatedly consulted by readers who at the moment want information only on one particular point. And the headings should be phrased so that these readers can find what they are looking for with a minimum of effort.

General Suggestions

The question of how many headings to use cannot be answered by rigid rules; but if you observe the frequency of headings in this book and in other material where headings are used systematically, you will be able to form a good idea of what is normal. Your aim should be to place headings close enough together to keep the reader constantly aware of what is being discussed, but not so close together as to reduce continuity.

You will also notice that the presence of headings somewhat reduces the need for transitional sentences and paragraphs. And finally, you will notice that the text does not depend on the headings for completeness. That is, words such as *this,* and *these* are not used in a manner that forces the reader to turn back to the heading in order to discover their meaning.

All in all, headings have an effect that is hard to overestimate. They encourage a reader to start reading a report by opening up its contents, they help him constantly as he is reading, and they subject the writer to a discipline that improves the orderliness of his writing.

SUMMARIES AND ABSTRACTS

The problem of discussing summaries and abstracts is complicated by the lack of anything approaching a general agreement about terminology. In a set of instructions for use within a single organization, the problem can be solved by the use of arbitrary definitions; but in a book for general use, this simple solution is unsuitable. The basic definitions that probably are most widely accepted in industry are as follows.

An abstract is a brief section—rarely more than six or eight lines long—the purpose of which is to tell the reader what points are covered in some piece of writing without attempting to tell him what is said about them. It may be written for a report or various other kinds of writing. If it is written as a section of a report, it is one of the first parts that a reader encounters. It is an independent element rather than an integral part of the body. Such abstracts sometimes appear by themselves in publications devoted entirely to abstracts of material in some particular field.

A summary—at least the summary of a report—is a section that presents in a highly condensed form the most important information that the report contains. It is brief, but not so brief as the type of abstract described above. It is placed, as an abstract would be, extremely early in the report (one rarely finds both an abstract and a summary in the same report); and like an abstract, it is an independent element rather than part of the body. Its functions will be described in more detail later.

Unfortunately, the situation is not so simple as it would be if the distinction established above were universally accepted. If one examines a substantial number of reports from different sources, he quickly discovers two additional facts:

1. The term *abstract* is often used before a section that would be a summary according to the preceding definitions.
2. The term *summary* is used widely and legitimately in reference to a section at the *end* of some piece of writing—a section that is an integral part of the piece and is used to summarize what has already been said. (In a working-level report, however, a summary of this kind is not often used, and even when used it is likely to appear without a heading rather than to be headed *Summary*.)

The problem of terminology is further complicated by the fact that in an effort to facilitate discussion, at least three other terms have been coined for use in books of general circulation. The term *descriptive abstract* is sometimes used to identify what was called merely an abstract in our original definitions. The term *informative abstract* is used in reference to what our definitions called a summary. (In fact one sometimes finds that the definition of *informative abstract* is sometimes followed by the statement that such an abstract is sometimes called a summary.) The term *introductory summary* is sometimes used in reference to what we called merely a summary, and sometimes in reference to a "summary" at the beginning that emphasizes such matters as an explanation of the problem or a review of the background. All these terms are used primarily in discussions and rarely appear as headings in reports.

The discussion of terminology might be further expanded to cover the occasional use of at least two other terms—*foreword* and *digest*—but we are concerned with other matters of more importance than mere terminology. The first major purpose of the present discussion is to emphasize the importance of the section at the beginning of a report that presents its most important contents in a condensed form—which for the sake of simplicity we shall henceforth call a summary. The second is to tell what a writer can do to make such a section, whatever it is called, serve its purpose effectively. It is unnecessary to deal with descriptive abstracts in the same manner, not because they are unimportant but because they are relatively easy to write.

If a report contains a summary, no other section with the possible exception of conclusions and recommendations is more important. This is the case because readers farther up the line than the writer's immediate superior may read the summary and little else. Also, readers who read the entire report on a single occasion may later refer to the summary repeatedly to refresh their minds. Still others will use the summary to decide whether to read the full report. In many great corporations, summaries of reports are widely distributed so that those whom they interest may request copies of the full version if they so desire. Also, by looking at summaries that are kept in carefully indexed files, people can tell what information is available from other departments than their own and avoid needless duplication of effort. In brief, when you prepare a summary of your report you may expect that it will be read more frequently, read by more people, and read by more important people than will the full report.

The condensation that is characteristic of a summary is achieved by leaving out preliminaries, details, illustrative examples—everything except the main facts and ideas. The relationship of the summary to the full report is roughly similar to the relationship of a topic sentence to a fully developed paragraph. (There is a difference, however. The topic sentence is an integral part of the paragraph, whereas the report is independent of its summary.)

To write a good summary you should not approach the job too casually. If you try to read a few lines and then condense them, read a few more and condense them, and continue this process to the end of the report, the result will be ineffective. Instead, you should go over the entire report—or an entire section if the report is extremely long—and make notes of the matters you consider essential for inclusion. Next, write the rough draft of your summary by using these notes and by referring to the full version whenever necessary. Then polish what you have written until it reads as smoothly as it can be made to read without the waste of space on introductory and transitional material.

Make a special effort to avoid wordiness. (A summary should rarely exceed 10 percent of the length of the full version, and it is unfortunate to waste words when you are constantly being forced to cut out facts and ideas you would like to include.) If you can preserve grace and variety of style in a summary, well and good; but if you must sacrifice those qualities for the sake of brevity, there is no reason for undue concern. In a summary, conciseness outweighs all other qualities of style except precision, clarity, and readability.

In writing a summary you will need to take special pains in order to make your work accurate. Figures must be double-checked to avoid error. Facts and ideas must not be distorted by the cutting out of words.

Care is needed in the selection of material for inclusion, for if its contents are unrepresentative, a summary can be misleading even though it contains no statement that was not in the original. In matters such as emphasis—in the implications as to what was important and what was less important, a summary should give the same impression that the reader would receive if he read the full report.

Still another consideration in writing a summary is the question of whether the reader or readers will be the same people who read the full report. When a report contains a good deal of technical material, it is very likely that some of those who read the summary will not have the technical background of the writer or the original receiver. Consequently, a summary will often be more useful if technical information and technical vocabulary are held to a minimum. You might do well to keep in mind three types of reader: the original reader, who wants a full picture but may not check details (he reads the introduction, summary, body, conclusions, and recommendations); the person whom the original reader depends upon to check everything for him (he studies the appendix, with all its data, calculations, tables, and figures to make sure that there are no errors or discrepancies); and the executive in a higher position, who may not be a specialist in the area that the report concerns, but who wants to know what is going on and who may say "Yes" or "No" to the entire project (he reads the summary and whatever is offered in the way of conclusions and recommendations).

To indicate how a summary compares with a full treatment of the same material, the following summary is presented. It covers all that was said in Chapter 7 about what reports are, their importance, and the difference between professional and school reports.

Example of a Summary

Though reports vary widely in form, length, and other respects, they are a distinct species of writing because they possess one characteristic in common: the responsibility of the writer of the report to the reader.

Reports are extremely important in government and industry because they are a major source of information for those who need it. Their importance to those who receive them makes skill in producing them important to the writers.

The merit of a report written on the job depends on the extent to which it meets the needs of the reader. This test of merit does not exist in the classroom. When a student graduates and takes a job, it is important that he realize this difference and form the habit of constantly asking himself, "Does the reader need this?" as he decides what information to include in a report and what to omit.

THE INTRODUCTION

Naturally, a report that is extremely short and simple needs little or nothing by way of an introduction. All that is necessary is to say what is required for the sake of self-sufficiency and to start smoothly enough that your reader feels at ease as he gets into the subject.

When a report is longer, it may be necessary to choose between an ordinary introduction and what we will call a formal introduction. The ordinary introduction in a report is no different from the introduction in any other kind of writing. It may consist of general comment on the subject as a whole. It may give background information on whatever the report concerns. It may explain the general plan that the report is to follow. It may serve, without being so labeled, as a summary of the contents. It may attempt to arouse interest. (The presence or absence of a table of contents and a summary will of course affect the need for some of these types of material.) In general, the normal introduction may say anything that needs to be said about the subject of the report before the first main division of the material is taken up.

Example of a Normal Introduction

The Klamath National Forest in northern California contains one of the most landslide-prone regions of the world. High rainfall, steep slopes, and weak foundation materials have combined to produce hundreds of landslides. The greatest instability is found in the western part of the Forest in a zone 20 miles wide centered along the Klamath River. Graphitic slates and serpentinized-ultrabasic rocks are especially unstable in this zone.

Every year on the Klamath Forest about 70 miles of new roads are built for timber harvesting and recreational access. One of the jobs of the engineering geologists is to make foundation studies for road route planning, location, and design. The foundation studies provide data for building roads which will have the least impact in terms of triggering new landslides or reactivating dormant landslides. The following deals with one aspect of these foundation studies: Making a preliminary estimate of slip-surface geometry based on the field expression of head scarps and lateral scarps.

In contrast, a formal introduction to a report is much like the introduction to a thesis or dissertation. That is, it deals not so much with the subject of the report as with the report itself. This type of introduction is likely to be used in a long formal or semiformal report that is written for some nonroutine purpose—often for the purpose of presenting the results of a special investigation. It will be discussed in more detail in Chapter 11, which deals with the formal report. Suffice it to say, at present, that in nonformal reports a formal introduction may be used to tell the reader whatever he needs to know about the reason the report is written, the function it is intended to perform, the scope of the problem it concerns, and the name of the person or organization that authorized it. A specimen of a formal introduction is included in Chapter 11.

Such an introduction may be followed by an ordinary introduction if one is needed; but in this case the ordinary introduction is placed under a heading such as "Background Information," "General Comment," "History of . . . ," or any other appropriate phrase.

CONCLUSIONS AND RECOMMENDATIONS

Though some reports, because of their purpose and content, do not need to include conclusions or recommendations, such material is almost always called for in a report that presents the results of an investigation. When it is included, it should usually be drawn together in one place and headed in a manner that identifies its special nature. In this connection, the term *conclusion* or *conclusions* means "convictions arrived at on the basis of evidence" rather than merely "the section that consists of an ending."

Though conclusions and recommendations sometimes merge naturally into a single section, their nature should not be confused. *Conclusions* is a term used to identify mere convictions. *Recommendations* refers to actions that are called for on the basis of such convictions. We might reach certain convictions without being in a position to recommend action because the wisdom of possible actions may depend on matters beyond the scope of our particular report. It might be possible, for example, to conclude that a certain site would be physically suitable for a factory, without being in a position to recommend that a factory be built there.

Logically, conclusions and recommendations should follow the body of the report because they depend on its contents. Often, however, they are placed before the body on the grounds that the reader wants to read them immediately. Most large organizations will have a standard practice regarding placement; but a writer who is not bound by a prescribed format may place them in either location with the assurance such placement is not unusual.

In a section labeled *Conclusions,* regardless of where it is placed, every statement should grow out of facts given elsewhere. A reader has a right to assume that you are not stating conclusions except on the basis of evidence. Overstatement should be carefully avoided. When some of the evidence seems opposed to the conclusions, that fact should be conceded; and in explaining away such evidence, the report should be scrupulously fair. If the evidence points toward some conclusion but does not definitely establish it, you should frankly say that the evidence is not conclusive. There is nothing wrong with offering an opinion if it is openly identified as such.

This advice to use restraint does not mean you should resort to evasion. In a conclusion, as elsewhere, such statements as "It is believed that . . ." are obviously meaningless unless we know who does the believing.

When conclusions are placed at the end of a report, the facts and reasoning that they are based upon may be pointed out briefly; but when they appear before the body, this support is not usually offered. Wherever they are placed, conclusions should be numbered if there are more than one, but the use of numbers is not a substitute for careful grouping of material into unified elements and careful arrangement in a logical sequence. Use of numbers does not achieve organization, but merely reveals it.

Even though the nature of a report sometimes makes it natural to offer a conclusion at the end of each major division, it is advisable to draw all the conclusions together in a single section unless the report is short and simple. It would be inconvenient for a reader who wished to study the conclusions and nothing else to be forced to hunt for them on several pages.

When recommendations are called for and do not merge naturally with conclusions, they should be keyed to the conclusions (that is, given similar numbers), when such treatment is possible. Like the conclusions, they grow out of the facts and reasoning offered in the body of the report. (Sometimes the recommendations grow out of the conclusions as the conclusions grow out of the body.) The occasional tendency to treat *Recommendations* as a section consisting of a few helpful suggestions on incidental questions is unsound in any report based on an investigation; rather than being incidental, the recommendations are the focal point of everything else in the report—the reasons for its existence.

One final point should be heeded. Like so many other tasks in producing a report, writing conclusions and recommendations is a matter of judgment, not merely compliance with rigid rules. In the preceding discussion, we have attempted to demonstrate that these materials should be presented in a manner that helps the reader to find what he is looking for as easily as possible. The organization and the choice of headings should be whatever is necessary to accomplish that result.

Following are the conclusions and recommendations arrived at in a report by the Technical Extension Service of a state university to a legislative fact-finding committee.

Conclusions and Recommendations

Conclusions

The survey of mobile home courts in Carmine County indicates that though the proposed regulations may be desirable and should be applied to any court opened in the future, they should be applied to existing courts with some restraint.

Fourteen courts are operating in the county at the present time. Details about each of them are included in the body of the report. If the new standards were in effect at present, only six of them would fully comply or could be made to comply by means of minor adjustments. These six, the newest and best, really present no problem. We have divided the others into three groups, as follows.

A. Three courts comply with all the standards that relate to health and sanitation and also with most of the other standards. Forcing them into absolute compliance would put them to such expense that it would not seem justified. If minor variations were permitted in individual cases, these courts could continue to provide needed service without being changed.

B. Four courts, which have been operating for periods varying from six to fourteen years, fall far short of meeting the proposed standards. In some instances, full compliance would be impossible. In others, though not impossible, it would be so costly that the courts in question would almost certainly be forced out of business. Closure of these courts would cause real hardships. It would not only mean heavy financial loss to the owners but would also dislocate the occupants of a total of 65 mobile homes. Since space in other courts is scarce, it is hard to conjecture what would happen to these units—in which, incidentally, many of the occupants have substantial equities. In this connection we should not overlook the fact that if continued occupancy of these homes in their present locations is prevented, their occupants may find it impossible to find other places of residence in this area, the housing shortage being what it is.

C. One court, in which only six of the twelve spaces are occupied, falls so far short of acceptable conditions that it probably could not be salvaged. It is over twenty years old, has been allowed to run down, and has been losing occupancy each year. It would be unlikely to survive much longer even if the new standards were not a factor.

Recommendations

1. The courts referred to in A, above, should be granted the variations necessary for their continued operation. (These courts and those in other groups are identified in a list in the appendix.)

2. All mobile courts more than five years old (in B above) should be exempted from proposed regulations except for those concerned with health and sanitation.

3. The owner of the court referred to in C should be notified that he will be forced to close within one year after the regulations become effective unless he submits an acceptable plan for improvements within 90 days and begins to implement it as soon as it has been accepted. (He is entitled to this opportunity even though it is unlikely that he will utilize it.)

THE APPENDIX

An appendix is often desirable in a report, as in a book, because it permits the writer to make material available without placing it in the main text. In a report, a typical example of this material would be the full results of a series of tests. The average of these results might be significant, but the results of the individual tests would often serve no function except to make it possible to confirm the average and perhaps to see the extremes. Other material sometimes relegated to the appendix includes tables, figures, calculations, lists (for example, lists of equipment), specifications, and even supplemental reports.

Various reasons can make it desirable to put something in an appendix. The appended material might be technical information that most readers would not use, though they might want to have it available for examination by specialists. It might consist of detailed or supplementary information that would be consulted on special occasions but otherwise would only be in the way. It might be material to be studied separately rather than during normal reading. It might be material such as photographs, that would make the report physically hard to handle if scattered through the body.

Obviously, anything placed in an appendix is not so closely integrated with the rest of the report as it would be if placed in the body. Sometimes, indeed, it does not need to be closely integrated. But when material in the appendix is closely related to something in the main report, it should be referred to at the appropriate point in the text.

Ordinarily the appendix is preceded by a title page bearing the single word *Appendix;* but if the appendix is brief and if the first item is not of full-page size, the title *Appendix* may be placed at the top of the first page rather than on a page by itself.

When the material appended is of more than one kind and when its amount is great enough to justify the extra complication, two or more appendixes can be used. In this case, each is identified by a letter, and its title indicates the nature of its contents. (For example, *Appendix A—Tables, Appendix B—Photographs.*)

In brief, use of an appendix is often a handy method of presenting material that should be available in support of the main report without causing the main report to become needlessly cluttered and complicated.

FINAL COMMENT ON SPECIAL ELEMENTS

Before the subject of special elements is dismissed, two more points should be brought out. The first is that in the report-writing world, uniformity is sadly lacking in the labels attached to these elements and in the details of their use. For example, what is called an *abstract* in one company may be a *summary* in a second and a *synopsis* in a third. The term *references* usually means a list of books and articles, but in some companies it means conversations, letters, and memoranda that are related to the subject of the report and are listed at the beginning. In short, though most of what has been said in the preceding discussion will apply in any position you may hold, it would be impossible for any book to give instructions that are entirely suitable for every occasion, because terminology and details of procedure vary.

Second, we should not lose sight of the fact that special elements of reports are merely tools. We should know how to use them, but we should decide in each case which ones are needed, and not use all of them just because they are available. In other words, planning the layout of a report is not just a matter of following hard and fast rules. On the contrary, it is a process of designing a product that will do its job effectively—and its job is simply to make sure that the reader can grasp the contents of the report as accurately and as easily as possible. Each of the elements is merely a means to that end.

EXERCISES

Following are excerpts from outlines made with the expectation of using the points of the outline as headings within the completed report. Revise the phraseology, when necessary, so that each point is worded as a topic and so that the principles of parallel form and self-sufficiency are applied.

(1)	(2)
Suggestions for reducing accidents	Advantages of electric heating
Reduction of speed limits	Cleanliness
Prevention of drunk driving	Quietness
Start driver education program	Saves space

(3)

Objections to drilling deep wells
Doubtful quality of water
Expensive to drill
Availability of surface water

(4)

Recommendations for applying
herbicide
Choice among types
Method of application
Fall better than spring

(5)

Reasons for moving highway off Main
Street
To improve flow of through traffic
To remove congestion from Main
Street
To reduce noise in business district

(6)

Danger of flooding
In the region east of town
In the region of White Avenue
In the Clover Creek drainage

(7)

Reasons for choice of Vanguard
antenna
Adequacy of gain
It is easy to assemble
Sturdiness of construction

(8)

Conditions that make cars dangerous
Defective brakes
Defective steering mechanism
Tires are badly worn

(9)

Claims made for the fabric
Water-repellent
Durability
Color fastness

(10)

Warnings concerning Kleenit Paint
remover
Need of good ventilation
Advisability of wearing neoprene
gloves
Guard against exposure to flame

10

PROCEDURE IN WRITING

A REPORT

The procedure described in this chapter is suitable when the complexity of a situation or the probable length of a report creates uncertainty about how to tackle the problem. It is offered not as a formula to be followed rigidly, but merely as a systematic approach that may be modified as circumstances require. It would call for the following steps:

1. Analyze and plan the job as a whole, so that you have in mind a picture of the kind of report you must produce and what you must do to produce it.
2. Form a general, tentative plan for presenting your material. This can include a preliminary decision about the entire layout of the report (what elements to include) and a highly tentative plan or outline for the substance.
3. Gather the facts that you need and interpret their significance.
4. Revise your tentative plan so that your outline reflects your final ideas about organization.
5. Decide so far as possible what use you will make of tables and figures if it seems likely that they will be helpful.
6. Write the first draft of the body of your report, and of your conclusions and recommendations.

7. Make your final decision about what elements (introduction, summary, and so on) you will use as a framework for the body, conclusions, and recommendations, and write those that you have not yet written.
8. Revise and polish all parts as necessary.
9. Care for such final details as providing the title page and table of contents. Go over everything carefully to be sure that it contains no errors, and then go ahead with the mechanical business of copying and fastening it together.

The occasions when you will follow this procedure exactly will be few and far between. When a report is long and complicated, preparing it is a matter of pushing forward on more than one front as conditions permit. It may be necessary, for example, to gather and analyze facts on one point before deciding whether another should be included at all. Again, you may produce a rough draft of one section, complete with tables and figures, before the investigation for another section has gone beyond its opening stages. Nevertheless, the procedure suggested is sound whether applied to an entire report or to a single section of a longer report. It at least provides a systematic sequence of action so that you do not need to start work with no plan at all and improvise as you go.

One more point should be mentioned in connection with the procedure as a whole. The order in which a writer produces the various parts of a report is not the order in which a reader encounters them. Such parts as the title page, the summary, the introduction, and the conclusions and recommendations are brought into existence as the work nears completion; but most of these parts are placed early in the report. A person who expects to produce a report of any length and complexity should grasp this essential point at the outset.

ANALYZING THE JOB

Analyzing the job really consists of analyzing two problems—the problem that the report itself is intended to solve, and the problem of what you must do to produce a report that will serve its purpose.

To analyze the first of these problems you will need to center your attention on the subject and purpose of the report. It is often the function of a report to answer some large, comprehensive question. As you think about the question that the report as a whole concerns, you will usually find that the answer depends on the answer to various other questions—many of which themselves depend on the answers to even smaller questions. For example, to answer the question of whether to produce some product by a new method it would first be necessary to answer questions about the quality of product that would result, the equipment needed, the space needed, the skills needed by the operators, the cost, and countless others. When such a list of questions to be answered has been drawn up, the scope of the original problem becomes clearly understandable.

But the second problem still remains: what you must do or have others do in order to produce your report. That is, you will need to figure out what methods of investigation will be necessary. You will need to decide when and by whom each portion of the total investigation will be performed. You will need to decide what kind of report (how detailed and what elements it should consist of) is necessary in presenting the information after you have gathered it. In brief, you will need to form a clear mental picture of the finished product, to anticipate the jobs that the production of such a product involves, and to lay plans that will result in orderly progress.

Admittedly, the two aspects of analyzing the job are not sharply distinct. The second is to some extent based on the first, for it is obviously impossible to decide where and how to get information without first deciding what information must be obtained. But in spite of this overlapping, the analysis of a report-writing job usually goes better if you think of it actually as the analysis of two separate problems.

MAKING A TENTATIVE PLAN OR OUTLINE

It is almost impossible to analyze the subject of a report without forming a general idea about your outline, at least in its larger aspects. As you think of various questions, you will consider their relationships toward each other, and these relationships will point the way toward main points and subpoints of an outline. And as you think of all your questions in relation to the function of your report, a general plan for presenting your material will begin to take shape. It is a mistake, however, to think of these early ideas on organization as final. Only after you have gathered enough information to be dealing with answers instead of just with questions can you be sure that any plan you form will be satisfactory. Then, some matters that looked important will dwindle into insignificance; points that seemed likely to be minor will assume importance; facts will emerge that you had not realized would call for consideration. All in all, though forming a tentative idea about organization is natural and useful, the principle urged in Chapter 3 is still sound. Only when you have gathered information and interpreted its significance can you decide definitely about the best way to organize it.

GATHERING INFORMATION

In a book dealing mainly with the process of writing, a discussion of gathering information for reports must be extremely general. The fact is that gathering information is not really a writing problem. Still, a few general comments about the sources of information apply widely enough to justify their inclusion. These comments will apply first to the reports you will write on the job and then to reports you may produce in a course such as the one based on this book.

Reports Written on the Job

When you are actually on the job, regardless of your field, your first step in gathering information will probably be to check the files of your own company in order to locate reports or other records that might answer some of your questions. Many companies have extensive indexes of reports and abstracts of their contents. Where no formal system for retrieval of information has been developed, personal inquiries as well as checking the files will often open leads that will enable you to avoid duplicating the work that others have done.

Published material may be the next source to try. A systematic survey of published literature is often the starting point when an extensive investigation of a scientific question is to be undertaken.

Another common method of obtaining information is by correspondence. Sometimes this involves the use of questionnaires, but a mere letter of inquiry is more frequently appropriate. Such letters are discussed in the section on letter writing. To a person on the job, the occasions when a letter is the best means of obtaining information are so obvious that there is no need to discuss the subject in detail.

In place of writing letters you may sometimes find it better to get information by personal interviews. Such interviews are of course more expensive than letters and are sometimes slower, but the flexibility of a personal conversation can bring out facts that would not come to light in an exchange of letters.

When the information needed for a report cannot be gathered from other people, personal observation and inspection is a possible method. A typical use of personal observation would be a visit to a factory to watch some process in operation, a field trip to observe the performance of a piece of farm machinery, or a trip to a mine to investigate its condition. Personal inspection might be the method used in an attempt to determine why accidents were frequent on a construction project, or what conditions were causing an epidemic.

Scientific tests and experimentation are of course the main source of information for many reports. Sometimes as a report writer you may do such work yourself and sometimes you may have it done for you by others whose special services are available. The contributions of scientific research to modern industry and government—contributions that are made available by means of reports—are so vast that it would be impossible to do more here than recognize their importance. However, we should not overlook the fact that reports based on scientific research are not always related to practical problems of the kind we have been discussing. Many a scientific investigation is made for no further purpose than to increase the fund of human knowledge, and many a report is written just because new knowledge exists and should be reported. The purpose and probable use of this book justify a focus of attention on practical rather than purely scientific problems; but many of the techniques recommended can be applied in reporting on scientific research that is not intended to solve a specific and immediate practical problem.

One final and general means of gathering information must be mentioned because, though it overlaps the others, it will often be the main method you will use on the job. It consists of merely doing the work that your profession calls for—sometimes with the help of others whose services are available. For many who use this book, such work will mainly consist of the scientific tests and experiments referred to above; but these and the other possible activities are so numerous and varied that no list could cover more than a small proportion of them. Actually, the work done in gathering information for a report can include any of the technical work done in our complicated industrial civilization.

It should also be noted that the information in a report does not necessarily come from a single source, and conversely, not all the sources are likely to be used on a single occasion. The fact is that though the information needed for a report may sometimes be far from easy to obtain, it will not necessarily be hard to decide what sources are promising and what method to use. The nature of the problem is likely to make the type of source and the method fairly obvious.

Reports in a Writing Course

Most of the sources of information described above may be utilized for a course in technical writing. But the problem of gathering information is complicated by the fact that resources available to a person holding a position in government or industry are often unavailable to a college student. On the job, the person who is asked to report on a problem has access to relevant files, opportunity to make necessary visits and inspections, and the time, facilities, assistance, and financial support that he needs in seeking a solution. In the classroom, finding a real or at least a realistic subject based on specific facts may in itself be a major difficulty

—especially since the problem should not only be realistic but must also be practicable for a student to handle. Consequently, instead of starting with a problem and then figuring out the sources of information and method of investigation, a student must often discover a subject he can obtain information about and then try to set up a problem that this information will be useful in solving.

Information available in books and magazines may unavoidably be a larger element in a report for a writing course than is usually the case in a report written on the job. To prevent a situation in which mere reference papers masquerade as reports it is well to remember that a school itself may provide suitable subjects. These would include such matters as parking, lighting, utilization of land, arrangement of laboratory equipment, condition of buildings, sanitation, equipment in kitchens, and many other questions that constantly arise in the physical operation of even a moderate-sized institution.

Also, a job past or present might provide information or access to information that can form the basis of a suitable problem and its solution. In any event, unless a subject is assigned or suggested by an instructor, it may call for considerable ingenuity for a student to figure out where information can be obtained and to develop a subject that permits him to use the facts available.

INTERPRETING THE FACTS GATHERED

A report writer's job is not limited to gathering facts and stating them. He must often interpret them for the reader's benefit. The nature of this interpretation is comparable to what happens when a patient goes to a doctor. The patient describes his symptoms, the doctor administers tests, makes examinations, and discovers facts about metabolism, blood pressure, and other physical conditions. Then the doctor must interpret these facts about the patient's condition in order to diagnose the ailment. Other examples would be as follows:

FACT: When a slide prepared by a medical technician is exposed to a certain stain, the smear upon it retains the color of the stain.
INTERPRETATION: A colony of bacteria has developed.
FACT: When the concentration of elk in a national forest is greatest the browse gets scantier each year.
INTERPRETATION: The herd is too large for the food supply.
FACT: The water above an industrial plant contains more oxygen than the water below, and the industrial wastes contain organic matter.
INTERPRETATION: The oxygen is being used up in the decomposition of the organic matter.

FACT: A Geiger counter ticks faster than usual.
INTERPRETATION: There has been an increase in radioactivity.

To be sure, the statements labeled *interpretation* in the preceding examples are themselves intended to be statements of fact, but the facts stated are a different kind. They are discovered not by observation alone but by observation plus reasoning, and they reveal the significance of the observed facts on which they are based.

Another kind of interpretation is the interpretation of the meaning of written material, as, for example, when the meaning of a statute is interpreted by a court decision. This kind is not so common in report writing as the kind previously described, but it is sometimes called for. For example, when there is doubt whether certain work is covered by a contract, interpretation of the meaning of the phraseology of the contract would be necessary, and the results of this interpretation might affect the recommendations of a technical report. To be sure, everything should originally be written in a manner that reduces the possibility of a dispute about interpretation. Nevertheless, in your technical writing, you will sometimes have to demonstrate the fact that a general or abstract statement covers a specific, concrete situation. And in doing so you will be interpreting the meaning of the statement.

The process of interpretation may call for any of the generally recognized processes of reasoning, such as generalizing on the basis of individual cases, applying a general rule to a specific case, ascertaining the cause of incidents or conditions, deciding what results will follow if certain actions are performed, or ascertaining the specific applications of a rule or regulation that is phrased in general terms.

Likewise, the errors in interpretation resemble the errors in reasoning in general. They include generalization on the basis of insufficient or unrepresentative evidence, or failure to consider alternative possibilities (that is, the possible existence of undiscovered facts or other possible interpretations of facts discovered). Also writers sometimes fail to realize that an entire line of facts and reasoning may be based on assumptions that might themselves be subject to question. Much of what was said in Chapter 7 under the heading of *Thoroughness* deals with these errors.

Whether to explain in detail the reasoning by which you reach your interpretation of the facts is a matter of judgment. Certainly it is not desirable to prolong a report by explaining the self-evident. There are times, however, when a reader will probably be unable to see why you interpret the facts as you do unless you tell him. And there are also times when it is desirable to tell him even if he might be able eventually to think his own way through, for with your assistance, he will be able to see the connection faster than he could figure it out for himself. Also, he is more likely to accept your interpretation if you show him how you reached it.

One more point needs to be cleared up: the distinction between interpreting the facts in your report and the process of forming your conclusions. Actually, there is no *essential* difference between the two. When you interpret a fact or a group of facts, you are drawing a conclusion, and when you settle upon the conclusion or conclusions for the report as a whole, those conclusions amount to an overall interpretation of all the facts. Nevertheless, at a practical, working level, a distinction is justified unless the report is short and uncomplicated. The difference lies in the scope. In stating the conclusions of a report you will be telling the convictions arrived at in connection with the major question or questions the report is supposed to answer. But in interpreting your facts you will be trying to find and reveal the significance of individual facts or small groups of facts. When you eventually draw your main conclusions, you base them on the interpretations that have occurred throughout the report.

MAKING THE DETAILED OUTLINE

As mentioned earlier, you are almost certain to form a general, tentative plan for a report before you gather all the information it will contain. But when you have gathered the information and analyzed it to ascertain its significance, you will usually need to revise the plan so that it will be a good organization for the facts as they actually turned out, and to develop it into a detailed outline. In most respects, making an outline for a report is not essentially different from making an outline for any other purpose. True, you may sometimes be required to place your material under a standard list of major headings; but even then it will still be necessary to plan the presentation of the material within each main division; and there will be innumerable occasions when you will be free to plan the entire treatment. Therefore, you should not overlook the techniques of outlining presented in Chapter 3.

In that chapter it was pointed out that making an outline is primarily a matter of organizing your material. As you organize the material of a report, one consideration needs special emphasis. A report, more than most kinds of writing, is written in order that it may perform a specific function when it reaches the attention of specific readers who read it for specific reasons. Consequently, it is especially important, in planning it, to refrain from concentrating on the subject matter only. It is usually possible to organize the same subject matter in more than one way, and the best of these possible organizations is the one that makes it easiest for the reader to find the answers to the particular questions he is likely to have in mind.

Consider, for example, a study made by a highway research laboratory to determine whether subgrade soil might be improved by mixing certain materials into it. Tests were made to determine, in regard to each admixture, its effect on the strength of the soil, the ease of compaction, the resistance to penetration of moisture, and the tendency of the soil to swell when moisture penetrated. Two of the possible organizations that would accommodate the facts are as follows:

(1) An organization in which main points were based on the admixtures and the subpoints were based on their effects. This organization would make it easy for a reader to learn all the results that would follow when a given substance was added; but in order to discover what substance to use if a certain result were desired he would have to look in as many places as there were substances.

(2) An organization with main points based on effects. This would make it easier to see what material might be used if a certain result were desired, but information about all that would happen if a certain substance were used would be scattered in several places.

Both of these treatments are logical, but on a given occasion one or the other might be decidedly superior.

PLANNING THE USE OF TABLES AND FIGURES

If a report calls for the use of tables and figures, you must decide whether to scatter them through the text, place them all at the end, or divide them between the two locations. It is usually best to make this decision before you do much writing. In fact it is better if you not only decide where figures and tables will be placed, but also decide what specific figures and tables you will use. If you cannot make this latter decision for the report as a whole, you can at least make it in connection with each section before you begin writing it. The reason for settling this question early is that your text itself will be affected by your decisions in regard to these nonverbal forms.

The technical aspects of using tables and figures in a report is no different from that of using them in other technical writing. Therefore the discussion in Chapter 6 makes a detailed explanation at this point unnecessary.

WRITING THE FIRST DRAFT

If the preliminary work has been done properly, the actual writing of a report is much the same process as writing anything else and should not be abnormally difficult. A few special suggestions, however, might be of value.

The first of these concerns the question of getting started. How to get under way may be a puzzle, of course, in any kind of writing. In writing a report, the preliminaries that must often be cared for present an additional complication. Consequently, if you find it hard to get a report started, the sensible thing to do is to postpone writing the opening and go to work on the earliest section that you feel ready to handle. If this is not feasible, start at the beginning in the simplest manner possible. Don't worry about smoothness. Just write the bare facts that are necessary if the report is to be self-sufficient, and then plunge into the first main section of the body. Later, you can revise your opening or write a new one; but it may surprise you by turning out to be better than you had expected.

In writing a report on an extensive project, you will probably not even consider writing the opening first. Rather, you will write up each section as soon as the information it will include becomes available. Your work will go faster and its quality will be better if you write while your information and ideas are fresh in your mind. Under these circumstances the preliminary portions of your report may be among the last to be written.

In any event, when you start writing, push ahead without worrying about details. If your flow of words is slowed down by an effort to improve your style, the flow of ideas may also slow down and it will be harder to preserve continuity. If something worries you as you put it down, make a note in the margin and settle the question later.

As you write or revise you should be especially alert for spots where ordinary text is ineffective. When you have a list to present, for example, you may be able to help your reader by presenting it with each item on a separate line. A typical use of such a list (sometimes called an informal table) would be as follows:

Gradation of Samples

Percentage passing ¾-in. screen and retained on ½-in. screen	16.7
Percentage passing ½-in. screen and retained on ⅜-in. screen	16.6
Percentage passing ⅜-in. screen and retained on No. 4 screen	16.7
Percentage passing No. 4 screen	50.0
	100.0

The use of nontextual form may often be carried even further with good results. To show how this helps the reader, the same facts are presented below in two forms, the second of which makes them easier to grasp than the first.

(1)

The savings in labor cost that we can realize by manufacturing the newly designed bobbin rather than the present bobbin will be in the neighborhood of $158,000. This figure is calculated as follows: We hire three crews to manufacture bobbins, each crew consisting of 45 men. The size of each crew, if we make the change, will be cut to 35, a reduction of 10. Thus the labor force will be reduced by a total of 30. As we now operate, a laborer normally works 1,920 hours per year. If we made the change, the elimination of 30 laborers would mean a saving of 57,600 man-hours. Our present wage to members of these crews averages $2.75. Thus by cutting the crews as indicated we would reduce cost of labor by a total of $158,400.

(2)

The savings in labor cost that we can realize by manufacturing the newly designed bobbin rather than the present bobbin will be in the neighborhood of $158,000, as shown by the following calculations:

Number of crews		3
Size of each crew		
Present	45	
Proposed	35	
Reduction	$\overline{10}$	
Total reduction of labor force		30
Hours worked per man per year		1,920
Man-hours saved per year		57,600
Average wage per hour	$2.75	
Savings in cost of labor		$158,400

REVISION

Unless you write with unusual skill, your work will probably call for considerable revision. If your schedule permits, you should not do your revising until you can look over your work with the feeling that you are a reader rather than a writer. It is possible, however, to let your work cool off too long. You should revise while your subject is still fresh enough in your mind that you will notice omissions and recognize errors.

The amount of time devoted to revision depends partly, of course, on what the project justifies, partly on the amount of time you have available, and partly on your degree of success in producing an acceptable version at the first attempt. If a report includes complicated ideas and deals with an important subject, there is no reason you should feel frustrated if it must be worked over three or four times. Most of us must face the fact that writing is not easy.

Since revision is tedious, one point needs to be emphasized. When you have revised a passage for what you think should be the final time, read it over once more from beginning to end to make sure that the changes you have made do not call for additional changes elsewhere. Many times an improvement at one point makes it necessary to change something that precedes or follows.

CARING FOR FINAL DETAILS

Caring for the final details of a report may include preparation of a title page, a summary, a table of contents, and a letter of transmittal or covering memorandum. Suggestions on the purpose and treatment of most of these elements have already been provided.

Also, one final task must always be performed: the close examination of the entire report for errors in manuscript, in form, or even in such simple matters as the order of pages. Everyone dislikes to make this final check, but it is so easy for errors to pass unnoticed that a final scrutiny is well worth the time it takes. Errors in figures are especially likely to slip in during the process of copying, and can easily escape detection. A typist can accidentally change *now* to *not*. *Dejected* can be miscopied as *ejected,* or *grout* can become *trout.* A writer who intends to change *not conspicuous* to *inconspicuous* may add the *in* and forget to remove the *not.* Such errors do not catch the eye, and close scrutiny is necessary if you are to succeed in actually writing what you intend to write.

In closing, it should be re-emphasized that the foregoing procedure for writing a report cannot and need not be followed exactly. Projects that call for reports develop in so many ways that it would be not only undesirable but probably impossible for you to follow a rigid set of instructions. But at least the steps suggested are definite and their sequence is logical. Though deviation may be necessary from time to time, you need not feel uncertain about how to tackle the job of writing a report if you will regard the recommended procedure as normal.

ASSIGNMENTS

Assignment 1

Analyze the problem one of the following subjects would present if it were to be reported on. That is, draw up a list of questions that would lead to a decision about what facts were needed. Break large questions down into smaller questions that would need to be answered first. Make your chosen subject more specific, if possible, by limiting it to some particular local situation that you know about or can learn about.

1. The feasibility of establishing a recycling center in your community.
2. The problems involved in setting up bus routes and schedules to reduce the need to drive to your school.
3. The question of the kind and amount of insulation a builder should use in the construction of several houses in the same area.
4. The question of converting certain streets or parking areas on campus or in the writer's town into a pedestrian mall.
5. The problems involved in deciding whether or not to construct an outdoor pool in some small community or in a park in the residential area of a larger town.
6. The feasibility of leasing rather than buying cars for a business.

Assignment 2

As directed by your instructor, write one or more reports of unspecified form resulting from an investigation of one of the following subjects, a subject listed in Assignment 1, or a subject of your own choice. Like all other subjects suggested, any of these subjects would have to be defined and adapted by setting up a hypothetical situation in which the report would be submitted to some person or organization that needs the information. The scope of your subject may also need to be broadened or narrowed so that you can use the information obtainable and have the size of the subject suitable to the assigned length of the report.

1. Ways to reroute traffic to make your campus totally pedestrian.
2. Steps involved in a civic beautification project.
3. The effects of a food additive, including both intended effects and possible unintended effects.
4. The relative merits of a Pass-No Pass grading system and the traditional five-letter grading system, particularly as regards student motivation and the use of grades in seeking jobs after graduation.
5. Traffic and pedestrian survey to determine need for a stoplight at an intersection in your community.
6. Survey and report on your average daily activities with a view toward reorganizing them to use your time more effectively.
7. Report on the relative merits of clear-cutting and select-cutting a specified forest area. Include, of course, a description of the area.
8. Analyze the geologic setting of some area near your school with a view toward building a house, recreation center, or business on it.
9. Report on the various causes of forest fires in your area and what is being done to prevent them.

10. Report on the market potential for a product new to your area. An example might be a Wankel-engine car or an artificial leather shoe.
11. Report on the advisability of a curriculum change that might make a field of study more responsive to current needs of its majors.
12. Compare the relative merits of two methods of testing a metal to be used for a specific purpose, which you should describe.
13. Report on a soil analysis of a local soil and detail what changes should be attempted to make the soil more productive for a list of crops that you specify.
14. Compare the relative merits of two design types for a local bridge construction job.
15. Report on the safety of nuclear-powered power plants to determine the feasibility of locating one near your campus.
16. Compare the merits of constructing a culvert or a bridge on a minor country road, considering especially cost and expected lifespan of each.

11

FORMAL AND SEMIFORMAL REPORTS

Though formal and semiformal reports have been mentioned briefly in earlier chapters, they give rise to so many special questions that they call for separate treatment.

DESCRIPTION OF FORMAL AND SEMIFORMAL REPORTS

The term *formal report* is somewhat misleading. Actually, the formality of reports can vary widely; a report can be quite formal in tone, and still not be classified as a formal report under most systems of nomenclature. Still, the term is widely used, and anyone concerned with reports should know what people are most likely to have in mind when they use it. The distinctive characteristics that the term implies, as gathered from observation of reports from many sources, are as follows:

1. A formal report has a title page.
2. It is usually accompanied by a letter of transmittal.
3. Unless it is unusually short it has a table of contents that includes or is followed by a list of nonverbal materials if any are used.
4. It has a summary (sometimes called an abstract) near the beginning.
5. It contains the kind of introductory information found in what we have called a formal introduction (see Chapter 9).

6. If it contains conclusions or recommendations, it identifies them by suitable headings.
7. It has internal topical headings, usually of more than one rank.
8. It is formal in tone and impersonal in style.
9. It includes tables and figures if they can increase its effectiveness. These may either be placed in the text that concerns them or gathered together at one place at the end.
10. It is documented by some conventional system if published information is used to any significant extent.

In this book, as mentioned earlier, the term *semiformal report* means a report that has the general qualities of a formal report but omits some of the elements. As you read the remainder of the chapter, remember that use of the elements of a formal report is not an all-or-nothing matter. On the contrary, if you are free to plan a report exactly as you think best, you can profit by following the suggestions about each element that you decide to use, but can dispense with any element that you feel can be spared. Though the term *semiformal report* has not been widely used, reports of that kind have been widely written and you need have no misgivings that they will seem strange to those who read them.

Formal reports are considered a distinctive type on the basis of form and tone, but they may be used for varied purposes. A formal report may be a progress report, a periodic report, a report on the completion of a project, or a report on the results of a special investigation. Obviously, however, it is suitable only when a report must be long enough to justify the somewhat complicated mechanism involved and when the occasion calls for dignity in treatment. Rarely would a formal report be written for a single reader. On the contrary, it is justified only when the contents will be of interest—in fact of more than temporary interest—to a number of people.

The general nature of a formal report becomes clearer when we note that its pattern is much like the pattern followed in most books. Like a book, it may have a cover and always has a title page. Just as a book has a preface, a formal report has a letter of transmittal. Like a book, it has a table of contents that includes or is followed by a list of nonverbal materials. Like a book, it starts its sequence of Arabic numerals on the first page of text that follows the table of contents or the list of nonverbal materials. Like a book, it may have an appendix and a bibliography or list of references after the main text. It might be an overstatement to say that every element in a book has its exact counterpart in a formal report; but certainly a writer who is puzzled by some question of form can often solve his problem by recalling or observing what is done in the books in his own field.

The resemblance of reports to books is worth remembering because it would probably be impossible to set down a rigid set of rules about form which would enable us to label each possible way of doing things *right* or *wrong*. But by following the conventions used in books we can at least use forms that our readers are accustomed to. The sensible attitude on questions of form is to settle them not so much on the basis of right and wrong as on the basis of effective versus ineffective communication. And we communicate more effectively when we do not depart from what is customary unless we have good reason to do so.

ARRANGEMENTS OF THE PARTS

The most important variation in the arrangement of the parts of a formal report concerns the placement of the conclusions and recommendations. Sometimes these materials follow the body, and sometimes they precede it. The first of these alternatives is logical in that it presents evidence before conclusions; but some users prefer the second on the grounds that readers want to know the conclusions at once. This variation in the placement of materials is shown in the two following arrangements:

Title page	Title page
Letter of transmittal	Letter of transmittal
Table of Contents	Table of Contents
Summary (sometimes called Abstract)	Summary
Introduction	Conclusions
Body	Recommendations
Conclusions	Introduction
Recommendations	Body
Appendix	Appendix
Bibliography	Bibliography

When the conclusions and recommendations follow the body, the other parts are almost always arranged as shown above; but there is no set order of arranging the introduction, summary, and conclusions and recommendations when they precede the body. The writer may arrange them as he thinks best unless he works for an organization that has standardized upon a single system. Whatever it may be, the arrangement has some effect on the contents of the various sections. The reasoning behind the conclusions, for example, is more likely to be explained when they follow the body than when they precede it. And the summary will cover the introduction when it precedes that section but not when the reader encounters the introduction before the summary. A writer's good judgment should enable him to adapt his treatment to the order he is following.

COMPONENT PARTS—ADDITIONAL COMMENT

Since many of the parts that comprise a formal report appear in other reports also, they have already been discussed in a general manner. But there is more to say about the treatment of three of these parts—the letter of transmittal, the table of contents, and the introduction.

Letter of Transmittal

The letter of transmittal in a formal report is more closely integrated with the report than is true of the covering letter or memorandum sent with reports of other kinds. In fact it is bound into the report, usually following the title page. It may vary widely in length and contents. If the report is entirely routine, it may consist of only one or two sentences, as for example: "Following is the annual report of the Kinman Soil Conservation District for the year ending December 31, 1960." At the other extreme, it may replace the summary and present the essence of the report in condensed form. Most of the time, however, its length is within the normal range of business letters and it presents at least the minimum of facts necessary to orient the reader to the report that follows. For example, it may identify the project the report concerns and also, if necessary, the specific subject of the particular report. It may indicate the scope and limitations of the contents if the title leaves any doubt about them. It may tell where and how the information was obtained and how the report came to be written. And finally, by its signature it may identify the writer. Of course any of these facts that is obvious is omitted, and other facts can be included if they are needed.

Since the letter of transmittal is by its very form a personal communication from the writer to the immediate receiver, it provides the writer with an opportunity to acknowledge assistance, to express his hope that the information will be useful, to offer to be of future service—in fact to say anything that he considers appropriate to the occasion.

There is no reason, unless it replaces a summary, that the letter of transmittal must provide a preview of the contents of the report, but if the contents are likely to be pleasing, there is no reason that they should not be indicated.

Being an integral part of the report, the letter of transmittal should not be conspicuously informal in style, but its style need not be as formal as that of the report proper. For example, it is not necessarily limited to the third person.

In general, the letter of transmittal in a formal report permits the writer to address his original reader personally and submit his report gracefully, and it enables the reader to start reading with the assurance that he understands the occasion for the report.

Its form is basically the normal letter form, but in case of doubt "Dear Sir" rather than use of the receiver's name is desirable as a salutation. In governmental reports the salutation may even be the single word "Sir," which is the most formal of all salutations, and the complimentary close is usually "Respectfully yours."

Following is a typical letter of transmittal.

Mobray, Washington
May 17, 197–

Mr. K. L. Keller
Head of the Transmission Design Section
Northern Empire Utilities System
Random City, Washington

Dear Mr. Keller:

As you requested me to do in your letter of November 18, 197–, I have investigated one of the possible causes of unexplained outages that have occurred on the Northern Empire Utilities System.

My investigation concerned the possibility that the outages have been caused by the melting of ice on the transmission line suspension insulators. To check on this possibility we negotiated a contract with the Central University Technical Institute, arranging to have them test the insulators under icing conditions in a low-temperature laboratory and compare their wet-flashover resistance with the standards established by the A.I.E.E. Since the resistivity of water varies considerably, samples with water of varying resistivity were used. The variation was produced by adding sodium chloride to the water used in these laboratory tests.

A second line of investigation was also followed. We arranged for the collection of rain water samples at 26 substations. The number of samples collected was 339. The resistivity of all these samples was tested in order that we might be able to judge the significance of the laboratory tests of the insulators in view of the actual qualities of the water comprising the precipitation in this area.

The data secured in all the tests mentioned above are presented and discussed in the following report. Our studies must be carried further before we can definitely conclude that the outages resulted from the conjectured cause; but the data obtained thus far make it apparent that further investigation is warranted. We are asking the A.I.E.E. Lightning and Insulator Subcommittee to review this report, and will submit a proposal for further investigation when we receive its comments.

Very truly yours,

John Doe

John Doe
Principal Technical Assistant
Transmission Design Section

Table of Contents

What has been said earlier about the table of contents applies with special force to the formal report. That is, a table of contents is a list of the headings that appear on the pages of the report; headings except those of highest rank are indented to indicate their rank; they are phrased as topics, and are worded exactly as they are worded where they appear in the text; and as a result of all this they make it easy for the reader to see what ground the report covers, how it is organized, and where each point is taken up.

Since a formal report is longer and more complicated than other kinds, unusual care must be taken so that the table of contents will be easy for the readers to use and so that it will observe the conventions that they are accustomed to. The danger of going astray is greatest when tables and figures must be listed. In this connection, two conventions should be borne in mind.

1. The titles of tables and figures are not mixed in among the headings that indicate topics discussed.
2. The items in any single list are arranged in the order of their occurrence. That is, an item with a smaller page number never comes later than an item with a larger page number.

In order to comply with both of these conventions, it is necessary to use a different system when the tables and figures are placed at the end of the report than is used when some or all of them are placed in the text. When tables and figures follow the text, they may be listed as part of the regular table of contents. But when tables and figures are scattered through the text, it becomes necessary to list them under a new heading, equal in rank and hence similar in form to the heading *Table of Contents*. The heading of the list of nonverbal materials can be anything appropriate: *Figures*, or *Tables*, or *Tables and Figures*, or *Exhibits*—anything that lets the reader see what kind of material is being listed.

The treatment can of course vary in detail; but the results will be acceptable if each kind of material is listed separately, if the items in each list follow the order in which the material referred to appears, and if the lists of nonverbal material are not treated as part of the main table of contents unless the report proper is complete before the first such material appears.

The terminology used to identify various kinds of material is not rigid, but comment on some of the terms should be useful. *Exhibits* is the broadest term used. Being practically all-inclusive, it can be used to head a list of specimens from outside sources or as a general heading under which figures and tables are separately listed. Tables and figures are of course different from each other in nature and would not be mixed together in a single list. The term *figure* includes graphs, charts, photographs—any of the kinds mentioned in Chapter 6. The term *plates* is sometimes used for full-page illustrations on paper different from the text pages. All the figures are usually listed in a single numerical sequence, but if there is a reason for doing so, each kind can be listed separately and given its own set of numbers.

The term *appendix* can be used to refer to material that follows the body of the report. (Separate appendixes, for example "Appendix A— Tables" and "Appendix B—Figures," can be used when the material is varied enough and extensive enough to warrant such treatment.) If the term *appendix* is used in the table of contents, it should of course appear also at the point in the report where the appended material begins. If the material in the appendix is extensive, a page bearing the single word *appendix* is useful because it makes the beginning of the appendix easier to find.

In spite of the numerous points made in this discussion, the listing of nonverbal material is not primarily an exercise in obeying rules. Rather, the main consideration is to make it easy for a reader to see what material of this nature the report contains and to find what he is looking for. The best system to use on any occasion is the simplest system that will accomplish that result. In order to increase your resourcefulness in doing this task and at the same time become better aware of the conventions that prevail, you should constantly notice the methods used in the books that you encounter.

Chapter 9, on Special Elements in Reports, contains a specimen table of contents, but to illustrate how the listing of visual materials can vary, some additional specimens follow. Each consists of the last heading preceding the list of visual materials (in order to show what page it appeared on) plus the titles of tables and figures. Where a centered heading appears, it may be assumed that its form matches the earlier heading *Table of Contents*. The examples do not show every conceivable situation, but they are adequate to illustrate the principles upon which a treatment for almost any situation can be devised.

LIST OF ILLUSTRATIONS

TABLES AND FIGURES

Specimen 1 shows an acceptable method of listing visual material when it consists entirely of figures, all of which follow the written material and are placed under the heading *Appendix*. It would have been equally appropriate to have replaced the heading *Appendix* with *Figures* and to have dispensed with the separate page on which *Appendix* appeared.

Specimen 2 varies from specimen 1 in that two kinds of visual material were used, each being treated as a separate appendix. Also, the writer placed his headings *Appendix A* and *Appendix B* at the tops of the pages where the appended material began rather than on separate pages.

Specimen 3 illustrates the form used when figures (presumably the only kind of visual material used) are placed in the body of the report. Whether the list of illustrations appears on the same page as the regular table of contents or on a separate page depends on the space available.

Specimen 4 is like specimen 3 except that both tables and figures were used and were placed in the body of the report.

Introduction

The introduction of a formal report is not the usual type, in which the writer begins his discussion of the *subject* of the report. Rather it concerns the report itself. (The introduction to a thesis or dissertation is this same type.)

In order that the report may be self-sufficient, the introduction answers such questions as the following to the extent that the writer thinks they might come to the mind of the readers: What is the purpose or function of this report? What circumstances caused it to be written? Who asked, or ordered, or decided that it should be written? How much territory (assuming that the title does not precisely show the scope and limitations) does it cover? Where and how was the information obtained? Until the reader knows the answer to these questions, he is not ready to receive the facts and discussion that the report offers him.

There is no single form for the treatment of this introductory material. One system is to use the general heading *Introduction,* and then place the facts under specific subheadings such as *Purpose* (or *Objective*), *Scope* (or *Limitations*), *Source(s) of Information,* and *Authorization.* When it is not top-heavy, this system is good because it reduces the likelihood that any of the various kinds of information will accidentally be omitted. Another system is to use the heading *Introduction* without any subheadings. A third possibility is to dispense with the heading *Introduction* and make main headings of the above-mentioned subheadings. Yet regardless of the headings under which it is placed, the introductory information is the same. For convenience in our discussion, we will assume that a main heading *Introduction* is used, plus whatever subheadings are needed.

Under *Purpose* or *Objective* the introduction explains as fully as necessary the conditions or events that created a need for the report and tells the exact function that the report is expected to perform.

The material under *Scope* or *Limitations* settles any possible doubt about what the report covers and what it omits. When the title itself performs this function, a section on *Scope* is superfluous; but it is often impossible to indicate the exact limits in the title without making it cumbersome.

The discussion under *Source(s) of Information* tells where and how the facts were obtained. If their source is obvious, this section may be omitted. If some of the information comes from printed sources, they may be named in this section provided there are not more than three or four; but if numerous, they should be placed in a list of references at the end, and this list should be mentioned in the introduction.

The section headed *Authorization* is extremely brief, merely telling who authorized the report, when, and how (letter, memorandum, personal conversation, or whatever other means). In a large organization this section is often replaced by something such as a project number, which would let the reader know under what authority the work was done.

The introductory material under each of these headings should usually be held to one or two paragraphs. If a long discussion on any of these points is necessary, it can appear under an appropriate heading in the body of the report, and the coverage under *Introduction* can be reduced to a brief general statement and a reference to the full discussion that will come later.

All in all, the formal introduction merely orients the reader. Rather than discussing the subject of the report, it tells about the report itself. Consequently, there may sometimes be a need for what we referred to in Chapter 9 as an ordinary introduction, which actually opens the discussion of the subject. If so, this ordinary-introduction material can be placed in the body of the report under a heading such as *General Discussion of* . . . , or *Background Information,* or *The General Problem of* . . . , or *Historical Review of.* . . . But whenever there is a formal introduction, the ordinary introduction is identified by some other heading than *Introduction.*

Some question may arise over why it is necessary to include parts of the introductory material mentioned. Since a report is usually submitted to the person who authorized it, it would seem that he might already possess most of the introductory facts. This is true; but the passage of time may have caused many of the details to fade from his memory. Also, the report may reach readers who are not familiar with the problem, or it may be filed and used again after a long time has passed. Therefore the report should provide all the information any likely reader may need at any time, and thus be self-sufficient.

It is true, also, that some points in the introduction may have been touched upon in the letter of transmittal; but there, they will probably have been barely mentioned rather than treated in full. The slight repetition is justified because there will be times when the letter of transmittal will be read yet the report itself may be set aside until later.

Following is a formal introduction of the kind recommended:

Introduction

Purpose

In local areas, highway engineers have found it necessary to use deposits of basalt rock as a source of aggregate for the base course of highways. Some of these deposits yield an aggregate that is entirely satisfactory for such a purpose, but others yield an aggregate that deteriorates rapidly, causing failure of the highway base within a year or two.

To cope with this condition the Highway Materials Testing Laboratory has developed tests intended to indicate in advance whether basalt aggregate from any particular source will degrade rapidly. Also, it has carried on experiments to see whether treating aggregates with asphalt emulsion would slow down the deterioration of those known to be unsatisfactory in their natural condition. The purpose of this report is to present the results of these tests and experiments.

Sources of Information

Information on the sources of aggregates that have not held up as should be expected was obtained from the State Highway Department. The techniques of testing were based in part on work done in other states, as indicated by the citation of references. Most of the facts presented were obtained, however, by the direct testing and experimentation done in the laboratory.

Authorization

This report and the investigation on which it is based were authorized in a letter from Hugh D. Lockhart, Chairman of the State Highway Commission, dated April 3, 197–.

STYLE IN A FORMAL REPORT

In Chapter 7, the difference between formal and informal styles was discussed. Obviously, preparing a formal report is one of the occasions when the language should be formal and impersonal. The desired effect is secured, however, not so much by aiming at formality as by avoiding noticeable informality. Slang is obviously ruled out. Colloquialisms—including contractions—are avoided. Technical jargon, handy though it may be, should be used only as a last resort.

The formality and impersonality prevail to such an extent in a formal report that little if any use is made of first and second persons. "I" and "you" are avoided, and "we" is used only when it means the organization in which both the reader and the writer are included. One reason for avoiding "I" and "you" is that they simply seem out of place in any piece of writing where the occasions for their use are few and far between. (Notice how seldom you see them in technical books and articles.) A second reason is that the meaning of these two words depends on who is writing and who is addressed. Consequently, when material by several writers is included in a single report, as often happens, the need of revision is reduced if each contributor avoids "I" and "you."

To be sure, formality and impersonality can be misused. Such a sentence as "It is believed that . . ." dodges the significant question of *who* believes, and overuse of *the writer* in place of *I* sounds stilted. But the real objection to these forms goes beyond the formality and impersonality. Basically, in the first instance the defect is vagueness and the desire of the writer to promote certain ideas yet leave himself an escape hatch if it turns out that he is mistaken. In the second, it is probably sheer awkwardness, for it is not usually difficult to construct a sentence in which neither *I* nor *the writer* appears. Formal style should not be condemned because it is bad when it is handled awkwardly, for if informal style is handled awkwardly it is equally bad.

In summary we need only say that the purpose of formal, impersonal style is merely to secure a reasonable degree of dignity when dignity is called for. Well handled, it is not stiff, pompous, indirect, or excessively ceremonious. It is a tool that a writer should learn to use, for informality is not always appropriate despite its many good qualities.

THE EFFICIENCY OF FORMAL REPORTS

A person who has never had a chance to see how reports are actually used may wonder why some of the features of a formal report are necessary. To understand their value we must remind ourselves that a report is not always written with the expectation that all of those who receive it will read it from beginning to end. On the contrary, there are many more occasions when a long report will be read in part than when it will be read in its entirety.

A formal report meets whatever demands are placed upon it. The reader who cannot remember or has never known the circumstances that caused it to be written can learn them as soon as he picks it up, and can decide how soon he must read it. When he reads it in full, he will find that it unfolds smoothly and that he can study the technical details, skim them, or ignore them as he desires. The executive farther up the line, who wants only a general picture, including the conclusions and recommendations, can get exactly what he wants without wasting time on details. The person who is interested in only part of the subject, or who wants to refresh his mind on some particular point, can easily find what he is looking for. The person who wants to study the full technical details thoroughly finds them easy to locate and identify.

All in all, experience will show that a well-designed formal report is an extremely effective instrument that performs not one but a variety of functions smoothly and efficiently.

(An illustrative specimen follows.)

SPECIMEN FORMAL REPORT*

The following report was one of five that were eventually incorporated into a single report. Each of the five covered a different aspect of the same basic situation. All were the work of authors with special qualifications in the areas that they dealt with.

The basic situation was as follows: A corporation, The American Metal Climax, Inc. (AMAX), had holdings in a Wyoming area—holdings in which it might eventually develop into an open-pit copper mining operation of moderate size. No immediate operation was contemplated when the reports were written, but an environmental inventory and impact analysis was considered desirable in case of future need. Hence the Kirwin Environmental Study was prepared for the company by the Rocky Mountain Center on Environment (Denver) and The Thorne Ecological Institute of Boulder, Colorado.

The five reports that were eventually combined covered wildlife, fisheries, limnology, recreation and scenic resources, and vegetation. The report used here as a specimen concerns limnology.

The cover of the report served also as a title page. Because of the close working relationship of the writer and AMAX and because—though it can stand independently so far as it concerns its own area—it was to be incorporated into a larger report covering other aspects of the same situation, a letter or memorandum of transmittal was not included.

* Courtesy of Rocky Mountain Center on Environment.

Limnology

Kirwin Environmental Study

Robert W. Pennak, Ph.D.
Boulder, Colorado

July, 1973

Following the table of contents the report had several elements that formal reports often include. It has both an abstract and a summary. The former is the type that mainly indicates what points are covered (sometimes called a descriptive abstract, as mentioned in Chapter 9). The summary, as you may observe, is a condensed version of the actual substance of the report.

Other special elements that follow include sections on conclusions and recommendations, under separate headings. For ease of reference the items in each of these sections are numbered. If a reader were to examine only the elements mentioned thus far, he would obtain a substantial though not a detailed picture of the contents of the entire report.

TABLE OF CONTENTS

LIST OF TABLES

ABSTRACT

The Wood River, in northwestern Wyoming, is an unproductive foothills and mountain stream that has a characteristic spring and summer runoff from the Absaroka Range. Chemical and physical conditions are typical and similar to those of other Front Range streams. The bottom fauna consists of a relatively small number of clean-water species except in the lower reaches near the water intake for the village of Meeteetse where the bottom fauna is negligible. Possible effects of mining developments on the chemistry, physics, and biology of the stream are discussed.

SUMMARY

This investigation of the Wood River in the Kirwin - Meeteetse area of Wyoming fixes "baseline" aquatic conditions and biological productivity during the open season of 1971. Five equally spaced stations were established between the altitude of Kirwin (9300 feet) and the Meeteetse water intake (6200 feet).

Wood River is a typical small Rocky Mountain stream with rubble substrate, 80 to 95 percent riffles, few pools, and torrential waters in the upper reaches. Temperature, dissolved oxygen, and free carbon dioxide concentrations are consistent with these features in other regional streams. According to local residents, there was a heavy spring and early summer runoff from the Absaroka Range. Nevertheless, Wood River had a surprisingly low silt load (less than 10 ppm) at stations 1 to 3 in June and July. Bound carbon dioxide readings ranged from 16.0 to 47.5 ppm, indicative of waters slightly harder and more productive than those of similar small Rocky Mountain streams. Hydrogen-ion concentrations ranged from pH 7.0 to 8.4; more typically the range is pH 6.8 to 7.4.

The bottom fauna is impoverished and typical of small Rocky Mountain streams, being dominated by clean-water species of Ephemeroptera (mayfly) nymphs and Trichoptera (caddis) larvae. At the lowest station (Station 5) near the Meeteetse water intake the fauna was negligible, possibly because of about two miles of upstream agriculture and pollution by cattle. The total bottom fauna list for this investigation consists of only about 40 species, in contrast to 80 to 100 species in streams east of the Mississippi. During August and September the bottom fauna made a rapid and remarkable recovery from the heavy load (50 to 250 ppm) of runoff silt during June and July. It is postulated that the fauna may undergo similar rapid recovery after short periods (one to five days) of accidental industrial silt pollution.

The periphyton film on the surface of rubble was typical and consisted mostly of organic materials, and especially algae. Inorganic silt formed a negligible portion of the periphyton, in contrast to the situation in other Rocky Mountain streams.

246

Possible ecological impacts of open pit mining, milling, pipelines, tailing ponds, roads, and reservoirs are discussed.

CONCLUSIONS

Future ecological changes in the Wood River flowage, such as those that could be triggered by road construction, forest clearing, and mining activities may alter, more or less, the following limnological conditions in Wood River: temperature
suspended organic and inorganic materials
bound carbon dioxide
pH
dissolved organic and inorganic materials
density of bottom organisms
species of bottom organisms present
  (certain species may be eliminated)
quantity and quality of the periphyton

More specifically, we might infer the following impacts on the Wood River when mining and associated construction projects are begun:

1. An open pit mine, mill site, waste dump or leach dump, as such, would have no direct effect on the stream, providing surface drainage were controlled.

2. Pipelines carrying tailings, etc. sometimes break at inopportune times. Since they are often laid alongside or near a stream, the danger of a gross siltation and stream kill cannot be overemphasized.

3. Tailing pond overflow can cause stream pollution, especially when the pond is not adequate to permit proper settling of solids and/or when it has outlived its projected efficient usefulness. A settling pond should be adequate to provide a clear water effluent.

4. Reservoirs, if properly designed, can minimize the deleterious effects of the annual runoff and summer "gully washers." Further, reservoirs have a well-known but poorly understood fertilizing effect on the streams below their dams, sometimes resulting in an increase in bottom fauna and fish populations. It might be excellent public relations if the reservoir(s) could be stocked and opened to public fishing.

RECOMMENDATIONS

1. Surface drainage from roads, mill site, waste dumps and leach dumps should be controlled so precipitation runoff would not result in toxic and excess silting of Wood River.

2. Settling ponds should be of adequate size so as to provide a clear water effluent.

The Introduction is clearly intended for the reader who expects to read the full report. It tells why a study of the stretch of the stream concerned was desired, identifies the three fundamental kinds of information within the scope of the report, and comments on the nature of the stream itself. It does a good job of orienting the reader to the discussion that follows.

The full substance of the report comes under two headings—*Methodology* and *Data and Results*. The noteworthy feature of the first of these two parts is its thoroughness. It refers to certain exceptional circumstances, tells the dates of observations, identifies the locations where observations were made and sample materials gathered, and explains the reasons for the choice of these locations. Techniques used in gathering samples are described. The significance of the kind of information sought is pointed out so clearly as to be easily understandable to a reader who is not a specialist in the scientific field concerned. Tests performed in the laboratory are identified.

The section on Data and Results is understandably the longest in the report. It includes six pages of discussion supported by several tables. (Two of the latter are included as examples but the others are omitted, not being essential to the intended function of the specimen in this book.)

The discussions of both methodology and data and results are clearly written and easy to follow—as are the other materials that precede them. Throughout the report, the contents demonstrate a high degree of scientific objectivity.

3. It might be an excellent public relations gesture to consider if suitable reservoirs could be stocked and then opened to public fishing.

4. The stream survey information in this report should be kept available for comparison with future stream conditions in the Wood River flowage. This procedure provides these benefits: protection of the company against possible unjust accusations or litigation; and providing background data against which to evaluate any future effects of operations.

5. All water effluents from the projected mining operations should be monitored frequently for copper and other toxic and/or heavy metals.

6. A brief biological reconnaissance of Wood River conditions should be made once or twice per year after mining developments begin in the Kirwin area. Observations from future studies should be compared and assessed against the data contained in this report.

INTRODUCTION

Early in 1971 the investigator was asked to conduct a series of comparative biological and physico-chemical determinations on Wood River, Wyoming, from the area of Kirwin downstream to the vicinity of the town of Meeteetse. The American Metal Climax, Inc. (AMAX) has holdings at Kirwin and is interested in the future regional ecology of the upper Wood River flowage. The fundamental purposes of the study were to: (a) establish present "baseline" conditions in Wood River; (b) assess the biological productivity of Wood River, especially as compared with other streams in the Rocky Mountain region; and (c) assess the possible effects of mining developments on the stream. Very little fundamental information is available in the literature about the biology of small streams in Wyoming. The investigator welcomed an opportunity to study undisturbed conditions.

At the outset, it would seem appropriate to comment on the general physical nature of Wood River. In most respects it is similar to other small streams of the eastern ranges of the Rocky Mountains. It is relatively narrow, with the width averaging about 4.5 meters at Kirwin and about 18 meters west of Meeteetse. The substrate is chiefly rubble (the most productive type of bottom), but unfortunately the river bed consists of only 5 to 15 percent pools and it therefore cannot support a good fish population. Indeed, in four trips to the Wood River only two fishermen were observed.

METHODOLOGY

Four visits were made to the study area during the open season of 1971. The first visit was not made until 9 June, chiefly because of the very heavy runoff and the late spring near Kirwin. In addition, field operations in the autumn were cut off early because of exceptionally early snows and cold weather. However, the four visits were made at evenly spaced intervals; collectively they represent a good picture of open-season

4

conditions in the Wood River. Data were gathered at five stations
on each of the following days in 1971: 9 June; 8 July; 5 August; and
11 September. The five stations were distributed as follows and may
be easily located on the proper U.S.G.S. quadrangle maps:

Station 1 - 9300' (3120 m), near Kirwin, about 100 yards above the
           mouth of Cascade Creek.

Station 2 - 8600' (2620 m), above midway between Kirwin and the site
           of a proposed AMAX tailings pond; 1.3 miles above the log
           bridge over Wood River.

Station 3 - 7600' (2315 m), 50 yards above Brown Mountain camp ground.

Station 4 - 6500' (1925 m), just above the first state-maintained bridge
           across Wood River.

Station 5 - about 6200' (1890 m), just above the Meeteetse water intake.

These stations were selected with possible future AMAX operations at
Kirwin in mind and with a view toward similarity of substrate and
current so that biological comparisons might be as significant as
possible. All substrates studied intensively were chiefly (70%) rubble,
composed of particles ranging from 1 to 9 inches in diameter. An
additional 20 percent was predominantly unproductive boulders (more than
9 inches in diameter), and the remainder (10%) consisted of unproductive
gravel, sand, and debris.

Bottom fauna samples were taken with a standard square-foot Surber
sampler at each station on most sampling dates. The usual practice
for sampling stream bottom organisms involves taking only one or two
square-foot samples at each station, but the present investigator
took five random samples at each station so that results would be more
significant statistically. The data in this study were derived from
81 separate samples, each representing essentially all of the food
organisms in one square foot of rubble bottom. Quantitative and qualitative studies of these organisms were made by the investigator in
the laboratory at the University of Colorado. In the interest of time,
identification to the species level was not considered critical for an
investigation of this kind.

A second type of biological sampling involves the thin film of dead
organic matter and associated microscopic plant material which covers
each stone, pebble, and boulder in a stream. This material is the
basic food source for the bottom insects. It is a feature that is
highly sensitive to the presence of polluting materials in the water.
This thin film, or periphyton, is scraped from exposed surfaces of rubble
with a sharp scalpel. Five or ten minutes of scraping from three to
eight pieces of rubble constitutes a "standard" sample. The periphyton
material is preserved in vials of formalin and brought back to the
laboratory. Sixteen such samples were taken during this investigation.

Measurements of certain standard physical and chemical parameters
were made either in the field or on water samples brought back to Boulder
by the investigator. Temperature and turbidity were measured in the
field. The following were done in the laboratory: dissolved oxygen;
free carbon dioxide; bound carbon dioxide; suspended organic material;
suspended inorganic material; dissolved organic material; and dissolved
inorganic material. Data on stream flow and detailed water analyses
are being supplied by other participants in this study.

DATA AND RESULTS

Inventory

A. Physical and Chemical Features

Table I shows some general features of stream temperature, suspended
materials, turbidity, and general flowage conditions during the sampling
periods. According to local residents, the summer of 1971 was characterized by an exceptionally heavy snow melt and runoff in the Wood
River drainage basin; the water at stations 4 and 5 was turbulent and
more than two feet deep in June and July. It was therefore impossible
to take periphyton and bottom fauna samples.

Temperature conditions were typical of those found in most Rocky Mountain
Front Range streams (Pennak, unpublished), except that the spring
of 1971 came later than usual according to local residents; as a result,
the June, July, and August water temperatures at stations 1, 2, and 3
were about three degrees lower than would be expected in a "normal"
year. Construction of reservoirs and settling ponds would have the
effect of warming the stream in summer and keeping it warm later in the
autumn. Biologically, this situation should improve the productivity
of fish-food organisms.

The lower part of Table I illustrates turbidity conditions in Wood River
in terms of parts per million of suspended organic and inorganic materials.
(These were considered more accurate measurements than the usual turbidi-
metric determinations.) It was notable that the upriver stations had
sedimentary loads that were surprisingly light (less than 50 ppm) in
spite of the heavy runoff in June. At stations 4 and 5, however, in June
and July the turbidity was high (up to 240 ppm), presumably because of
the contributions of tributaries and because of the more erodible nature
of the shorelines at lower elevations. Natural runoff data as given in
Table I can be compared with data derived from pollution conditions such
as silting caused by road-building and some types of mining operations.
Efficient settling ponds and reservoirs should cut down on silting.

The waters of Wood River had no appreciable true color, except at Station
5 in September where the color was 10 (slight) on the platinum cobalt
scale. Such colorless water is normal for mountain streams that do not
have marshy or boggy areas along their courses.

Various chemical conditions are summarized in Table II. Several parameters
need only brief comment. Dissolved oxygen, for example, was always near
saturation or slightly above 100 percent saturation. This is the usual

Table I. Physical conditions in the Wood River near Kirwin, Wyoming, 1971.

	Date	Sta. 1	Sta. 2	Sta. 3	Sta. 4	Sta. 5
Temp., °C.	9 Jun	2.5	4.0	7.5	10.0	11.1
	8 Jul	3.9	5.0	7.0	12.0	13.5
	5 Aug	6.4	7.9	8.6	17.7	18.6
	11 Sep	4.2	5.6	10.4	14.8	16.8
Visible turbidity	9 Jun	light	moderate	high	high	high
	8 Jul	0	trace	trace	high	high
	5 Aug	0	0	trace	medium	high
	11 Sep	0	0	0	0	0
General remarks	9 Jun	water level high	water level high	water level high	full flood	full flood
	8 Jul	moderate	moderate	moderate	water level high	water level high
	5 Aug	average flow	average flow	average flow	average flow	average flow
	11 Sep	low flow	low flow	low flow	low flow	low flow
Suspended inorganic material, ppm	9 Jun	29.71	30.63	90.90	120.90	220.15
	8 Jul	3.82	5.46	5.13	44.13	54.13
	5 Aug	4.62		2.80	8.18	9.40
	11 Sep	8.18	6.41	9.73		4.59
Suspended organic material, ppm	9 Jun	2.11	2.54	5.03	13.89	20.55
	8 Jul	0.58	0.39	0.50	4.20	6.20
	5 Aug	0.68		0.48	1.75	1.96
	11 Sep	1.31	0.64	0.45		0.50

Table II. Chemical conditions in Wood River near
Kirwin, Wyoming, 1971.

	Date	Sta. 1	Sta. 2	Sta. 3	Sta. 4	Sta. 5
Oxygen, % saturation	9 Jun	106.3	102.8	100.2	100.4	101.0
	8 Jul	106.4	106.6	106.0	104.2	103.0
	5 Aug	110.2		104.7		108.7
	11 Sep	103.2				126.4
Free CO_2, ppm	9 Jun	1.5	1.5	1.4	1.3	1.5
	8 Jul	1.1	1.5	1.5		
	5 Aug	3.5	1.5	1.0	1.0	-2.0
	11 Sep	1.0	1.0	1.0	0.2	-0.5
Bound CO_2, ppm	9 Jun	18.5	20.0	21.0	25.1	28.2
	8 Jul	16.0	18.0	24.5	30.5	33.0
	5 Aug	26.5	25.2	27.5	35.5	47.5
	11 Sep	29.0	29.0	28.0	38.0	44.4
pH	9 Jun	7.3	7.2	7.1	7.0	7.0
	8 Jul	7.3	7.3	7.3	7.2	7.2
	5 Aug	7.3	7.4	7.9	8.1	8.3
	11 Sep	7.1	7.7	8.0	8.2	8.4
Dissolved organic matter, ppm	9 Jun	54.45	27.74	25.76		51.52
	8 Jul	42.70			45.70	
	5 Aug		23.96		36.60	42.62
	11 Sep	29.02		35.34		46.57
Dissolved inorganic matter, ppm	9 Jun	154.55	52.42	47.22		67.34
	8 Jul	41.84		49.66	59.74	
	5 Aug		53.10		70.17	84.82
	11 Sep	68.61		88.21		122.45

situation in unpolluted mountain streams. Supersaturation readings are possible because of entrapped bubbles in the field samples. Incidentally, all preliminary analytical reagents were added to samples in the field, so that the final laboratory titrations would be accurate.

Most of the free carbon dioxide readings ranged from 0.2 to 3.5 ppm. These again are close to the equilibrium values with air where water is unpolluted. The -0.2 and -0.5 ppm readings at Station 5 are a reflection of a high photosynthesis rate by the periphyton algae but are of no special ecological significance.

Most rapid mountain streams in the Rocky Mountains have bound carbon dioxide values ranging from 10.0 to 25.0 ppm, a reflection of the fact that most of the substrates are poor in carbonates and other soluble salts. In general, the results for Wood River were somewhat higher with 14 readings in excess of 25.0 ppm. It would appear, therefore, that the waters of this flowage are slightly harder than average. As a consequence, the general productivity of the bottom fauna and fishing potential should be slightly better than for the "average" small mountain stream, since there is a general (but poorly understood) correlation between productivity and water hardness.

The slightly harder waters were further shown by the fact that hydrogen-ion values were higher than usual. Most similar mountain streams have pH values ranging from 7.1 to 7.4, but a few readings for the lower stations of Wood River were more alkaline than pH 7.4, and up to pH 8.4.

Similarly, both dissolved organic and dissolved inorganic materials were about 50 percent higher than is customary for Rocky Mountain streams (Amitage, 1958; Cordon and Kelley, 1961; Gaufin, 1959; McConnell and Sigler, 1959; Pennak, unpublished). The former averaged 38.25 ppm and the latter 73.86 ppm. These are of no special significance other than to indicate a somewhat higher level of potential productivity. Indeed, they are actually considerably lower than similar determinations made in Michigan and New England trout streams, where general productivity levels are notably high.

Bound carbon dioxide, pH, dissolved organic materials, and dissolved inorganic materials are useful baseline parameters. In the event of chemical pollution and settling pond construction, all of the values for these features will change. On the other hand, reservoirs along the course of Wood River should have no important effects on the water chemistry.

With reference to physical and chemical factors, then, in summary Wood River is a "typical" Rocky Mountain stream with somewhat higher than average biological potential.

B. Bottom Fauna

Table III shows the detailed results of the bottom fauna study, a critical item in assessing stream conditions. It illustrates particularly the necessity of taking samples several times during the open season because

of the wide variations in the density of the various species from one
month to another, owing to variations in their life histories. Many
species, for example, were not taken by the Surber sampler early in the
season (May and June) because they were still in the egg or first larval
stage and consequently were too small to be retained by the net. Similarly, sampling in August and September often failed to take certain other
species because they had already emerged from the stream as adult, flying
insects.

Unfortunately, samples could not be taken at stations 4 and 5 in June and
July because of the extremely high runoff and water at these stations.

All of the genera listed in Table III are typical of small Rocky Mountain
streams. None of them are especially rare or unusual. Station 5, which
is well out toward the plains, had a poor insect fauna from the standpoint
of number of species. In part, this situation is due to the fact that
Station 5 is not a true mountain stream habitat, and ecological conditions
are not favorable for cold water species. Of greater significance,
however, is the fact that Station 5 had the appearance of some organic
pollution. It is possible that conditions in the several miles of farm
land and gross disturbances by cattle between stations 4 and 5 were
responsible for the restricted fauna found at Station 5. For example,
large quantities of cattle droppings were found along the shore upstream
from Station 5. Although little specific information can be found in
the published literature, stream biologists believe there are deleterious
effects on stream organisms from large amounts of cattle droppings.

The very large population of mayfly nymphs on 5 August at Station 4 is
remarkable (Table III). It is an excellent example of the ability of
Ephemeroptera to withstand and recover from heavy (up to 135 ppm)
runoff silting such as that experienced in this part of the stream in
June and July. Although mayflies from mountain streams are particularly
sensitive to chemical and sewage pollution, they appear to have developed
an effective resistance to temporary inorganic silting from runoff.

A total of about 40 species of macroscopic bottom organisms were found
during this study. This list is similar to those found by the investigator (Pennak, unpublished) in other Rocky Mountain streams. Chemical or
physical pollution would cut this species list to perhaps five or ten.

As shown in the summarized bottom fauna data in Table IV, stations 1,
2, and 3 had similar population densities of bottom insects. Station 4,
however, had an exceptionally high standing crop in August (4046 per sq m),
chiefly because of large numbers of very small mayfly nymphs and midge
larvae, as noted above. Presumably these insects emerged as adults
before the last samples were taken a month later. Station 5 had a negligible bottom fauna, perhaps for reasons already mentioned.

Total numbers of bottom organisms ranged from 0.8 to 378.0 per square
foot for various dates and stations. Omitting these two extreme figures,
the mean standing crop for all five stations was 23.5 organisms per square
foot. Thus Wood River must be considered a poor producer. The most
widely used stream productivity criteria in the United States have been

promulgated by fisheries biologists:
  Poor streams   - 0 to 100 organisms per square foot.
  Average streams - 100 to 200 organisms per square foot.
  Rich streams   - more than 200 organisms per square foot.
A similar, gravimetric system of grading streams classifies "poor"
streams as those having less than 1 gram of organisms per square foot.
Thus it is further obvious that Wood River is a poor producer. There is
no specific reason for this situation unless it is the generally short
Wyoming growing season, the relative softness of the water (16.0 to 29.0
ppm bound carbon dioxide at stations 1 to 3), and the small size of the
stream (4.5 m at Kirwin). Table V contains comparable data for other
Rocky Mountain streams; all of these have higher productivities than Wood
River; except for biomass, the data for these streams are similar to
those for Wood River. Additional data (Pennak, unpublished) show that
two small Idaho streams have productivities even lower than that of Wood
River, while data for a small New Mexico stream show a standing crop more
than 200 percent greater than that for Wood River.

Frequent intermittent silting will knock out most of the species listed
in Table III. Effluents from carefully constructed settling ponds should
not have serious effects on the stream fauna provided that the effluent
is less than 50 percent of the flow of the Wood River at the effluent
entry point. Indeed, there may be an actual "fertilizing effect" if the
settling pond effluent contains no toxic ions in excessive quantities.

C. Periphyton

Table VI contains data derived from the quantitative study of the periphyton.
Periphyton serves as an important food source for the bottom insects,
most of which are grazers over exposed surfaces. Comparable data are not
yet available in the scientific literature but on the basis of unpublished
work (Pennak, literature citation) the summer standing crop of periphyton in Wood River at the first four stations is similar to the variable
conditions found in other unpolluted streams in the Front Range of the
Rocky Mountains. In general, the data in Table VI indicate major rapid
productivity of periphyton following the spring runoff. Essentially
all of the periphyton consists of living algae with little inorganic
silt. The genera of algae represented are common in unpolluted swift
streams everywhere. If Wood River were to be subjected to continuing
mine tailing waste, etc., the periphyton picture would change quickly.
The rubble would lose most of its coating of algae, and surfaces would soon
become covered with inorganic particulate material.

Impact

Upper Reservoir. This body of water should have no deleterious effects
on the biology of Wood River. The altitude is such that temporary storage
of stream water behind the dam can have only a slight warming effect on
the flowage, not enough to influence the species and abundance of downstream organisms. On the other hand, proper manipulation of the water
level could improve biological conditions in the stream, especially if

greatest storage could be attained during spring runoff. The more constant the flow in a stream, the better the biological conditions; even a small mediation of annual high and low water extremes would have beneficial biological effects. The reservoir could become an excellent fishing area.

Lower Reservoir. The same comments apply to both reservoirs. Limnologically, there is no preference between these two reservoir sites.

Tailing Pond. Much of the biological impact of the tailing pond depends on the way it is constructed and its relation to the Wood River flowage. If, for example, all of the water in the thalweg flows through this pond, there might be problems with the efficiency of the settling process. It would be likely that excessive tailing silt would wash downstream in the spring and early summer, to the detriment of the biology of Wood River. If, however, a large conduit were laid in the present thalweg so that much of the stream water (especially in spring and early summer) would pass under the tailing mass, then the situation would be much less serious. It would mean that the tailing pond water overflow would join the stream proper just below the dam. Runoff from the tailing might actually have a beneficial effect on downstream biological productivity because of the higher level of dissolved nutrients in the water (Pennak, unpublished).

Waste Dump, Leach Dump, Open Pit, Mill Site. None of these areas should ordinarily produce an impact on Wood River. However, there is always the danger of washouts and silting from heavy rains. Care should therefore be used in laying out these sites, with special emphasis being placed on local drainages, especially since the protective vegetative cover will presumably be gone. A serious problem might develop if chemically toxic substances are washed into the stream from the operations areas.

Services and Transportation. These facilities can ordinarily affect Wood River only when they (1) are built too close to the stream and (2) are not well maintained. When a gravel road is bulldozed alongside a stream, much of the streamside herbs, shrubbery, and trees are destroyed. This has the effects of (1) increasing erosion and silting, (2) decreasing those stream fish-food organisms that rely on streamside vegetation for breeding areas, and (3) destroying shade over the stream and thereby raising stream temperatures.

Chronological Sequence of Impacts. When the Kirwin operation is under construction, most of the stream impacts would be centered around the silting produced by road building and the erection of plant facilities.

During the 15+ years of plant operations major potential impact areas will be related to: settling ponds, leach dump drainage, waste dump drainage, and roads and service areas.

When the Kirwin area is eventually abandoned, residual major impacts will be the settling pond and the danger of a break in its dam, the leach dump, and the waste dump. An intensive revegetation program will be necessary.

Sphere of Influence. Most of the items mentioned in this impact section are of primary importance to the Wood River biology only in the Kirwin property and perhaps two or three miles below the tailing pond. This would apply unless, of course, there were to be a major plant breakdown or severe precipitation, in which case there might be stream impact extending much farther downstream. Because of the long distance and because of the diluting effect of tributaries, however, it is doubtful that there would be measurable stream impacts as far east as the Meeteetse water intake.

ACKNOWLEDGEMENTS

The investigator would like to acknowledge the patience, help, and understanding extended to him by the administrative personnel of the Thorne Ecological Institute, the Rocky Mountain Center on Environment, and American Metal Climax, Inc. He is especially indebted to James J. Cooper who made special field work arrangements in the Cody and Kirwin areas, and who extended much practical advice.

SELECTED REFERENCES

Armitage, K.B. 1958. Ecology of the riffle insects of the Firehole River, Wyoming. Ecology 39:571-580.

Cordone, A.J., and D.W. Kelley, 1961. The influence of inorganic sediment on the aquatic life of streams. Calif. Fish and Game 47:189-228.

Fisher, S.G., and G.E. Likens. 1970. The annual detritus budget of a small stream. (Abstract). Paper submitted at 33rd annual meeting of the Amer. Soc. Limnol. & Oceanogr.

Gaufin, A.R. 1959. Production of bottom fauna in the Provo River, Utah. Iowa State Coll. J. Sci. 33:395-419.

Hynes, H.B.N. 1960. The Biology of Polluted Waters. 202p. Liverpool, England.

_____. 1970. The Ecology of Running Waters. 555p. Toronto, Canada.

Logan, S.M. 1963. Winter observations on bottom organisms and trout in Bridger Creek, Montana. Trans. Amer. Fish Soc. 92:140-145.

McConnell, W.J., and W.F. Sigler. 1959. Chlorophyll and productivity in a mountain river. Limnol. & Oceanogr. 4:335-351.

Mecom, J. 1969. Productivity and feeding habits of Trichoptera larvae in a Colorado mountain stream. 148p. Ph.D. thesis, Univ. of Colorado.

Muttkowski, R. 1929. The ecology of trout streams in Yellowstone National Park. Roosevelt Wildlife Ann. 2:155-240.

Needham, P.R. 1940. Trout Streams. 233p. Ithaca, N.Y.

Pennak, R.W. 1953. Fresh-water Invertebrates of the United States. 769p. New York, N.Y.

_____. Unpublished data.

Pennak, R.W., and E.D. Van Gerpen. 1947. Bottom fauna production and physical nature of the substrate in a northern Colorado trout stream. Ecology 28:42-48.

Slack, K.V. 1955. A study of the factors affecting stream productivity by the comparative method. Invest. Indiana Lakes and Streams 4:3-47.

ASSIGNMENTS

Write a formal or semiformal report (length to be specified by the instructor) on one of the following subjects or on a subject of your own choice.

1. The advantages and disadvantages of establishing a free day-care center on campus. Consider such factors as building space, personnel, availability to people other than students, insurance, and other factors that might apply at your particular school. If there is already such a facility at your campus, examine it with a view toward improving it or changing the services to make it more responsive to student needs.

2. The artistic, economic, and psychological factors affecting the selection of specified child or adult clothing by a department-store buyer.

3. Devise a program of drug education for use in grades one through six. Such a program should include education about all types of drugs, not just illegally used drugs, and must consider parental reaction as well.

4. Report on the current intramural recreation program at your school, and make suggestions for areas that could be expanded or changed and give your reasons.

5. Examine the performance of your school's student government and make recommendations for whatever changes you feel are necessary.

6. Determine some area in your field of specialization that you wish to learn more about. After getting your instructor's approval, investigate the area and write a report on what you find.

7. Analyze the atmospheric conditions surrounding your campus and determine which conditions might aggravate pollution. Make recommendations that would minimize effects of pollution when the specified conditions exist.

8. Determine a pest that attacks trees or crops in your part of the country. Detail the types and amounts of destruction and describe biological, chemical, and/or other methods of control, giving advantages and disadvantages of each method. Recommend one for a specific location and give your reasons.

9. Consider the effects on your community of the current methods of financing government services and schools. Examine other possible methods of raising money and consider the possible effects; make suggestions for changes if you think they are called for or explain why you feel the current methods are best.

10. With the consent of the people involved, report on a research project currently under way at your school.

12

PROPOSALS

A proposal, as the word itself indicates, is a suggestion or request that some particular action be taken. Usually a specific person, business, or agency wishes to do a job or solve a problem for another person, business, or agency, and writes a proposal in an effort to secure the opportunity to do so; the proposal amounts to an offer to do the job for payment or at least with funds provided by the other party. For example, a corporation seeks, as a subcontractor, to develop and perhaps manufacture some part of a missile for an aeronautics firm which has itself, by means of a proposal, secured a government contract to develop the complete missile. Sometimes, to be sure, the person or organization submitting a proposal attempts to persuade those to whom the proposal is addressed to do some job—as, for example, a student group urging the municipal government to establish special bicycle lanes as a public convenience. Sometimes, too, a proposal involves joint participation in a project—as, for example, the proposal by a school board to a municipal recreation board, or vice versa, that the two bodies join in developing a tract of land for recreational use by both the school and the general public. But most characteristically a proposal is an effort by the proposer to be authorized to do a job, for payment, as a service to the organization to whom the proposal is addressed.

261

Proposals differ from other technical writing in one important respect. While most technical writing deals with things that exist—events that have taken place, projects that have been completed, operations of machinery—proposals are concerned for the most part with the future— with projects that someone merely wants to have undertaken. This major difference makes writing proposals particularly difficult. Most people in industry become accustomed to describing with precision things that exist; but how does one describe accurately something that may be little more than an idea in someone's head? A student is likely, sooner or later, to face this problem and indeed may find that some of the most important writing which he or she does involves proposals. In fact, the talent that major corporations assign to the preparation of proposals is just as high-caliber as the talent they assign to carrying out a job once the proposal has been accepted.

The importance of proposals is easily shown. All the projects that have been written about, after completion, all the changes in procedures, started as proposals. Thus it is that everyone in education, science, industry, or government at a decision-making level utilizes proposals as a basis of action.

CONVINCING THE READER

A proposal may come into existence either because those who receive it make known their desire or need for it or because those who submit it do so on their own initiative. In either case, the problem for the writer is the same: to convince the reader that the writer or his organization is capable of solving the problem. Until the reader is convinced that those who make the proposal have a reasonable chance of performing the task successfully, a go-ahead is unlikely. The reader's time or money is usually involved, and he doesn't wish either to be wasted.

Whether a proposal is long or short, simple or complicated, a writer can improve his chance of securing conviction by making sure that its contents cover the following questions:

1. What do we propose to do?
2. How do we propose to do it?
3. What evidence can we present to show that the methods we propose to use will actually get the desired results?
4. What evidence can and should we present to show that our way of obtaining the desired results is better than any other way?
5. How can we demonstrate our ability to do what we propose to do?
6. What evidence must we present to show that the cost will be acceptable and, perhaps, that we can meet a satisfactory time schedule?

In providing the information called for it will be necessary to explain what methods we propose to use, to show that we have or will obtain the personnel and facilities necessary to use these methods, and to offer enough information about costs to show that our estimates are realistic. Each of these points will be covered in more detail later in the discussion.

KINDS OF PROPOSALS

Proposals are infinitely varied. They range in size from a one-page letter or memorandum to many volumes; in subject matter from suggestions on such minor matters as better lighting for the sake of employee morale to suggestions for joint American-Soviet space exploration. Some of the main types can be described, however, under the following headings: Research, Research and Development, Sales, and Planning.

Research

Proposals for pure research are usually written by workers at universities, science centers, or in some cases, research departments of major industrial concerns such as General Electric or Dupont. Most teachers in the sciences at the university level are involved in projects that receive funding through proposals written to some funding agency. Such teachers often are happy to show interested students examples of the proposals they have written that have been successful in getting grants. A great deal of the work on such grants may be performed by students, usually working on graduate degrees. Sometimes, however, eager undergraduates help out as well. In fact, in psychology, a great many of the projects use undergraduates as subjects. You may already have taken part in a research study without being aware of the process by which the researcher received permission to proceed.

Examples of the kinds of projects that fall under the heading of research abound, but a few that have involved student help include an analysis of animal life in an Iowa cavern, the simulation of movement of fluid in the small intestine, and a study of the growth of bacteria in sewage. A proposal for a grander project was first unveiled at a New York Academy of Sciences conference: Three scientists from the Lawrence Livermore Laboratories, which is funded in part by the United States government through the University of California, proposed harnessing "black holes" within the solar system to provide for future world-energy needs.

Research of these types is frequently funded by philanthropic foundations such as the Carnegie Institute or the Rockefeller Foundation, or by government agencies such as the National Institutes of Health or the National Science Foundation, but much of it is funded through other channels as well. Most of these organizations have exacting formats for the proposals they solicit. For example, a National Science Foundation pamphlet, "Guide for Preparation of Proposals," lists four major sections: Narrative, Abstract, Budget, and Cover Sheet, with details about what should be included in each. Under *Narrative,* which contains the bulk of the explanation, are listed *Major Objectives of Project, Previous Experience, Project Description, Staff,* and *Selection of Participants.* It would appear that writing a proposal for projects of this type would be quite simple, but the problem of what to put in each section remains. In unsolicited proposals to these same agencies, the problem of what to include is even greater.

Research and Development

Usually referred to as R & D, research and development accounts for a great many of the proposals written in industry, as well as some written in universities. Proposals in this area usually have as a goal the actual development of a project or process of some kind. Examples include the development of a more effective crabgrass herbicide, the development of better seals for the Wankel engine, or the development of a preliminary treatment of waste from a food-processing plant so that it can be discharged into a municipal sewage system.

Proposals for R & D projects are frequently solicited by government departments such as the Department of Defense or the Department of Agriculture, or by industrial firms such as drug or chemical companies. Once again, when requests are made for competitive proposals, the form is usually specified for ease of comparison, but the problem of content remains, even if the format is clearly given.

Sales

Sale of products and services forms the backbone of American commerce. Decision-making officials throughout industry agree that the formal written sales proposal is the most important factor in the total sales effort. A sales proposal is an attempt to persuade a customer that a product or a service will fill a particular need. It frequently describes a product already in existence, but it must also convince the customer that the product will outperform any other product in a task yet to be done. Although sales proposals often include a great deal of what is called "boilerplate"—printed material ready for inclusion in any report—successful proposals are generally tailored to the particular customer's needs and specifications. The care taken in the tailoring is one indication to the customer that the company really wants the business.

In addition, sales proposals frequently offer to develop new products to fill a need for the customer. They may be made in response to inquiries or bid requests, or they may be submitted on its own initiative by an aggressive firm as a way of expanding business. Proposals of this type resemble the R & D proposal more than they resemble the traditional sales proposal.

Planning

Planning proposals are frequently written for governmental bodies or by organizations that are in the process of change and are trying to determine the best changes to make. Examples might include plans for new school districts, a new division within an organization, or a new procedure in a business. Frequently a person or committee is chosen to study the case and write up the results as a proposal. The organization then decides on the basis of the information contained in the proposal whether the changes will be advantageous.

CONTENTS

It might seem foolhardy to attempt to give a list of ingredients for such varied types of proposals as those described above, but despite their differences, all contain similar information and are governed by the same general rules.

The first rule concerns the audience. Perhaps more than any other form of technical writing, proposal writing is audience-oriented. A reader who is at all confused about the content of the proposal will be reluctant to approve the project. Consequently, almost all large organizations insist that the material be organized to their own specifications and be aimed at their own situation. Generally, too, they want the presentation of each section to move from major points to details so that readers interested in different aspects of the proposal may all start at the same place but stop where they wish. To make this possible, sections of the proposal must be clearly identified, usually with headings, so that readers may quickly turn as they please from one section to another.

Another consideration that cannot be overstressed is language. Many students, particularly those in science, feel that they will never need to write for anyone but other experts, and those "experts will know what I'm talking about." As a U.S. Department of Commerce pamphlet on proposal writing points out, however, "The reviewer may be assumed to be knowledgeable. . . . He may also be assumed to be a very busy person who has little patience with verbosity or fuzzy language." The concern for language is even more important in business, where proposal readers generally want the same proposal to serve managing, engineering, purchasing, and operating staffs, and thus want as much of the proposal as possible written in plain, clear language that can be understood by people of different backgrounds. All proposal readers are suspicious that the person who cannot write out the idea in plain language will not be able to carry out the idea either. Basically, if the idea cannot be understood, it cannot be approved.

INFORMAL PROPOSALS

While the student might wait years before writing all or part of a major formal proposal, he will find that he will be writing informal proposals much sooner. In fact, the informal proposal is a good way to advance in a job. Most jobs that students get following college tend to be routine in nature. Although it is no longer the general practice it once was, some firms used to stockpile engineers—hire more than they needed because having large numbers of engineers available looked good when writing proposals for government contracts. These engineers, usually right out of college, would be set to work on drafting or other nonprofessional tasks simply to keep them busy. Those who did only these jobs tended to stay longer at lower levels than those who looked around and found something more important that needed doing. It is still true in any field, not just engineering, that the person who looks around to find more work advances faster than the person who does only what he's told.

Ordinarily, when a person discovers a particular problem that he would like to work on, he brings it up orally with his immediate supervisor. If the supervisor finds it to be a task that requires time away from other work, or requires extra money or equipment beyond his authority to provide, he'll ask for a memorandum on it which he can take to his superior. Such a memorandum is an informal proposal.

Sometimes an informal proposal is a letter rather than a memorandum. This is usually the case when the proposer is not an employee of the organization he would like to perform the task for.

Since the proposal is informal, the contents tend to vary more than those of formal proposals. In some cases, sections are labeled, in some cases not. Nevertheless, the following elements need to be included:

Problem

Somewhere close to the beginning, a statement of the problem should be presented. The first reason is to identify the subject matter of the proposal in detail. But beyond this, the situation is often one that no one else sees as a problem. If anyone had, he might have solved it. The writer of the informal proposal must indicate that the situation is in fact a problem, and one that can be solved. He might do this by describing the *background* to the situation—that is, how it developed into a problem. Or he might describe the *significance* of the problem, showing that it is important enough to justify spending the time to solve it.

Proposed Solution

Obviously it is not enough simply to point out the problem, so the proposal must next offer a solution. If a clear solution cannot be given, then the writer concentrates on his *proposed method* toward finding a solution. In this section the writer can show that he has given some real thought to the problem and not just written a memo "off the top of his head." In informal proposals, just as in formal, the reader wants to be sure that the writer is genuinely interested in doing the work, and not just suggesting one thing to get out of doing another.

Statement of Request

Near the end of the proposal the writer requests permission to perform *specific tasks* that should solve the problem or provide information that may lead to a solution. At any rate, it is necessary to indicate in a proposal exactly what the writer wishes to do, since it may not be evident from the description of the solution. When the case is put in terms of a specific request, the recipient can respond easily and quickly and need not puzzle over what to do with it. Whenever the writer can make the reader's job easier, a positive response is more likely.

While informal proposals are often used on the job, they are also used as forerunners for formal proposals. When a company or organization has an idea that it thinks another company or organization might be interested in, it will often write an informal proposal to determine whether there is enough possibility of acceptance to warrant producing a formal proposal. Because of the great amounts of time and money required to produce formal proposals, some government agencies, such as the National Oceanic and Atmospheric Administration, recommend using the informal proposal first as a standard procedure. When a great deal of detail is included, the informal proposal is sometimes called a "pre-proposal." If the reader is interested, he states that he would like to see a full proposal on the subject, which the writer then proceeds to produce.

FORMAL PROPOSALS

When proposals are solicited, the sections for inclusion are frequently specified, but the problem of what information to include remains. The following discussion is intended as a guide to the kinds of information formal proposals usually include. Although those who solicit proposals do not all want the information to appear in the same order, the sections are discussed here in an order that is often used. Whenever appropriate, sections are condensed or combined. It is always better to have the shape of the material dictate the pattern rather than to try to force the material into some preconceived format. Nevertheless, knowing what form to put the information into often helps the writer understand what is needed. (Many of the sections have already been discussed in Chapter 11 on formal and semiformal reports, which should be consulted as necessary for further details; but there are some details to add in connection with proposals.)

Front Matter

Letter of Transmittal.   A formal proposal is almost always accompanied by a letter or memorandum of transmittal. Usually the letter identifies the proposal, gives the highlights, and expresses interest in doing the work. It also will usually refer to the circumstances that gave rise to the proposal, such as the letter or meeting at which permission was granted to submit it.

Cover Sheet/Title Page.   The cover sheet always identifies the work as a proposal, usually in large type. It also gives the title of the proposal; the name, formal title, and address of the person or organization to whom the proposal is submitted; and the identifying number, if as usual there is one, to link the proposal to an announcement or bid request. Usually the date of submission is given, along with the date by which the work will be completed, and perhaps the date after which the proposal will no longer be in effect. When money is requested, the total sum is sometimes included on the cover sheet. Finally, the formal name, title, and address of the person or organization submitting the proposal are given.

Summary or Abstract.   The summary is a brief statement of the total offering. In commercial proposals, the summary indicates that the proposal meets all specifications or clearly states the exceptions. It also identifies the problem and briefly states the proposed solution. It should be no longer than three percent of the total length of the proposal.

Table of Contents.   Any proposal (or report) of more than five pages should probably have a table of contents.

Statement of Request.   Although the statement of request is sometimes combined with the summary, in larger proposals it is a separate section. In it, the writer states clearly who wishes to do the work, what the work is, and what the cost will be. Frequently this section contains only a few sentences, but it is often used by evaluators to help rank competing proposals.

Body

Statement of the Problem.   In both solicited and unsolicited proposals the section defining or describing the problem is important, even though the writer may feel that the reader understands his own problem. In solicited proposals, it shows that the writer, too, understands the problem and has the reader's concerns in mind when he sets out a solution. In unsolicited proposals, the writer often needs to convince the reader that the problem exists. In either case, the writer might spend a paragraph to several pages defining the problem, its ramifications, its significance, and its relation to larger problems, issues, activities, or missions. Often, in his statement of the problem, he includes the material covered in this discussion under the heading *Background.*

Background.   The background, history, or, in proposals for pure research, the state of the art concerning the problem is given in this section. Usually the conditions leading up to the problem are described, indicating why the problem is now being considered. If previous attempts at solution have been made, they are described along with their results and shortcomings. A review of the literature on the subject is often given as well. What the writer needs to show is an understanding of the total context of the problem and an awareness of previous work in the area.

Scope.   The section on scope sets the limits within which the proposal writer will work. When contracts are involved, this is a vitally important section, since the writer does not wish to be held accountable for more than he has agreed to produce. Even if for no other purpose than preserving good will, a clear understanding of the limits of the work to be done is important. For instance, if a writer promises to sample opinion on a stated subject, he would not want the reader to think he was going to sample every person affected when he planned to sample only 5 percent. In this section, then, the writer states exactly what he plans to accomplish, and often, what he does not. The scope section is usually separate from other sections because of its importance in protecting both the writer and the reader from misunderstandings.

Methodology.   The section explaining how the work is to be accomplished is perhaps the most important in any proposal. The reader wants to know that the methods used will in fact produce the results promised. It is not enough simply to find a solution; it has to be a solution that will withstand all kinds of attacks. When competing proposals with similar costs are compared, the one offering the more comprehensive methodology is often the one chosen. For these reasons, the methods used to solve the problem or do the job are always given in detail. When the methods are standard and are understood by both writer and reader, they need only be identified, but any prospective deviations from them are fully described. When the methods are innovative, they are described step by step, with reasons for each step included, in enough detail to convince the reader that they will work.

Facilities.   Usually the section describing facilities follows the section on methods, for it amplifies that section. It does little good to promise work in sterile conditions, for instance, if the proper laboratories are not available. Equipment to be used is frequently described in this section, although sometimes it is listed separately. Equipment might range from normal laboratory equipment to massive roadbuilding machinery, but it must clearly be capable of doing the job. If the writer promises to gather certain information from published sources, for example, he would do well to indicate that the library he plans to use is likely to contain the information he will need. In major proposals, one of the further reasons for including this section is the description it contains of what the contractor will be getting for the overhead charges, which often range from 40 to 60 percent in addition to the cost of actually doing the work.

Personnel.   The people who will be doing the work are listed in the personnel section. When they are known, their names are given, and frequently a page or two is included giving a biography of each person complete with his pertinent publications. The purpose, of course, is to convince the reader that the people can do the work proposed. When the personnel are not yet hired, the grade or degree of competency of those to be hired is described. Part-time personnel are listed, and the percentage of time they will be devoting to the project is indicated. When major personnel will be working only part-time on the project, their other commitments are listed as well.

Advantages and Disadvantages.   Often the section listing advantages and disadvantages of a proposal is omitted because it is thought that the first is obvious and the second ought not be pointed out. Both reasons are shortsighted. This section can sum up a proposal in an honest manner that goes a long way toward convincing the reader to grant permission to proceed. Listing all the advantages in one place lets the writer summarize the total offering. He might list the products that will result, possible side benefits, improved methods of production, savings, or simply the knowledge that will be accumulated. The argument for listing the disadvantages of a proposal is even stronger. The writer, of course, is concerned to have the current proposal accepted, but of more importance is the proposal after the current one. He does not want the reader to accept something that will turn out badly and result in suspicion and ill feeling. A dissatisfied customer not only will refuse to give repeat business to the source of his dissatisfaction; he will also warn others of his experience. Nothing spreads through a field faster than a reputation for questionable practices. Therefore, in plain, clear language, the disadvantages of the proposal should be mentioned.

Duration.   The section headed Duration tells how long it will take to complete the project. If the project is divided into segments, then firm time schedules for completing each are given. Delivery dates are also listed where appropriate.

Costs.   The section on the costs of the proposed project is crucial. Sometimes the writer concludes with a statement of costs in hopes that after going through everything else, the reader will be convinced that the expense is justified; sometimes costs are detailed in a section separate from the proposal so that they will not influence other deliberations; sometimes they are presented first on a special budget sheet. In any case, all costs should be itemized under headings such as salaries, capital equipment, expendable equipment, miscellany, and overhead. Often only estimates are possible, but they obviously should be made with the greatest care. In industry, at least, expensive cost overruns are rarely tolerated.

Reports.   Finally, there should be a section detailing what reports will be written, giving dates for progress and final reports. Often included, too, are the names or titles of the people responsible for the reporting, and the form that the reports will take.

EVALUATION OF PROPOSALS

It might seem that in a chapter on writing proposals, a section on evaluation would be inappropriate, but generally writers produce better products if they understand how their work will be judged. The following set of questions was devised by an agency within the Department of Health, Education, and Welfare to aid readers in preliminary review of proposals. By giving a scale of values to the questions, the reader can make a point comparison between competing proposals. Although the questions were intended for particular kinds of research and development proposals, they apply to proposals of all kinds. Writers would do well to ask the questions themselves before submitting a proposal. If any question receives an inadequate response, the section concerned should be restructured.

1. Understanding of purposes, objectives and tasks—30 points.
    a. Does the bidder demonstrate clear understanding and acceptance of the requirements presented in the rfp (request for proposal)?
    b. Are the tasks outlined in the proposal clear and well defined?
    c. Are there important omissions in the specified tasks?
    d. On a scale of 30 points please assign a rating to this area.

2. Technical quality of methods proposed—30 points.
    a. Are sufficient time and manpower resources specified to accomplish the quality outlined in the proposal?
    b. Does the proposer emphasize quality as an important criterion when presenting his methods?
    c. Will the quality of the proposed methods be monitored throughout the contract period?
    d. Is this monitoring sufficient to insure quality?
    e. On a scale of 30 points please assign a rating to this area.
3. Quality of management plan and planning—10 points.
    a. Has a management plan been designed to insure receipt of materials at certain specified times?
    b. Does the proposal clearly identify working relationships within the contractor staff and with this agency's staff?
    c. Is sufficient technical management assigned to the task to insure production and quality of output?
    d. Will information be available to this agency sufficient to permit analysis of cost and effectiveness?
    e. On a scale of 10 points please assign a rating to this area.
4. Qualification of staff—20 points.
    a. Have individuals assigned to the task had prior experience in the required technical areas?
    b. Have key personnel been assigned to the project for large enough percentages of time?
    c. Have project directors and those assigned management roles been experienced in similar management positions?
    d. Is there sufficient depth in the staff to provide backup and overload capabilities?
    e. On a scale of 20 points please assign a rating to this area.
5. Corporate capability and experience—10 points.
    a. Has the organization had previous experience in planning and managing efforts of this type?
    b. Has the organization previously managed efforts of the size and complexity of this agency's?
    c. Is the organization of sufficient size and stability to undertake the responsibility called for?
    d. Is there any "track record" of performance available, indicating consistent meeting of schedules, with quality output, within fiscal limits, or the inverse?
    e. On a scale of 10 points please assign a rating to this area.

Although none of the preceding questions deals directly with expenses, costs are a primary consideration. In commercial proposals, the first consideration is cost, followed in order by performance, reliability, economy of operation, and early delivery of the product.

Beyond these considerations are many others, but one of the most important is the tone of the proposal. Those addressed are extremely concerned that the responsible officials of the proposing organization are genuinely interested in doing the work and are committed to providing complete satisfaction, even beyond normal guarantees. In other words, proposal readers are likely to pass up a proposal if they feel that they would have to put up with sharp dealing—or with indifference.

As a final word on evaluation, all proposal writers should heed the warning given by the National Science Foundation concerning a major reason for rejection of proposals:

> It cannot be emphasized too strongly that the reviewing panels must base their recommendations largely on the information contained in the proposal. The most frequent fault of proposals is vagueness.

(An illustrative specimen follows.)

SPECIMEN PROPOSAL*

Full-scale formal proposals often are so long that using one as a specimen would be impractical, but the following preproposal illustrates the manner of their construction. The additional material in a full, formal proposal on the same subject would consist mainly of much more detail and a more extensive analysis of costs.

The page opposite to this comment was printed on heavy paper and served both as a cover and as a title page. It shows to whom the proposal is submitted, and it names the three bodies that would cooperate in the project—assuming that a full proposal was encouraged.

The title page is followed by the letter of transmittal, which orients the reviewers to the project, provides evidence of the care with which the material submitted had been prepared, and makes the actual request. Note that even at this early point attention is directed to the fact that the value of the project could extend far beyond the benefit to the particular location involved.

---

* Courtesy of William A. Smith, City Engineer, Moscow, Idaho.

A Preproposal To The
U.S. ENVIRONMENTAL
PROTECTION AGENCY
For

A RESEARCH &
DEMONSTRATION
STUDY ON THE
REMOVAL OF
IRON & MANGANESE
FROM POTABLE
WATER SUPPLIES
USING OZONE

SUBMITTED BY
CITY OF MOSCOW, IDAHO

IN COOPERATION WITH
UNIVERSITY OF IDAHO,
MOSCOW, IDAHO

AND
STEVENS, THOMPSON & RUNYAN, INC.
BOISE, IDAHO

OCTOBER 1972

CITY OF MOSCOW

MOSCOW, IDAHO
83843

122 East Fourth

Telephone: 882-5553—Area Code: 208

October 31, 1972

Dr. Gordon G. Roebeck, Director
Water Supply Research Laboratory
Taft Environmental Research Center
Environmental Protection Agency
Cincinnati, Ohio 45268

Dear Dr. Roebeck:

Enclosed for your consideration are copies of a preproposal for a research and demonstration study on the removal of iron and manganese
from potable water supplies using ozone. Professor Calvin C. Warnick,
Director of the Idaho Water Resources Research Institute, talked with
you last summer with regard to the possibility of EPA participating in
the proposed water treatment system.

Reviewers are respectfully reminded that this document is a preproposal and therefore does not contain detailed information on equipment,
budgeting, study programming, manpower requirements, etc. The intention would be to include such detail in the formal proposal and grant
application.

However, the preproposal is based upon a comprehensive literature review. Particular emphasis is directed to the reference by F. L. Evans,
"Ozone in Water and Wastewater Treatment." This recent publication
is a compendium of the published literature on ozone, and offers a thorough review and report of previous work with ozone.

As demonstrated herein, the City of Moscow is committed to improving
the water quality of the City and will commit a substantial amount of the

276

City's resources to the proposed research and demonstration plan. We hope that the Environmental Protection Agency will view this study as being of significance to water supply systems throughout the nation and will assist in funding of the project.

Sincerely yours,

U OF I CHEMICAL ENGINEERING DEPT.

Robert R. Furgason, P.E.
Professor and Chairman

CITY OF MOSCOW

W. A. Smith
Director of Public Works

STEVENS, THOMPSON & RUNYAN, Inc.

Richard O. Day, P.E.
Boise Manager

RRF/WAS/ROD:sm

Encl.

The preproposal proper, as can be seen from the table of contents, follows a pattern in which, first, the Ozone Water Treatment System is explained. (Two figures are provided to make the general nature of the process easier to grasp.) Then, the development of such a system is justified. Next, the capabilities of the cooperating agencies are established.

This pattern provided answers, in a logical order, to the questions most likely to arise in the minds of the reviewers.

CONTENTS

FIGURES

279

PROPOSED WATER TREATMENT SYSTEM

PROJECT OBJECTIVES

The purpose of the proposed water treatment demonstration program is to investigate the feasibility of using ozone to remove iron and manganese from potable water supplies. Results from the operation of the ozone treatment demonstration will be used to establish the practicality and cost effectiveness of ozone treatment for removal of iron and manganese from domestic water supplies. Specific objectives are to:

1. Determine in a pilot plant operation the optimum operating conditions and design criteria for ozone oxidation of iron and manganese, using the water from wells No. 2 and No. 3 of the Moscow, Idaho, water system. Water from these wells typically analyzes 2 - 10 mg/l iron and 0.1 - 2.0 mg/l of manganese. Pilot plant control variables will include ozone dosage levels, residence time, pH and ozone injection method.

2. Determine the optimum filtration rates possible for satisfactory removal of the iron and manganese precipitates.

3. Utilize the design criteria and optimum operating conditions derived from the pilot operation to design, construct and operate a full-size demonstration plant of approximately 3.7 million gallons per day capacity.

4. Establish actual construction and operating cost data from the full-scale demonstration plant in order to make an accurate economic comparison of ozone treatment to other accepted iron and manganese removal methods.

PROPOSED TREATMENT SYSTEM

Features of the proposed treatment system are as follows:

1. Full-scale, pressurized filters utilizing chlorine-potassium permanganate oxidation will be initially installed on the combined discharge from City wells No. 2 and No. 3. This oxidation process is temporary and is intended to remove only the iron, not the manganese.[9][10]

2. An ozone pilot unit will be constructed simultaneously in parallel with the above full-scale filtration system. The purpose of the pilot plant is to obtain accurate scale-up design data for the full-sized demonstration plant. Wherever possible, the pilot plant equipment will be designed to be incorporated in the final plant equipment, for example, the ozonator.

1

3. Upon successful completion of the pilot studies, the full-scale ozone oxidation equipment will be installed replacing the chlorine-potassium permanganate oxidation facilities, thus providing complete removal of the iron and manganese. At this point the full-scale demonstration plant will be complete so that accurate operating and cost data can be obtained and documented.

The above operational plan facilitates iron removal from the water supply during the initial development of the demonstration system. This is desirable from the City's standpoint because it removes most of the aesthetically objectionable material as soon as possible with only a minor amount of additional cost involved. Figures 1 and 2 illustrate the proposed system operation.

FINANCIAL OBLIGATIONS

It is requested that the Environmental Protection Agency fund one-half of the cost of the total project by supporting the full cost of the ozone research-demonstration facilities. Thus the EPA grant would include: (1) a pilot plant designed for a capacity in the range of 20 - 200 gal/min consisting of an ozonator, gas-liquid contactor and a precipitate removal system; and (2) ozone demonstration equipment consisting of full-size ozonators and gas-liquid contact system.

The City of Moscow will provide the pressurized filtration equipment, the chlorine-potassium permanganate feed system, personnel, building and other In-Kind services.

Costs are estimated as follows:

City of Moscow

Pressurized filter system and the $KMnO_4$ - Cl_2 system	$200,000
In-Kind services	25,000

EPA

Ozone research-demonstration system including O_3 pilot plant ($75,000)	225,000
TOTAL COST	$450,000

2

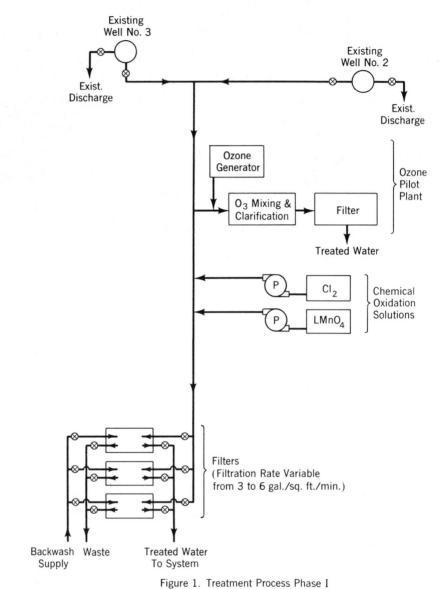

Figure 1. Treatment Process Phase I

3

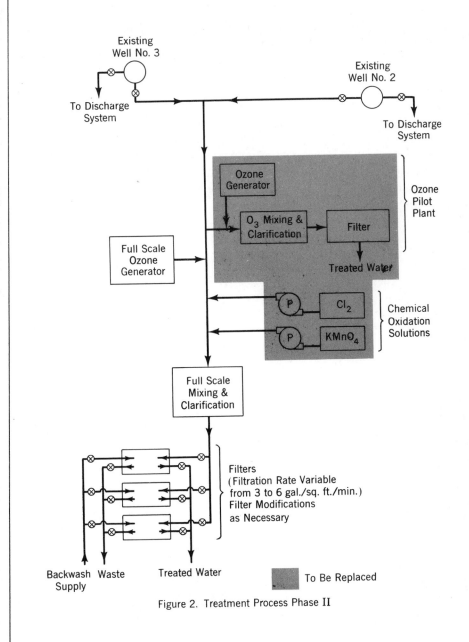

Figure 2. Treatment Process Phase II

4

The nature of the proposal, having been explained, the reviewers are prepared to evaluate its justification.

The first part of this section covers the widespread need for such a project and then tells why Moscow, Idaho, would be a desirable location for it. Thereafter, it provides more information about the Ozone Treatment—concentrating attention, at this point, not on the technique but on what the treatment should be able to accomplish, its use elsewhere (primarily in Europe, where it is common), the reasons that it has not been extensively explored in the United States, and the reasons that experience with it elsewhere does not cover the needs for the work proposed on this occasion.

PROJECT NEED

The problem of iron and manganese removal from domestic water supplies is not unique to Moscow, Idaho. A 1970 study by the Bureau of Water Hygiene, U.S. Public Health Service[8] analyzed community water supplies across the nation from Vermont to California. Nearly 1000 water supply systems were involved. This study revealed that iron and manganese exceeded the Drinking Water Standards (DWS) for the <u>delivered</u> water in 159 water systems. No other constituent exceeded the DWS more times than iron and manganese. Of the communities tested 32% of the groundwater supply systems had iron and manganese in excess of the DWS recommended limits for <u>delivered</u> water.

The Bureau of Water Hygiene report found that the smaller water systems had more difficulty in meeting water quality standards than did the systems serving larger populations. The recommendations of the report contain the following statement pertaining to iron, manganese, color, turbidity, etc.

> "Further research is necessary to understand what causes these problems as well as how to more effectively remove objectionable constituents."

Two other recommendations of the Community Water Supply Study are relevant to this proposal. They are:

> "The presence of coliform organisms at a level above the DWS limit in 120 systems indicates that better source protection and/or disinfection should be instituted as soon as possible,"

and

> "Inasmuch as tastes and odors remain a common complaint in public water supplies, more effort should be made to employ the present technology of oxidation and adsorption to control or remove these troublesome constituents. Further research is needed to reduce cost of such treatment to make it more practical and acceptable."

The Moscow, Idaho, water system is an ideal supply for a demonstration study of ozone technology applied to water treatment: first, the system is intermediate in size serving a current population of about 10,000; and second, the test system can be built and run without affecting the normal operation of the basic water supply.

The City of Moscow uses two main groundwater aquifers as sources for its domestic water supply. One aquifer serves as the source for two deep wells, No. 6 and No. 8 (approximately 1300 ft.), both of which are relatively free of iron and manganese. These two wells produce an average of 3.2 million gallons per day and can satisfy normal water demands during fall, winter and spring periods. However, during the heavy demands of the summer months, water must be drawn from wells No. 2 and No. 3, which are shallow wells (approximately 500 ft.) and are characterized by a very high iron and manganese content. Thus during the construction period of the demonstration plant, the normal water supply will not be disturbed and any unforeseen difficulties can be dealt with while the normal system is isolated from the proposed water treatment plant since the ozone treatment is needed only for wells No. 2 and No. 3.

Although the poor quality water from wells No. 2 and No. 3 is currently being used only for peaking purposes, it is projected that the population growth by 1975 will require these wells to be used on a regular basis.[10] Thus if a research demonstration project is to be done with a minimum amount of system disruption, the time to do it is now.

The City of Moscow has investigated alternate sources of water. One apparent solution to this problem would be the development of new deep wells which would draw water from the same aquifer as wells No. 6 and No. 8, providing the same high-quality water. However, the expense of developing deep wells requires an excessive capital outlay.[10]

Previous investigations first examined the feasibility of a surface supply. In 1970, Stevens, Thompson & Runyan, Inc., completed a study[11] for the development of a joint surface supply for the City of Moscow, the City of Pullman, the University of Idaho and Washington State University under an intergovernmental, bi-state agreement. The study concluded that development of a surface supply was feasible, provided conjunctive use of all existing wells (including wells No. 2 and No. 3) was a continuing source of supply.

Since the existing wells would still be necessary even if a surface supply were developed, it became apparent that consideration should be given to the concept of water treatment in lieu of development of new deep wells. This concept would consider the feasibility of treating the water from wells No. 2 and No. 3 to provide a suitable quality of water, and compare the expense of such treatment to the cost of developing a comparable supply through construction of new deep wells.

Accordingly, the City had preliminary pilot tests performed by Keystone Engineering & Products Co., Inc., in 1971.[9] These tests were limited to chemical treatment and filtration of water from the two wells. Generally, the results showed that this process would require uneconomic chemical costs to achieve complete removal of the iron and manganese. Keystone recommended that an ion-exchange process be tested as an alternate method of treatment.

In November of 1971, the Mayor and City Council authorized Stevens, Thompson & Runyan, Inc., to investigate feasible methods of water treatment to remove the iron and manganese from wells No. 2 and No. 3.[10] This report found ozone treatment to be one of two feasible methods of mineral removal from the sources involved. The intergovernmental group that sponsored the earlier surface water supply study, the Pullman-Moscow Water

6

Resource Committee (PMWRC), took action in August 1972 in support of a planned program to encourage the construction of a water treatment system that would remove the iron and manganese from the groundwaters. Professor Warnick, chairman of this intergovernmental group, made contact with Dr. Roebeck of the Taft Engineering Center, EPA, to seek advice and counsel on possible support of a research and demonstration grant.

The University of Idaho, located in Moscow, Idaho, maintains its own water system. The University's water is supplied by deep and shallow wells reflecting the same problems as those of the City. The results of the proposed demonstration plant will substantially influence the type of water treatment installation that the University would install to improve its own supply.

OZONE TREATMENT

The impetus to use ozone as a treatment method originates from the ability of this material to solve simultaneously several of the problems associated with potable water supplies. Ozone is a very powerful oxidant second only to fluorine. Due to its strong oxidizing power it is known to be able to deodorize, decolorize and sterilize water supplies in addition to oxidizing the iron and manganese.[5] If properly handled, it has few adverse environmental side effects since it decomposes readily into oxygen. The proposed treatment plant can be designed so that no free ozone can escape into the atmosphere.

The current use of ozone for water supply treatment is concentrated in Europe and to a limited degree in Canada. There are a number of operating systems that use ozone as a disinfectant and as a polishing treatment for odor and color removal. The only known system which utilizes ozone principally for iron and manganese removal is in Duesseldorf, Germany, with a capacity of 38.5 million gallons per day.[7]

There are several reasons why ozone has not been widely used in the United States. With the development of large and relatively inexpensive sources of chlorine, ozone at over twice the cost could not compete economically with chlorine as a general disinfectant. Also, chlorine escapes from a system quite slowly, thus providing a disinfectant residual that is carried throughout the water system. Ozone, on the other hand, has a half-life of only 20 minutes in water and its decomposition to oxygen is catalyzed by many metals such as iron. As a result, many regulations require a chlorine residual to be present in the water system which precludes the use of ozone. However, ozone is a more powerful disinfectant than chlorine and the presence of a chlorine residual is not necessarily good when organic materials are present. Also, the cost of ozone has decreased substantially in recent years.

Nonetheless, ozone treatment remains as a very viable alternative to other methods for iron and manganese removal and as an agent for eliminating odors, color and microorganisms. The fact remains there is little published in the literature on the design and operating characteristics of a treatment plant for iron and manganese removal using ozone. The available information on the Duesseldorf plant[7] which is many times larger than the proposed system for Moscow, Idaho, does not duplicate the effort herein proposed. Thus the objective of the research-demonstration plant is to generate usable data to evaluate the economic feasibility of such a treatment system and to make this information available to small municipalities across the country who share in the problem.

7

Having explained the system and justified the request, the preproposal takes up the question of whether the three cooperating agencies involved are capable of handling the job. It explains the responsibilities each would accept, providing a figure to illustrate the organizational structure. It presents a copy of the official resolution authorizing the request to show the firm commitment of the municipality to carry the project forward. It covers the experience in similar work of the university department involved, and its available facilities. It provides evidence of the qualifications of the engineering corporation that would cooperate in the work.

This material is followed by a list of references showing familiarity with the background of previous work in the field (one such reference had been mentioned earlier as particularly significant) as well as listing literature written by personnel who would participate.

CAPABILITY

The proposed organizational structure for the demonstration process is shown on Figure 3. The City of Moscow will act as grant applicant and administrator. The actual demonstration process will be a cooperative effort between the City of Moscow, the University of Idaho's Chemical Engineering Department and Stevens, Thompson & Runyan, Inc. Performance of the individual work elements to be conducted by the City, the University and the Consultant are shown on Figure 3. Division of the work elements is based on using the individual strengths of each of the three team members.

Summaries of the capabilities of the City, University and Consultant, and of key personnel who will work on this project also are included.

CITY OF MOSCOW, IDAHO

Public Works Department

The City is managed by a Mayor-Council form of government. Departmental supervision is administered with a full-time Administrative Assistant and Director of Public Works/City Engineer. Subdivision within the Public Works Department includes a Department of Sanitation, Department of Water, Department of Streets, Department of Building Inspection and a Department of Engineering. The coordinator of the Departments of Sanitation, Water, and Streets is the Superintendent of Public Works, who was recently promoted to that position from Water Department Superintendent. Each department operates at the direction of a superintendent. Currently there are two foremen and seven laborers working under the Water Department Superintendent, who perform the daily tasks of maintaining and improving the City water and sewer collection systems. The existing crew can maintain the proposed project equipment within the scope of its present duties.

Wells No. 2 and No. 3 are located adjacent to each other and are enclosed in a masonry structure which serves as storage, office and maintenance area for the Water Department. In a preliminary review of this proposed project, the building and site were found to be more than adequate for the proposed treatment and filtration facilities. The Water Department has a full complement of maintenance tools and equipment which would be available to the project, should it be funded.

The City Council, in their meeting of October 16, 1972, passed a resolution authorizing the City Engineer to pursue EPA participation in a research and demonstration grant (see following resolution).

City Financing

Should this grant be awarded, the City is committed to funding their proportionate share of the project as previously outlined in this pre-proposal. Currently the City is retiring $309,000 in both long-term revenue and general obligation bonds. The assessed evaluation

8

for taxing purposes is approximately $11,750,000 and the current assessment is at the rate of 32 mills. Consequently, bonding for this proposed project is fully within the City's capabilities. In addition, the current budget (1972) has set aside $103,000 toward capital improvement of the water collection and distribution system. Accrued water revenue will allow an increase in this figure in the 1973 budget.

The City is in a sound posture to accept the responsibility of funding, coordinating and administrating the contracts for the proposed facility.

9

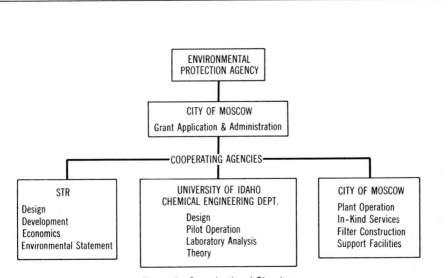

Figure 3. Organizational Structure

10

CITY OF MOSCOW

RESOLUTION

WHEREAS, the waters from City wells No. 2 and No. 3 contain quantities of iron and manganese which exceed current drinking water standards, and

WHEREAS, acceptable quality of water from these wells is necessary to meet future demands of the City of Moscow, and

WHEREAS, treatment of such waters to remove the iron and manganese utilizing ozone has been investigated and found to be possible, and

WHEREAS, ozone treatment for mineral removal is unique and possibly eligible for federal assistance,

NOW, THEREFORE, BE IT RESOLVED, that the City of Moscow make application to appropriate federal agencies for grant assistance for a Research and Demonstration Study on the Removal of Iron and Manganese from Domestic Water Sources using Ozone, and

BE IT FURTHER RESOLVED, that W.A. Smith, Director of Public Works, be authorized to act on behalf of the Mayor and City Council in making such application, and

BE IT FURTHER RESOLVED, that the City of Moscow participate in the demonstration process as a cooperating agency in conjunction with the University of Idaho and Stevens, Thompson & Runyan, Inc.

APPROVED this 16th day of October, 1972.

Larry Merk
Mayor

Councilmen:

Glenn Utzman

Larry Kirkland

Paul Mann

C. H. Bond

Cliff Lathen

George R. Russell

ATTEST:

Marvin Kimberling
City Clerk

11

Ozone Research

The Chemical Engineering Department at the University of Idaho has been engaging in research involving ozone for over 10 years. Projects utilizing ozone for pollution control work particularly in the pulp and paper industry are underway as well as studies that have involved the physical properties of ozone.[1][2][3][4][6] Dr. Furgason and Dr. Edwards have participated as consultants in a number of projects when ozone was utilized including odor control for a blood drying plant, purification of brewery water, waste water treatment from a chemical plant and ozone bleaching of paper pulp. Of particular significance from an equipment standpoint is the ozonation facilities currently available in the Chemical Engineering Department. A portable ozone test unit was developed and built by the Department through a grant from the Water Resources Research Institute. The portable unit contains a small ozonator, a stirred tank reaction system, pH level and temperature control, an ozone injection system, feed and recirculation pumps and other peripheral equipment. The test unit has been used at several locations throughout the West to examine the feasibility of ozone treatment of various materials. It was used on the Moscow water system and did demonstrate that ozone was capable of removing the iron and manganese (see Appendix). It is not, however, large enough to be used for design and scale-up purposes.

The Department also has a large 51-tube ozonator capable of producing 20 lbs/day of ozone which is currently used in several research projects. The ozone equipment of the Department are the only ozone generation facilities available at the University of Idaho or at Washington State University (nine miles away). Thus these facilities serve a regional need and are used by researchers on both campuses.

In the proposed research-demonstration program, the expertise of the University faculty primarily involving Dr. Furgason and Dr. Edwards would be an integral part of the project. Analyses of the plant's operation would be incorporated into the graduate program of the institution, thus providing for extensive study of the system. Publications of the results would appear when appropriate.

Chemical Engineering Department Facilities

The Chemical Engineering Department occupies extensive facilities in the relatively new $2,500,000 Buchanan Engineering Building on the University of Idaho campus. Approximately one-half of the laboratory building is available for Chemical Engineering activities, including 19 specialized laboratories (250 - 750 ft.^{2}) for research work, two large 3500 ft.^{2} and 2500 ft.^{2} general purpose laboratories, an analytical laboratory, a well-equipped shop and a full complement of study and office areas.

STEVENS, THOMPSON & RUNYAN, INC.

Stevens, Thompson & Runyan, Inc., was formed in 1920 as a partnership by J.C. Stevens and R.E. Koon under the name of Stevens & Koon. Upon the retirement of Mr. Koon in 1951, the name was changed to Stevens & Thompson. Mr. Stevens retired in 1954 and Mr. Runyan became a partner in 1955. To assure better continuity of service to clients, the firm was incorporated in 1961. STR is an independent consulting firm in which all the owners are active managing officers. It has no affiliation with any manufacturing, construction or other commercial interests.

The initial assignments of the firm were in the public works field, but through the years the firm has had a steady growth in personnel, responsibility and type of projects. Today the staff of approximately 200 is engaged in a variety of projects including water and sewerage planning and design, industrial waste treatment systems, urban planning, architecture, highway and airport planning and design. In addition, the firm participates in several joint ventures with other consulting firms.

STR has direct access to GE-235 and GE-635 computers as well as the IBM 360. Sending and receiving consoles are located in both the Portland and Seattle offices. This readily available tool with its numerous programs for solving engineering problems supplements other computer services utilized by Stevens, Thompson & Runyan, Inc.

Sanitary engineering laboratory facilities maintained in the Portland office are equipped for routine water and sewage analyses in conjunction with design and research projects conducted by the firm. Much of the equipment is portable to permit setting up field laboratories and pilot plant studies.

All projects undertaken by STR are under the direct management of one of the officers of the firm. Also, regardless of the size of the project, STR utilizes a team approach in which all the disciplines represented on our staff are available and responsible to the officer in charge. For a large, multi-facet project several teams may be established, each of which is responsible to the officer in charge. Since the credentials of a consulting firm are based equally on the experience and reputation of the firm and the capabilities of its personnel, STR has developed a professional staff of specialists who contribute a wide range of engineering and planning services to its clients.

STR has considerable experience in the field of municipal water supply and distribution. Recent experience includes the design of municipal water treatment facilities at:

Eugene, Oregon	100 mgd
Bellingham, Washington	24 mgd
Anacortes, Washington	24 mgd
Oregon City, Oregon	10 mgd
Anchorage, Alaska	8 mgd
Fairbanks, Alaska	4 mgd
Orofino, Idaho	2 mgd
Castle Rock, Washington	1 mgd

13

STR has provided engineering services to the City of Moscow for many years. The firm has completed the following projects for the City's water system:

Preliminary Treatment Report	1953
Distribution Analysis	1962
Reservoir, Pump Station	
Control System Design	1968
Water Supply Study[11]	1970
Water Treatment Analysis[10]	1972

Consequently, the personnel at STR are thoroughly familiar with the City's water system and its inherent problems.

REFERENCES

(1) Edwards, L.L. "Heat Transfer in Thermally Decomposing Ozone," Ph.D. Thesis, University of Idaho, 1966.

(2) Edwards, L.L. and Furgason, R.R. "Heat Transfer in the Thermally Decomposing Ozone System: Theoretical Investigation," CEP Symp. Ser. 61, 229, (1965).

(3) Edwards, L.L. and Furgason, R.R. "Heat Transfer in Thermally Decomposing Ozone: Experimental Investigation," IEC Fundamentals, 7, 440 (1968).

(4) Edwards, L.L. and Strecker, S. "Operation of Welsbach C-1-D Ozonator at Low Temperatures," Report to Welsbach Corporation, Philadelphia, Pennsylvania (1966).

(5) Evans, F.L. "Ozone in Water and Wastewater Treatment," Ann Arbor Science Ann Arbor, Michigan (1972).

(6) Furgason, R.R., Harding, H.L. and Langeland, A.W. "Use of Ozone for the Decolorization and Deodorization of Pulp Mill Wastes," submitted to TAPPI.

(7) Hopt, W., "Water Purification with Ozone and Activated Carbon (Duesseldorf Process) II: Uses of Ozone," Gas-Wasserfach, Wasser-Abwasser, 111, 156 (1970).

(8) "Community Water Supply, Analysis of National Survey Findings," U.S. Department of Health, Education and Welfare, Bureau of Water Hygiene, (July 1972).

(9) "Pilot Plant Test of Wells Nos. 2 & 3 for the City of Moscow," Keystone Engineering & Products Co. Report (April 1971).

(10) "Water Treatment - An Alternative Analysis for Moscow, Idaho," Stevens, Thompson & Runyan, Inc., Report (April 1972).

(11) "Water Supply Study," Stevens, Thompson & Runyan, Inc., April 1971.

14

The proposal proper was followed by an appendix consisting of two parts. The first (omitted) consisted of eleven pages describing the qualifications of the key personnel in the three agencies that would work on the project.

This was followed by a report on laboratory-scale tests already conducted—tests that demonstrated that the Ozone Treatment would actually produce the desired results. This report consisted of one page of discussion, a flow sheet illustrating the laboratory setup and process, and a table showing the results of the tests performed.

OZONE OXIDATION TESTS

ON MOSCOW WELL NO. 3

March 27, 1972

The principal objective of the tests was to determine the capability of ozone to oxidize iron and manganese from the water from Moscow Well No. 3. For three days prior to sampling, the well was intermittently operated for 2 - 3-hour periods by city personnel in order to obtain representative samples. A test sample of 60 gallons was withdrawn from the well and transported to the University of Idaho Buchanan Engineering Laboratory for testing.

Figure 1 demonstrates the flow process utilized in this test. Raw water was metered to a reaction vessel through an adjustable feed pump. Flow rate of raw water was set at one gallon per minute for all tests. From the reaction vessel the raw water was pumped through a venturi, at which point the ozone was introduced, and then returned to the reaction vessel. Retention time within the reaction vessel, after introduction of the ozone, was set at one minute by regulation of the height of an overflow weir. Ozonated water left the reaction vessel over this same weir.

Gas flow from the ozonator was regulated by a rotometer for application rates of 0.5, 0.75 and 1.0 scfm, providing ozone production of 0.1667, 0.201 and 0.206 mg O_3 per second, respectively. Ozone was produced by a Welsbach Model No. C-D-1 generator.

Samples were taken from the effluent for each different gas flow after five minutes of operation at that flow rate to achieve stabilization. Chemical analyses were run on raw water for iron and manganese. Chemical analyses for iron and manganese residual were run on treated water from the reaction vessel after filtration. All analyses were performed in accordance with the latest edition of "Standard Methods."

The test demonstrated the principal objective of the study in that significant oxidation of the soluble minerals could be accomplished using ozone without pH adjustment. For both iron and manganese the mineral concentrations were reduced over 90% but did not quite reach drinking water standards; however, no attempt was made to further reduce the concentrations or optimize the operating conditions.

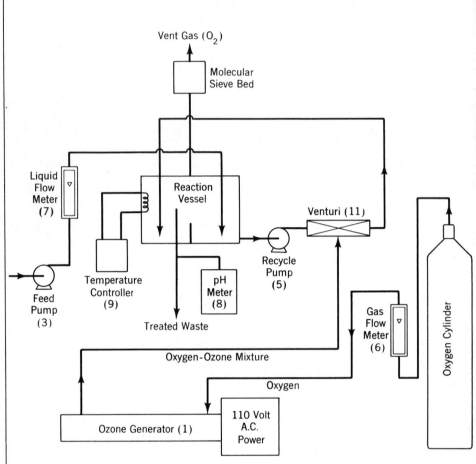

Figure 1. Schematic Flow Diagram of Portable Test Unit

28

Run No.	Gas Flow scfm	Titrant Sample	ml	PPM	mg/sec

OZONE

Ozone in Effluent - 100 ml Samples

Run No.	Gas Flow scfm	Sample	ml	PPM	mg/sec
No. 1 Unfiltered	0.5	1	26.35		
		2	26.40	.016	
No. 2 Unfiltered	1.0	1	18.5		
		2	16.5	.0106	
No. 3 Unfiltered	0.75	1	21.1		
		2	20.9	.0128	

Ozone from Generator - 10 ml Samples

Run No.	Gas Flow scfm		ml	PPM	mg/sec
No. 1	0.5		82.1	5.0	0.1667
No. 2	1.0		94.6	6.2	0.206
No. 3	0.75		92.2	6.04	0.201

Run No.	Gas Flow scfm	Sample	Unknown	Trans-mission%	PPM	Avg. PPM

IRON

Run No.	Gas Flow scfm	Sample	Unknown	Trans-mission%	PPM	Avg. PPM
Raw		1	Iron	25.0	5.0	
		2		25.0	5.0	5.0
No. 12 Filtered	0.5	1	Iron	66.0	0.40	
		2		76.0	0.25	0.325
No. 23 Filtered	1.0	1	Iron	76.0	0.25	
		2		79.0	0.225	0.2375
No. 34 Filtered	0.75	1	Iron	90.0	0.1	
		2		95.0	0.05	0.075

MANGANESE

Run No.	Gas Flow scfm	Sample	Unknown	Trans-mission%	PPM	Avg. PPM
Raw		1	Mn	82.5	0.93	
		2		(95)(4)	0.90	0.91
No. 1 Filtered	0.05	1	Mn	96.0	0.265	
		2		96.5	0.21	0.2375
No. 2 Filtered	1.0	1	Mn	97.00	0.17	
		2		98.0	0.13	0.15
No. 3 Filtered	0.75	1	Mn	98.5	0.09	
		2		98.0	0.13	0.11

ASSIGNMENTS

Assignment 1

In the list of assignments following the chapters on reports, find a subject you would like to report on, and write an informal proposal suggesting that you do the research and write such a report. Write the informal proposal to your teacher, or if he suggests, to a third person or agency that would be interested in receiving the information.

Assignment 2

After you have received approval of your informal proposal, research the subject adequately enough to write a formal proposal. The formal proposal should be aimed at the same audience as the informal proposal.

13

ORAL PRESENTATION
OF TECHNICAL INFORMATION

Anyone who is involved in a technical field will be called upon from time to time to present technical material orally. While a written discussion of this subject will, of course, be no substitute for a course in speech, the presentation of technical information presents special problems that often are not covered in a college speech course. The discussion in this chapter is intended to bridge the gap between the speech courses you may take as a student in college and the time when you must rise before a group of people and speak as a professional in a technical field.

When would such a necessity arise? One occasion would be a meeting of a few people to whom you wish to explain a project you propose to undertake. Such meetings are common in all proposal negotiations. Again, after submitting a report on a finished project, you might be asked to discuss aspects or ramifications of the project that were rightfully not included in the written report. It cannot be assumed that any report, however good, can cover every aspect of a problem that might interest people.

A typical occasion when you might need to speak to a larger group would be a public meeting at which you present the attitudes or intentions of your employer or profession concerning some project of general interest. For example, a forest ranger might speak to an environmentalist group explaining logging procedures in nearby forests; a home extension agent might speak to a group about methods to save money in shopping for food and in preparing meals; a health department agent might discuss the problems involved in following certain organic diets. The list could be prolonged indefinitely, but is long enough already to show that men and women in technical professions are constantly called upon to speak on matters in their fields as well as write about them.

Still another type of speaking occasion for those who write on technical subjects occurs in connection with technical articles. If you look over technical journals you will observe that many of the articles are papers that have been read at professional gatherings. A first reaction to the idea of reading a professional paper may be that this activity is peculiar to the academic profession, but such is not the case. For example, in a recent issue of a periodical in the field of paper manufacture, more than half of the articles were printed versions of talks at conventions and similar meetings by speakers employed in industry.

PREPARATION

Although preparing material for oral presentation is in many respects the same as preparing to write about it, there are important differences, because listening and reading are different processes. These differences are especially noteworthy when the material in question is a technical report, where a great quantity of specific detail needs to be conveyed accurately. What is said here, therefore, is especially applicable to report situations, but not limited to them.

When a person reads what you have written he does not necessarily read in the same manner from beginning to end. For instance, he may read some parts carefully, pausing to consider them if he feels so inclined; other parts he may skim because the topics lie outside his own particular interests or because he accepts what you say without needing to evaluate your supporting evidence. Sometimes he may turn back to refresh his mind on what he has read earlier, or he may skip a section and turn, with the help of headings, to material that comes later. In general, he is master of the situation, free to decide for himself how he wishes to approach the report.

But when a person, as a member of a group, is listening to a speech, options such as these are not open. He must accept what you offer. Unless you repeat, he hears each fact only once. Unless he wants to risk missing something that follows, he can think about what you say only while you are saying it. He can't skim; he can't turn back; the time he devotes to each aspect of the subject depends on the time you spend on it.

Of course, a listener may interrupt with questions, but unless the group is extremely small, many people are reluctant to interrupt out of consideration for others who might prefer to have you proceed. Thus when you speak to even a small group—actually speak, rather than engage in a general discussion—you have to some extent a captive audience. Captive, that is, to the extent that in most cases they won't stand up and walk out—but there is no way that you can hold their minds captive. You can not force anyone to listen who doesn't want to; you must persuade people to listen willingly by the skill with which you prepare your material.

Audience Adaptation

The first step in preparing your material does not deal with the material at all, but with the audience. You need to analyze the audience to determine the best method of presentation. There have been many methods and checklists suggested to aid in analyzing an audience, but the best method is sitting down with a paper and pencil and jotting down notes about the composition of the group that will be listening to you: What brings them together? How do they feel about the subject? How much do they already know? In addition, you need to decide what the purpose of your speech is: Is it, for example, to inform the audience of action already taken, to persuade them to approve action you wish to take, or to persuade them to engage in action themselves?

For instance, if you plan to speak before a civic group about the need for an improved sewage system to reduce pollution in the local river or lake system, there are many factors about the audience that you need to consider. Are they generally in favor of pollution controls, or do they feel that environmentalists are impractical idealists? If the latter, then your task would be to organize your material to show how an improved sewage system is in fact practical. On the other hand, the audience may well feel that a better system would be desirable, but too expensive for the town. Your task would then include showing how the system could be financed without raising taxes, or that it was more important to have a new system than to keep taxes low. The organization and emphasis of your presentation would reflect these considerations.

Another factor that would cause your presentation to differ from audience to audience would be the extent to which your audience is already informed on the subject. How much background information must you include before you can discuss the issue at hand? What terms must you define? In other words, it is not enough simply to decide that you understand your audience; your presentation must reflect that understanding if you are to be an effective speaker.

Arrangement of Material

The preceding considerations should, of course, affect the arrangement of your material. Before more is said on the subject, however, one point should be made clear: Nothing that follows is intended to advise that the demands of logic or continuity be ignored. Overall, the method for organizing information to be given orally should follow the suggestions on organization in Chapter 3. The suggestions that follow should be used after the demands of logic are met. For instance, the logic of a subject may require the consideration of a number of points, but the order of presentation may well be arbitrary. Thus you are free to plan your delivery to capture your listeners' attention so that they will more easily assimilate your information.

Listeners are likely to be fully attentive to what a speaker says at the beginning of his talk. Later, unless he is unusually effective or the subject is one of absorbing interest, their concentration becomes less intense. For this reason the skillful speaker provides variety from time to time to revive interest. As the listeners sense that the speaker is nearing the end, they once again become more attentive. It is sound strategy, therefore, to utilize these natural tendencies as you arrange your material.

At first it might seem obvious to arrange your material in the order of decreasing importance, but this method creates the same effect in speaking as it does in writing. That is, if the material becomes less and less interesting or significant as time goes on, people are not only naturally inclined to become less attentive, but justified in doing so. Consequently, the following pattern is one method used to overcome the difficulties of oral presentation. It should be remembered, however, that other methods may be more appropriate to a specific situation, and that this method should be followed only to the extent that it is compatible with logical, coherent treatment.

The first step is orientation: You need to let the audience know what to expect, and how you plan to present your information. Once they have a good idea of your plan of organization, they can fit the information better, as they receive it, into the overall scheme and can better judge its relationship to the problem or situation you are talking about. In other words, prepare your audience: Tell them what you are going to say and how you plan to say it.

Then, as you get into the body of your speech, take up some part of your material that will make a strong impression. When this has been covered, turn to something less vital and build from it with stronger or more interesting parts. Such a pattern will help capture the listener's attention by the strength of the first material, and then retain it because of the increasing interest or importance of the remaining information.

At the end, summarize the material to leave the listeners thinking of the entire picture rather than the last detail. In other words, tell the audience what you have said. Such an arrangement utilizes the natural tendency of listeners to pay closest attention at the beginning and also makes use of the fact that people remember best what they have heard last.

Preparing Notes

People vary so much in how natural they feel when they use notes that it would be undesirable to urge a specific manner to prepare them, just as it would be undesirable to insist on a single method of organization. As a person gains experience in public speaking he discovers what kinds of notes he prefers and prepares them accordingly. The following suggestions are intended for students who have not yet discovered their individual preferences.

1. How detailed and extensive your notes should be depends, of course, on how familiar you are with your subject. It is possible to have too many notes. When this happens, you may keep looking at them so often that you seem to be speaking to your notes rather than to the people present. Or perhaps, not needing them all, you may talk so long without looking at them that you will have difficulty finding your place when you need to refer to them. It is not the function of notes to encompass all the material to be presented. Rather, they should serve to remind you of the organization of your material. Another function of notes is to provide exact quotations or figures where these are required. No audience expects a speaker to have quotations or figures memorized, and will in fact have greater confidence in a speaker if he refers to notes at such times.

   In this connection, there is no reason your notes need be equally detailed for all parts of your presentation. Where you are especially familiar with your material, they may be sketchy. Where you are less familiar, they should be more detailed.

2. If time permits, you may find it helpful to make notes in considerable detail so that you will really think your speech through, and later make a new and shorter set for actual use—a set that has been cut down to what you find you really need as you run through the others. It is inadvisable, however, to attempt to use new notes unless you have time to become thoroughly familiar with them.

3. While an outline is always useful for organizing information, it may

not be the best form of notes for a speech. Using an outline to speak from sometimes produces a mechanical-sounding result—making it seem as if you are running down the points of an outline rather than thinking about what you are saying, particularly if you simply follow the points in the outline rather than use them as notes to jog your memory for more extensive discussion. All speeches are more effective if the remarks seem spontaneous rather than coming from a piece of paper. Many speakers find that they get the best results from notes consisting of key words and phrases which keep the ideas in the right sequence and launch the topic without further assistance. An example of such key words and phrases would be:

Sources of pollution: *Mines   Farms   Homes   Boats*

These few words might be enough to cover what would need to be said about the pollution of a lake—the facts about each source being so familiar that details would be unnecessary.

4. Although obviously your notes should be legible, don't worry about their appearance. Irregularities, once you are familiar with them, may serve as landmarks. The unique appearance of each one may help you find your place faster than you could find it by looking for typed or longhand words that look much the same as other words on the page. But by all means be thoroughly familiar with the physical appearance of your notes as well as with their content, regardless of what their distinctive features may be.

5. It is harder to judge in advance how long you will need to cover the material in notes than to read from a previously written text. Most people need to feel more concern about talking too long rather than talking too briefly. You can provide for this possibility by marking your notes to show what may be omitted if time runs short. For instance, suppose you asserted that fire was not being guarded against adequately, and were supporting your assertion by citing three examples. You might include all three in your notes but put two in parentheses or otherwise mark them to indicate that they could be omitted or only referred to briefly if necessary. Remember, too many simplistic examples bore an audience, but important detail should never be omitted.

6. In addition to notes that you definitely will use, supplementary notes, to be used only if needed, are frequently helpful. For instance, you might include additional statistical data in supplementary notes. Having them available would enable you to answer possible questions without cluttering your regular notes with figures you might not need.

LANGUAGE

Assuming that you have satisfactory notes, you still need to think about putting your facts and ideas into words. It would be a glaring omission if a discussion on this point failed to take up the subject of "standard" English, or "correct" English, as it is sometimes called.

Standard English and Dialects

Actually, except for some college professors, newscasters, and an educated minority, few people use standard American English as a matter of course. Most of us speak some variety of the three major dialects of American English in use in the United States. The three dialects are usually referred to as Northern, Midland, and Southern—terms which correspond somewhat to their original distribution. The dialects derive in part from the region of origin of the settlers in the area, in part from the period of settlement, and in part from naturally differing development of spoken language. For instance, settlers coming from England at an earlier period spoke a different dialect of English than those who came later. Thus, the dialect spoken in New England in some respects represents a development of the language from an older period than does the dialect spoken in Pennsylvania.

These dialects differ in grammar, vocabulary, and of course pronunciation. Thus in Southern, the construction "hadn't ought to" is not only common but acceptable, while in Northern the same thought is expressed by "ought not to." In vocabulary, although few Northern speakers would be confused by the Southern "bucket," they would tend to use "pail"; but not many Northern speakers would recognize that the Southern "pulley bone" is the Northern "wishbone."

While this is an immensely oversimplified discussion of American dialects, it does suggest the problems a person might have if he is speaking outside his own dialect area, or if he speaks a minor variety of one of the major dialects. In addition, on more or less formal occasions such as a speech, your audience will probably expect you to use some level of standard English, which is largely Northern and Midland in composition, and will view any dialect as funny, peculiar, or incorrect.

If you are unaware of the difference between your own speech and standard English, you may well have difficulties in communicating with your audience, for they may be distracted from your message by your language and miss the points you wish to make. Some people, of course, use reasonably standard English as a matter of habit because they have been exposed to it all their lives, have picked it up through extensive reading, or have had intelligent instruction in the classroom. If you are among their number, you are fortunate.

If you are not, however, you should face the fact that most of those with whom you will wish to communicate in your career know standard English reasonably well and will notice deviations that might pass unnoticed in your previous surroundings. Consequently, the sensible thing to do is become concerned about your speech and try whenever you speak to use the language and terminology you will encounter in your professional field. It will not be enough to be able to avoid deviations when you concentrate on doing so. When you talk to people about technical topics, you will need to concentrate on the substance of what you are saying without worrying about your English.

What kinds of deviations do you need to concentrate on? It would be unrealistic to maintain that everything that has ever been labeled "error" must be regarded as equally serious, especially in speech rather than in writing. Some deviations, in fact, are not very serious, and the distinctions are being dropped from American dictionaries. Many a literate person, in speaking, uses *like* when standard English calls for *as, who* at the beginning of a sentence when traditional grammar calls for *whom*, or "It was me" instead of "It was I."

At the other extreme are deviations so well known that you may be avoiding them already: *ain't*, for example, most double negatives, and changes in case such as "My colleague and *me* agree. . . ." The deviations that persist are those less well known. Typical examples are *come* for *came* in the past tense, "We *was*" for "We *were*," or *good* for *well* in adverbial constructions as in "The motor runs *good*." These deviations, if you use them, may cause some people to regard you as illiterate.

The above discussion should not be read as permissive: Deviations from standard English in either speaking or writing can gravely hinder a career. In addition, there are real errors in grammar and usage that do not arise from dialect differences. For instance, "It makes three days that I am here" is not correct usage in any American dialect. Correct language is largely a matter of convention; that is, a group of people have agreed, through custom, that certain combinations of words will mean certain things. Other combinations are confusing. If a technician is confused about a set of directions, or if an audience does not clearly understand what is said, action may be postponed, or perhaps worse, the wrong action may be taken. The Handbook sections *Correctness in Usage* and *Glossary of Usage* will help you identify the kind of language that calls for attention.

Formal versus Informal Style

Avoiding conspicuous deviations does not mean becoming pompous or excessively formal. Just how informal you should be depends largely on your relationship with those to whom you are speaking—your position, perhaps, in relation to theirs. It might be advisable to be less informal in addressing superiors than in speaking to subordinates, but a degree of formality that seems natural in writing may sound stiff in speech. For example, one might write, ". . . people with whom I talked," but it would be both natural and perfectly acceptable in speaking to say ". . . people I talked to." Or again, contractions would be natural and unobjectionable used orally, though it would be preferable to avoid them on some occasions in writing.

In general, unless you are frankly reading from a prepared text rather than putting your thoughts into words as you speak, a moderate degree of informality, if compatible with your personality, is well advised.

Personal versus Impersonal Style

When your language is informal, it is likely at the same time to become more personal. This is all to the good. It is entirely appropriate that speech should be less impersonal than writing. When a person is reading what you have written, he is in direct contact only with the printed page and is not necessarily thinking of the writer; but when you are speaking, what he hears comes directly from a person, and it seems only natural that what comes directly from a person should be expressed in an honest or candid personal manner.

Variety

Finally, there is no reason that during a talk your style need be the same at all times. Even when for the most part you use moderately formal language, an occasional injection of informal personal style may be not only unobjectionable but also highly effective. In fact, such a conversational style gives the listener a feeling of involvement similar to what he feels when you talk directly to him in a personal conversation. He feels that you are speaking to him instead of at him, and he is much more inclined to listen. A good opportunity for such a change is the time when you have finished one point and are about to take up another. A shift to informal, conversational style during the transition not only conveys the impression that you are relaxed but also gives your listeners a chance to relax. In written material, such a device not only would seem out of place, but it would serve no purpose because a reader can pause and relax for a moment whenever he desires, rather than depending on you, as a listener does, to provide the opportunity.

DELIVERY

Needless to say, no amount of advice about delivery can approach the value of coaching by a competent instructor; and even if you can obtain such coaching, you will derive its full benefit only as time passes and you gain experience. After all, you have been talking a long time and can hardly expect to change your abilities overnight. Still, you do not need to feel discouraged about improving your skills.

In discussing this subject we will first assume that the occasion is one on which you are speaking to a small group—eight or ten to perhaps fifteen persons. Later we will add a few ideas that would apply mainly when you speak to a larger group.

General Suggestions

1. Consult your dictionary in advance to check the pronunciation of any word you are likely to use and feel the slightest uncertainty about.

2. Talk fast enough that your remarks don't drag—which may be hard if you customarily talk unusually slowly—but not so fast that you find it difficult to enunciate clearly. If you speak clearly, don't worry about your audience's ability to comprehend the words themselves. Experts in communication estimate that the average individual can distinguish aurally 400 to 800 information bits a minute—much more than you can speak even at your fastest If you talk too slowly, you give the listener opportunity to let his thoughts wander, and once you've lost him, you have difficulty getting him back because he is no longer oriented to your information. Nevertheless, you should vary your rate of speaking from time to time. When an idea is harder than average to grasp, you should consider repeating the information at some point in your speech rather than slowing down and risking losing the audience. When a statement is especially significant, consider pausing a moment at its end to give the listener time to think about it.

3. Enunciate distinctly. Avoid dialect pronunciations such as *gonna* for *going to*, *dunno* for *don't know*, *yuh* for *you*, and *em* for *them*. When a word ends in *ing*, don't pronounce it as if it ends in *n*. On the other hand, don't go to the opposite extreme and pronounce words like *cupboard* as if it were *cup board* or *often* with a *t*. With this in mind, note that no objection is raised on the grounds of pronunciation to recognized contractions such as *can't* and *doesn't*, which are pronounced as they are spelled.

4. Speak loud enough that no one must exert a special effort to hear and understand you, but not so loud that you seem to be shouting and sound like a television with the volume too high. If some members of the audience seem to be talking among themselves and making a distracting noise, a trick many seasoned speakers try is talking *softer*, rather than louder, to force the listeners to concentrate on listening. Save the extra volume for moments when some point calls for extra emphasis.

5. Avoid a monotone, which as the word implies is a monotonous tone which fails to accentuate words that need to stand out. In varying pitch, however, do not vary it in the same manner in every sentence— which is just as monotonous as no variation whatsoever.

6. By varying speed, volume, and pitch as recommended above, stress words that call for special emphasis, either for their own importance

or for clarity. For example, compare the meanings between "Is *that* your only objection?" and "Is that your *only* objection?" The value of proper stress will become apparent if you listen to effective speakers, including the leading television commentators.

7. Keep most of your sentences relatively short and simple. A listener, unlike a reader, can not look back at what came earlier if he loses the gist of what you are saying.

8. Don't indulge in oratorical flourishes like those too often heard in political speeches. Your language should sound natural—especially when you are speaking to small groups. For example, "in the United States," or "in this country," sounds less affected than, "in this great democracy where we are fortunate enough to reside." Follow the advice of the critic who said that a major cause of bad writing (and bad speaking) is the writer's striving to be considered a man of genius rather than being satisfied to be regarded as a man of sense.

Transitions

Transitions call for a little more comment than the items in the brief recommendations above. They need a somewhat different treatment in speech than in writing—especially writing where headings are used. The presence of a heading often minimizes or eliminates the need for a transitional sentence or paragraph. A reader who sees a heading knows that you have finished one point and are about to take up another. When you are speaking, however, it is often necessary to use a sentence or two to introduce your new subject or point. If you have already told the audience in your introduction the major aspects of your talk, it is much easier to introduce a new point by reminding the listeners of that fact. Moreover, you can sometimes dramatize a transition by physical movement: If you are standing, by a change in position; if you are sitting, by an action such as setting aside a sheet or note card and picking up another. You must be careful, however, that such attempts at dramatization do not distract from what you are saying. It does no good to emphasize a transition if your audience is going to remain thinking about the transition device rather than the subject to come.

Visual Aids

Visual aids to a talk come in many shapes and sizes. There are slide projectors, movies, drawings, tape recordings (which aren't visual, of course, but are frequently used to aid a speech and so are mentioned here), objects of any kind—the list could go on and on. The primary point to remember, however, is they are *aids,* not substitutions for your talk. Visual aids should be used whenever you need to support or clarify a point with further evidence that can be shown more easily than told, and in talks about technical material, this is often the case. Likely aids include tables, bar charts, and graphs. Before preparing any, you should review the chapter on *Illustrations,* for many of the same rules apply. However, when illustrations are used in connection with a speech rather than with writing, further factors need to be considered.

Any visual materials must be large enough and clear enough for everyone present to see and understand easily. Visual aids should not be so numerous that your talk seems to be merely comments on exhibits. They should also be somewhat simpler than you might use in writing, since the audience can not study them as long as every member might wish.

Even though an exhibit may not be complicated, you should point out its salient features. This is advisable for two reasons: First, those present can discover what is significant to your talk faster with your assistance, and second, you lose contact with them if their attention to the illustration is not integrated with attention to your remarks. With this in mind, it is easy to see why you should not introduce a visual aid until you wish your listeners to consider it, and why, if possible, you should withdraw the visual aid from view after you have completed discussing the information it illustrates.

The factor of audience attention is also the reason why handouts should be avoided during a speech, despite the frequency of their use in technical talks. Few people can listen well while passing papers around, and fewer people can resist reading the handout instead of listening. The best solution is to avoid handouts altogether, but if they are necessary, pass them out after the speech. If you plan to use them as illustrations during the speech, label them clearly and pass them out prior to beginning. Then you can call attention to the proper handout at the proper time in your talk. Unfortunately, however, you will not be able to regain attention easily after you have discussed the handout.

SPEAKING TO A LARGER AUDIENCE

Everything that has been said about speaking to a small group also applies when the number of listeners is larger, but when it is larger, some additional suggestions may be helpful. Some of them, to be sure, may well be borne in mind even when the number listening is small, but are taken up in connection with the larger groups because that is when they become especially important. Others are of concern only when you speak in front of enough people to be regarded as an actual *audience.*

The Opening Moments

For your own benefit, when you talk to a sizeable audience, you should take a few moments to adjust to the situation. That is, you need to establish eye contact by looking over the audience, thus causing them to look at you, to form an impression about how loud you must speak in view of the size of the room and the number present, or to position yourself to use the microphone comfortably if one is provided.

For the most part, however, what you need to do is based on the needs of the audience. For the first few moments people are pre-occupied, at least subconsciously, in forming an impression about *you* —in reacting to your appearance and manner, and in getting used to your voice and way of speaking. Consequently, you should not be in a hurry to plunge into the substance of your talk. If you notice what is done by effective speakers, you will find that many begin by making a few preliminary remarks that do not constitute the real introduction. If you do this, the remarks should appear spontaneous rather than pre-pared. You might express your appreciation of the opportunity to speak. You can thank the person who introduced you. You can compliment the preceding speaker if there was one. You can talk for a moment in a general way about the occasion for the meeting and the importance of whatever it is concerned with. You must be careful, however, not to trivialize your own talk by attempting to be humorous if the event does not warrant it, or by calling attention to your lack of preparedness in a false show of humility, or by any other mannerism that will distract from your message. Remember, the audience came to hear what you had to say, and most of them want you to get on with it.

Speed

It is even more important when the audience is large to avoid talking so fast you fail to enunciate clearly. If you naturally talk slowly this caution is probably unnecessary, but if words come easily to you, don't let yourself, as you pick up speed, indulge in rapid-fire delivery and thus become hard to understand. If a formula for an ideal speed could be devised, it would include several ingredients, instinctively blended in proportions suited to the occasion. Among them would be what seems natural to you, the size of the room and the audience, the quality of your voice and the distinctness of your articulation, and the nature of your material—whether at any moment it is easy or hard to grasp. And during any talk an occasional change of pace is desirable to avoid monotony.

Articulation

Clear articulation, mentioned earlier as desirable even when only a few are present, becomes immensely important when your audience is larger. Whatever problems in understanding that might exist in a small group are intensified as the size of the audience increases. With more people in a room the possibilities for distractions are more numerous, the general noise level is higher, and the distance between the speaker and much of the audience is greater. Every speaker is aware that the hardest part of an audience to hold is that part in the back of the hall. With greater distance, the possibility for eye contact becomes less, and therefore the relationship between speaker and listener becomes more impersonal. In addition, if a listener can not easily make out the expressions on a speaker's face, he has more difficulty in understanding the intent of the speaker's words. If poor or sloppy articulation is added to noise, distractions, and distance, the problems for the listeners may make continued close attention more trouble than they feel it is worth. You can do little about the size of the audience, but a careful attention to articulation can minimize its effects.

Use of a Prepared Text

The nature of technical information is such that often the only way to be certain of the accuracy of what you say is to read it. When you prepare a text with this in mind, it is doubly important to remember the points made in this chapter. A written text is of necessity different from a spoken text, but if you plan on writing out what you say, it is easy to forget this fact and write as if it were to be read, not heard. The best way to make sure you have allowed for your audience's listening problems is to read your text aloud several times. If possible, read it to a friend or colleague, just as you would show him a draft of an article or report. If you cannot, read it aloud by yourself. Your own ear will tell you much about the presentation.

For instance, you will discover places where what you have written is unsuitable for oral presentation, although it would have been fine on the printed page. Sometimes you will find problems in making pronunciation distinct. Consider for example Pope's lines

> When Ajax strives some rock's vast weight to throw
> The words, too, labor and the line moves slow.

Pope wrote the lines to make a point about poetic style, and no adverse criticism is intended. They are quoted here to show how hard it is to pronounce some combinations of sound. You may also become aware, when you read your text aloud, that some word has been repeated so many times it becomes unpleasant, that you have accidentally written distracting rhymes, or that some sentences are so long and involved that it will be difficult for the audience to follow them easily, even though the sentences may have been easy to follow when read.

In addition, reading the text aloud will help you to become thoroughly familiar with it, so that you may look up more frequently and appear to be talking to the audience rather than to the lectern. Professional readers, to accomplish this, do not read at all, but deliver lines, apparently read, from memory. Charles Laughton told of a time he was supposedly reading Dickens when he glanced down at the page to further the illusion and his eye caught the actual words he was then supposed to be reading. They so surprised him that he forgot what he was saying, and was forced to actually read for a while until he regained his composure and could remember the lines again.

If your talk is to be printed after you have delivered it, you should edit your text before you turn it over to the printer. When you prepared it you had in mind, or should have had in mind, using inflection, vocal emphasis, and the many other resources available to the speaker for making his meaning clear. The same text might not be clear if the words alone had to be depended on.

There are times, however, when such editing cannot be considered. When what you are reading is a prepared statement or the report of a committee, written for the express purpose of becoming a matter of record, changing it after you read it, except for minor emendations, could not be justified. The general principle is: Sometimes what you read orally will be exclusively your own, written at least ostensibly to be read before an audience although it may also be published later. In this case, you should feel free to exercise your judgment in adapting it to the nature of the new medium. On other occasions what you are reading will have been prepared primarily to become part of the written record —in which case the oral version and the written should not differ.

Oratorical Style

Nothing said here is intended to recommend against being emphatic or, within reasonable limits, enthusiastic. But unless you are an accomplished speaker, you would be ill-advised to try to be oratorical— especially in any talk that a book on technical writing should concern itself with. Remember, science convinces not by rhetoric, but by content. Be sparing in the use of dramatic effects. Exercise restraint if you feel tempted to become emotional—even though your own emotions may be aroused by your message. You are more likely to be effective in a technical talk by making your facts pertinent and your ideas clear.

ATTITUDE

Perhaps most important of all is your attitude as a speaker. Any audience, of whatever size, will react quicker to your attitude than to anything specific you have to say. And your attitude is almost impossible to hide or disguise. It will show in your choice of words, in your posture, your gestures, your tone of voice. If you feel superior to your audience, or arrogant, they will respond by rejecting both you and your message. If you are bored by your material, your audience will be bored. If you show fear or lack of confidence, your audience may well interpret this as doubt about your material, and your message may lose credibility.

The subject of fear, or "stage fright," is one that every speaker should face. At some point, everyone has felt some degree of fear at facing a gathering of people. Accomplished speakers avoid fear in two ways. First, they learn their material so well they have total confidence in their ability to remember all the pertinent information. Their knowledge is gained by practice, by reviewing the material so many times they have no doubts about it. Second, accomplished speakers concentrate on the needs of the audience. They have no doubts about the value of the information to the audience. The confidence they gain also helps them avoid the distracting mannerisms many people have when they speak, such as putting their hands to their faces, slouching in an unnatural way, or swinging a leg back and forth.

The most important factor to remember in any speaking situation is your audience. Without an audience, you would have no speech. In a very real sense, the people in the audience have honored you by coming to hear what you have to tell them. Although they may appear to you to be a mass of people, they listen to you as individuals, and they will react to you as individuals. You are not, finally, speaking to a mass, but to John, and Jack, and Jane, and Mary. You owe them the respect to present your information honestly, in a clear fashion, so that they will feel they have spent their time well. If you do this, even though your delivery is not perfect, they will listen to what you have to say, for that is why they came.

ASSIGNMENTS

Assignment 1

Present orally to your class, in a talk of a length specified by your instructor, the information you would include in a formal proposal. The goal of your presentation would be to convince the class members to give you permission to proceed with the project you are proposing. You may find subjects in the list of assignments following the chapters on reports, or you may propose a subject of your own choosing.

Assignment 2

After you have completed a written report, present it orally to the class. Although the length must in part depend on the nature of the material and the length of the original, attempt to keep within the time limits specified by your instructor.

Assignment 3

Take an illustration that you have devised for a written report and redo it for presentation as a visual aid. Unless you plan to actually use it with a talk, you need not make it the actual size you would need, but you should indicate the dimensions that would be necessary for the size of the audience you have in mind.

PART III

BUSINESS CORRESPONDENCE

14

GENERAL PRINCIPLES

OF BUSINESS

CORRESPONDENCE

If we regard the term *technical writing* as broad enough to include whatever writing you will be responsible for doing on the job, there can be no doubt that letter writing is an important part of the subject. Anyone who cannot write effective letters will be seriously handicapped in his career.

Study of letter writing is especially desirable because of the difference between letters and most of the other material written by those trained in technical professions. In many professions a person will spend much of his time thinking not about people but about things—about steel, or oil, or concrete; about transistors, or chemicals, or geological formations; about smog, or insecticides, or machinery. And when much of his writing deals impersonally and objectively with such subjects, he can profit by a reminder that there are times when writing is done for the sake of personal communication, and when different techniques are therefore desirable. A study of letter writing constitutes such a reminder.

To make it easier to visualize the finished product, attention is first centered on letter form. Next, the substance of letters in general is discussed. In following chapters, specific types of letters are considered, in order to illustrate how the general principles of letter writing work out in a variety of characteristic situations.

CORRECT FORM IN BUSINESS LETTERS

This discussion of letter form is not exhaustive, for covering every possible question of form would occupy too much space and would make it harder to identify and remember the points that are essential. If you need additional information on some points, you may find it in a special section in the best collegiate dictionaries. The information that follows, however, will answer most questions that arise in ordinary correspondence.

Identification of Parts of the Letter

The parts that appear in *every* letter are (1) the heading, (2) the address, (3) the salutation, (4) the body, (5) the complimentary close, and (6) the signature. The heading is the section that tells the address of the writer and the date; the address is the section telling whom the letter is addressed to; the salutation is the "Dear Sir" or its equivalent; the complimentary close is the "Very truly yours" or its equivalent; and the signature includes the longhand signature of the writer plus the typed material that accompanies the longhand signature. These and certain other parts that appear with varying frequency will be discussed later in more detail.

Miscellaneous Mechanical Details

Stationery. The stationery should be good, unruled bond paper, preferably white, 8½ by 11 inches large. If the first page of a two-page letter bears a printed letterhead, the second should match it in quality and color but should be blank. Smaller stationery is sometimes used for extremely short letters, but paper of standard size is usually preferred even for short letters because of the greater convenience for filing.

Placement of the Letter on the Page. To place a letter on a page so that the margins are well balanced demands either long experience or careful planning. The result aimed at should be approximately as follows: (1) there should be more white space at the bottom than at the top unless the length of the letter and the size of the letterhead makes this impossible; (2) the side margins should be approximately equal and should ordinarily be no larger than the bottom margin—often being smaller; (3) regardless of the length of the letter, the body should fall partly above and partly below the center of the page. A short letter will look better if slightly more than half the body is above the center of the page.

In spacing a letter it is helpful to remember that there are six lines of type to the inch, and either ten or twelve letters to the inch (usually twelve) in the line. Most writing averages about six letters to the word, including space between words. By utilizing these facts you can reduce the uncertainty about the space a letter will occupy if you must do your own typing.

When a letter is written on a blank sheet, the minimum margins all around are 1 inch—preferably 1½ inches at the top and bottom. The maximum margin at the top is about 2½ inches, and at the sides is 2 inches. The space between the heading and the inside address may vary from a double space in the longest letters to six spaces in the shortest. In the shortest letters, the bottom margin is obviously controlled by the length of the letter.

When the stationery bears a printed letterhead, the point to consider when setting the top margin is the first line of the inside address. *In the shortest letters* this may be as much as 3½ inches (21 lines) below the top of the page. This would normally leave room for six or seven lines of the body of the letter above the center of the page. The side margins would, of course, be the maximum for such a short letter. *For the longest letters,* the inside address may begin only a double space below the date line, which would itself be only a double space below the bottom of the printed letterhead. For shorter letters, the date line would be placed either (1) two spaces below the printed letterhead or (2) halfway between the letterhead and the inside address.

Spacing Within the Letter. The best procedure, normally, is to single-space within the paragraphs of all business letters and to double-space between paragraphs. The space between the heading and the inside address varies. There should be a double space, however, between any other two parts of the letter, namely (1) between the inside address and the salutation, (2) between the salutation and the body, (3) between the body and the complimentary close, and (4) between the complimentary close and the signature if the first line of the signature is typed. More will be said later about spacing the signature.

Systems of Indentation and Punctuation

The heading, if a letter is on blank stationery, is in block style. (That is all lines begin the same distance from the left side of the sheet.) It may be placed so that all of its lines begin just to the right of the center of the page, or so that its longest line extends to the right-hand margin. The inside address is also blocked, with all lines beginning at the left margin. The salutation begins at the left margin. The beginning of each paragraph is indented either five or ten spaces. Terminology varies, but the form described is usually called the semiblock system.

Education Commission of the States
300 LINCOLN TOWER • 1860 LINCOLN STREET
DENVER, COLORADO 80203 • (303) 893-5200

April 11, 1974

Mr. Kenneth Plate
Counselor
Belpre Middle School
200 Rockland Avenue
Belpre, Ohio 45714

Dear Mr. Plate:

Thank you very much for your recent letter to Dr. Hazlett and for your
interest in National Assessment. Dr. Hazlett has asked me to reply to your
several questions.

As you probably know, NAEP assesses the knowledge, skills and attitudes of
young people in two learning areas each year and reports its findings on an
item-by-item basis and on several variables, one of which is geographic
location. Thus, we do have data on 9, 13, 17 and 26 to 35 year olds for the
central region of the nation (which includes Ohio); but these data cannot be
regarded as "norms." Rather, they are statistics showing performance on a
particular skill, piece of information, behavior, or attitude and can best be
interpreted by the reader as to whether they are good or poor.

I am enclosing a NAEP publication, which I think will be of real help in
understanding NAEP's method of procedure, entitled, "What Is National
Assessment?" and a copy of our "Publications List," which shows and describes
all of our reports, booklets, monographs, and similar publications. Possibly
after reading these, you will think of additional materials which would be of
help. In that case, feel free to write me again with your request.

Best wishes for the success of your reevaluation, and let me know if we can
be of further help.

NAEP

Sincerely,

Ken Seaman

Kenneth W. Seaman, Director
Utilization/Applications

KWS/lm

Encls: "What Is National Assessment?" Publications List

cc: Dr. James Hazlett

NATIONAL ASSESSMENT OF EDUCATIONAL PROGRESS

FIGURE 15. Letter of Approximately Average Length, Well Placed on
Letterhead. It shows use of a full block form.

741 Kearney Street
Colter Bay, Wyoming 83001
March 8, 197-

Mr. John Walters, Chairman
Springfield Environmental Council
2414 Walnut Boulevard
Springfield, Colorado 81073

Dear Mr. Walters:

Thank you for your letter of March 2 inviting me to participate as a panel member at the April 25 meeting sponsored by
the Springfield Environmental Council.

Much as I would like to accept this invitation I am unable to
do so. As you may know, I retired at the end of January, so
I am now free to do some travel that my work schedule formerly
prevented. Consequently, I am planning to leave Colter Bay
for a six week trip early in April.

Otherwise, I should enjoy attending your meeting, for I admire
your work and am sure that the meeting will be interesting.

Sincerely yours,

Arthur M. Perkins

Arthur M. Perkins

FIGURE 16. Extremely Short Letter, Correct in Form and Well Placed
on a Blank Sheet.

In the full block system, which some companies prefer, not only the paragraphs but also the date line (which is the only typed material in the heading of a letter on a letterhead), the inside address, the complimentary close, and the signature begin at the left margin. However, the most frequent deviation from the semiblock system consists of not indenting for paragraphs. This deviation is entirely acceptable if the letter is single-spaced but not if it is double-spaced.

The modern system of punctuating the heading and inside address is the open system. That is, no punctuation is used at the end of any line unless, of course, it is necessary to use a period after an abbreviation.

Figures 15 and 16 show business letters in correct form, well placed on the page.

Form for Individual Parts of the Letter

The Heading. If the stationery does not bear a printed letterhead, the heading contains from two to four lines, usually three. The first line, plus the second if necessary, tells the local address from which the letter was written; the next line tells the town and state; and the last line tells the date. The name of the writer does not appear in a typewritten heading. The heading is placed in the upper right-hand portion of the page. It may be placed so that its longest line ends at the right margin of the page, or it may begin slightly to the right of the center of the page, regardless of where this causes the lines to end. The top line of the heading is typed 1 inch to 2½ inches from the top of the page, the exact distance depending on the length of the letter.

If the letter is written on stationery with a printed letterhead, the typed heading consists of the date line only. This may be placed halfway between the bottom of the letterhead and the inside address, far enough to the right so that it extends to the right-hand margin, or it may be centered and placed two spaces below the letterhead. A printed letterhead is often designed so as to provide a natural space for the date line.

Examples of Typewritten Headings

```
1316 Twenty-seventh Street      The Associated Engineers
Alameda, California 94501       University of Idaho
May 15, 197-                    Moscow, Idaho 83843
                                February 20, 197-
```

Abbreviation, capitalization, and use of numbers are illustrated in the preceding headings; but since these are matters that apply to other parts of the letter also, they will be discussed as the final point under *Form.*

The Inside Address.   The examples of inside addresses given later illustrate usage in the most common cases. As indicated, considerable flexibility is necessary to prevent some of the lines reaching an unwieldy length or to prevent the number of lines from becoming so great as to be awkward. Usually, however, three or four lines of reasonable length are sufficient.

When a letter is addressed to an individual, some title or the equivalent must always be used. To omit it is discourteous. The title *Mr.* is of course the most common, but it may be replaced by *Doctor, Captain, Professor, Rabbi,* or any other term one might use before the person's name if one were actually speaking to him. The terms *The Honorable* and *The Reverend* are sometimes used in place of titles. *Honorable* is appropriate when one addresses an important governmental official. *Reverend* is appropriate when one addresses a Protestant or Catholic clergyman. (Full coverage of all the possible forms appropriate for all clergy may be found in many books.) It is also correct to indicate the reader's degree, for example, Ph.D. or M.D., after the name rather than placing a title before the name. There should be no duplication, however; *Dr. Ralph Morgan, M.D.* would be incorrect.

If your letter is addressed to a woman you may be uncertain about whether to call her *Miss* or *Mrs.*° The following suggestions may be helpful:

1. If you have received a letter from the woman you are addressing, look at her signature. If you see nothing to indicate that *Mrs.* is called for, address her as *Miss*. (If *Mrs.* were called for, she would presumably have indicated that fact unless she assumed that you knew it already.)

2. If you are addressing a married woman or a widow, the correct form would be *Mrs. John R. Brown* rather than *Mrs. Mary Brown,* unless she herself uses the latter form in business.

3. If you are addressing a divorcee, the correct form is *Mrs. Mary Brown.* (This suggestion is based on the assumption that *Brown* was her married name. If she has resumed her maiden name, she is addressed as *Miss Mary Brown,* just as if she had never been married.)

4. The use of some other title (*Professor* or *Dr.*, for example) rather than *Miss* or *Mrs.* is entirely appropriate.

The Salutation.   When a letter is addressed to an organization, the salutation should be *Gentlemen:* (or *Ladies:* if the organization is composed of women). *Dear Sirs* is out of date, except in a letter addressed to

---

° Another possibility is *Ms.*, but unless and until it becomes established as firmly as *Miss* and *Mrs.*, it cannot be recommended as a solution to the problem unless one knows that it appeals to the taste of the person addressed.

more than one individual. Though some companies omit all punctuation after the salutation, the vast majority use a colon.

For letters to individuals, note the following salutations, arranged in the order of decreasing formality:

```
Sir:  (too formal for most letters)
My dear Sir:  (rather ceremonious)
Dear Sir:  (very common)

My dear Mr. Jones:  (often appropriate)
Dear Mr. Jones:  (most widely used of all forms)
```

It is often appropriate to replace *Mr.* by some other title, such as *Doctor, Professor, Captain, Governor,* or *Mayor.* It is more complimentary to refrain from using any abbreviation for a title, except *Mr., Mrs.* and *Dr.* The terms *Honorable* and *Reverend* are not titles, and though they sometimes replace titles in the inside address, they should not be used in the salutation. Many other terms, such as *Superintendent,* are sometimes erroneously used as titles. When there is doubt whether some particular term should be used as a title, it is advisable to refrain from using it.

As for the choice between the reader's name or *Dear Sir,* it used to be customary to use the name only when one knew the reader or at least had received a letter from him. In recent years this convention has been widely disregarded, for business letters are becoming less formal and more personal. To use *Dear Sir* may make a letter sound cold and remote. The modern trend is to use the reader's name unless his high rank or the occasion of the letter would cause informality to be in bad taste. The same situation holds true in a letter addressed to a woman. If her name is not used, the salutation is *Dear Madam.* There is a strong tendency, however, to avoid *Dear Madam* and use *Dear Mrs.* (or *Miss*) *Jones,* or *My dear Mrs. Jones.* In the last analysis, your own sense of appropriateness must determine your choice of salutations. There is no reason not to use the first name of a reader if you would use it in speaking to him personally and if there is no special reason informality would be unsuitable to the occasion.

The salutation must harmonize with the inside address. If a letter is addressed to an individual, the salutation must be to an individual and the title must not be different. If the letter is addressed to an organization, the salutation must be to an organization. This holds true even when an "attention line" comes between the inside address and the salutation.

To show how inside addresses and salutations can be handled on many occasions, the following examples are provided:

Mr. David M. Jones, President
Western Testing Laboratory
Olympic Building
South Bend, Iowa 57087

Dear Mr. Jones:

Western Testing Laboratory
Olympic Building
South Bend, Iowa 57087

Gentlemen:

Mr. Walter S. Stevens
Secretary of the Chamber of
   Commerce
Box 101
Caldwell, North Dakota 58203

Dear Mr. Stevens:

Messrs. William Smith and H. R. Brown
1148 Harrison Boulevard
Twin Cities, Arkansas 92116

Dear Sirs: (or Gentlemen:)

Dr. Frederick C. Moore
1415 West Grant Street
Tulsa, Oklahoma 74199

Dear Dr. Moore:

Frederick C. Moore, M. D.
1415 West Grant Street
Tulsa, Oklahoma 74199

Dear Dr. Moore:

Dr. Walter C. Church
Professor of Mathematics
University of Kansas
Lawrence, Kansas 66025

Dear Dr. Church:

Professor Walter C. Church
Department of Mathematics
University of Kansas
Lawrence, Kansas 66025

Dear Professor Church:

Director of Personnel
Hammond and Haskins Milling Company
Fourth and Main Streets
Sandpoint, Utah 84336

Dear Sir:

American Association of University
   Women
Cascade Building
761 Main Street
Sandpoint, Utah 84336

Ladies:

The Honorable John R. Doe
United States Senate
Washington, D.C. 20515

Dear Sir:
   (or My dear Senator Doe:)

The Honorable John R. Doe
The House of Representatives
Washington, D.C. 20515

Dear Sir:
   (Or Dear Mr. Doe:)

The Honorable John R. Doe
Governor of (State)
(Capital City and State)

Dear Sir:
   (or My dear Governor Doe:)

The Honorable John R. Doe
Mayor of (City)
(City and State)

Dear Sir: (or My dear Mayor Doe:)

The Complimentary Close. The complimentary close is placed a double space below the final line of the body of the letter. It usually begins slightly to the left or right of the center of the page.

The forms used most frequently are *Very truly yours* and *Sincerely yours. Yours truly* is more formal and somewhat less friendly. *Cordially yours* is a gracious form if one is doing the reader a favor or in a position to do him favors or merely wishes to be friendly, but it seems inappropriate in a letter soliciting a favor. *Respectfully yours* is used mainly when the reader's position entitles him to special respect or when one is formally submitting material to a person who outranks him in his own organization. It is too formal and too subservient for ordinary use. Participial endings—that is, *hoping, trusting,* and similar terms ending in *ing*—are out of date. The best attitude is to use *Sincerely yours* or *Very truly yours* unless there is reason for deviation.

The Signature. The place and contents of the signature are shown in the various examples following. Whenever a letter purports to be personally written, the name of the person who wrote it must be signed in longhand, even if the primary signature shows the letter to come from an organization. The name of the writer must always be typed, however, as well as signed in longhand.

The first example below indicates that the writer is acting merely as an individual; the second and third both show that the writer acts in behalf of an organization; the fourth shows the letter as being primarily from an organization, even though the name of the person who actually wrote the letter is also given. Note that in Numbers 2 and 3 the organization is not named, the assumption being that it was named in the letterhead. Otherwise, the name of the organization would appear in the signature. In all of the examples, the complimentary close is shown above the signature in order to show the relative positions of these two parts of the letter. At least a triple space would be left for the writer's longhand signature. If the first line under the complimentary close is typed, as in Number 4, a double space is used between the two.

The signatures of women call for additional comment. When a woman who is signing a letter feels that the person whom she is writing to needs guidance about whether to address her as *Miss* or *Mrs.* in an answer, long-established conventions call for the use of one of the following forms.

> For a single woman: *Mary L. Brown*
> For a divorcée: *(Mrs.) Mary L. Brown* (It is assumed that her ex-husband's name is Brown.)
> For a married woman or a widow: *Mary L. Brown,* followed by *Mrs. John J. Brown* on the line underneath.

(1)

Very truly yours,

John M. Jones

(2)

Sincerely yours,

John M. Jones, Secretary

(3)

Sincerely yours,

John M. Jones
Production Manager

(4)

Very truly yours,

THE REED PAPER COMPANY

Robert L. Brown
Sales Manager

By using parentheses around the material where they are used above, the woman signing the letter provides the information needed; but since the parenthetical material is not regarded as part of the actual signature, she is conforming to the convention that one should not place a title before his or her signature. The conventions are often disregarded, however, by women who write in a business or professional capacity, prefer not to use *Mrs.*, and sign as if they were unmarried—except that they use their husbands' surnames. Also, in some professions, especially the theatrical, a woman who goes by her maiden name professionally uses it as a matter of course as her signature.

The Attention Line. In a letter addressed to an organization, an attention line is sometimes used in addition to the inside address. By using it one can indicate the name, the position, or both, of the individual from whom the letter should receive attention. Though there are several places

The H. C. Capwell Company
Broadway and Twentieth Streets
Oakland, California 94610

Attention of Mr. H. L. Klarnet

Gentlemen:

The H. C. Capwell Company
Broadway and Twentieth Streets
Oakland, California 94610

Gentlemen: Attention of the Credit Manager where such a line can be placed, the two places illustrated above are probably the most common. Note that the salutation matches the inside address in spite of the attention line.

The Subject Line. Only a very small proportion of business letters contain subject lines, for a good opening will identify the subject without a special line being necessary. Sometimes, however, a subject line is desirable, as, for example, when you have been requested to refer to a file number. It is most frequently placed to the right of the salutation and on the same line or a double space below. When it is not on the same line as the salutation, it either may be centered or may align with the paragraph openings. Note the following illustration:

```
The Prudential Insurance Company
The Paulsen Building
Spokane, Washington 99201

Gentlemen:                    Subject:  Policy No. 6,499,313

As you requested in your letter of July 23, I am sending you more
complete information on . . .
```

Stenographic References, Enclosures, Postscripts. When a letter has been transcribed by a stenographer, her initials and sometimes those of the person who dictated the letter are placed at the left margin, either on line with the bottom of the signature or slightly lower. If enclosures are to be sent with the letter, that fact is indicated either one or two spaces below the initials. The following illustration shows these details in one of the many acceptable forms.

```
            We shall hope to receive your consent in the near future.

                            Very truly yours,

                        THE CENTRAL ELECTRICAL COMPANY

                            W. I. Bowers

WIB/lc
Enclosures: 2
```

If a postscript is added, it is placed a double space below everything else in the letter. It is typed in the same form that has been used for the paragraphs in the body of the letter. It may or may not be preceded by *P.S.* Obviously, it should be short.

As used in modern business letters, a postscript is not a means of adding some statement that has been omitted by accident, for letters should be written well enough that such accidents do not happen. Rather, it is a device used to throw emphasis on some matter of special importance, or else to add something that it seems desirable to mention but that is not really related to the main message of the letter. Since such a device loses its effectiveness if overworked, postscripts should be used sparingly.

Additional Pages. When a letter must consist of more than one page, the stationery used for all pages should be exactly alike except that as mentioned previously, only page 1 bears a letterhead. The side margins of extra pages should be the same as those on the first page. The top margin should be equal to the left margin. Often, in order that the second and later pages can be identified, the name of the person or organization addressed, the page number, and the date are typed at the top of the second and later pages. (This precaution, however, is not mandatory when common sense indicates that it is unnecessary.) The first line of the body of the letter comes three or four lines lower. Several forms may be used for this reference information, one of the commonest being as follows:

```
Dr. Irvin B. Steadman          -2-        October 3, 197-
```

Abbreviation, Capitalization, and Numbers

Abbreviation. In the body of a letter, the usage in abbreviation is the same as in any other writing. In all other parts of the letter, the main point to remember is to refrain from overabbreviating. The first rule is: In case of doubt, don't. Do not abbreviate the name of a person or organization unless the person or organization does so. Do not abbreviate months. Many organizations do not abbreviate the names of states (except that *N.Y.* and *D.C.* are customary), nor do they abbreviate such words as *Street, Avenue, East,* and *West* in addresses (*2315 West Walnut Street*). Some latitude is permissible if abbreviation is necessary to prevent a line being so long as to appear awkward. Yet even then you should use a consistent style in any given letter. That is, it would be wrong to abbreviate a term such as *Street* in the heading and write out the same term in the inside address.

As mentioned in the discussion of the salutation, it is more courteous not to abbreviate any title except *Mr., Mrs.,* and *Dr.* In the inside address, the terms *Honorable* and *Reverend* are sometimes seen abbreviated, but it is more courteous to write them out. The compliment that you imply by using one of these terms is cancelled when you attempt to save time by reducing the term to an abbreviation.

Capitalization. In the body of the letter, capitalization is the same as in any other writing. In other parts of the letter, the examples that have been given are a sufficient guide.

Use of Figures or Words for Numbers. In the use of figures or words to express numbers, the body of a letter is no different from any other writing. One point however—the writing of dates—calls for comment. The best form is as seen in the example, *June 17.* The year would be added only if needed. You should not use such terms as *the seventeenth* or *the 17th* unless the month has already been named in the body of the letter.

In other parts of the letter than the body, the best usage is as follows: A house number is written as figures. The number of a street is written out unless it consists of more than two words, in which case it is written in figures. Examples are: *Eighteenth* Avenue, *Forty-ninth* Street, but *114th* Street. (Some authorities permit the omission of the *th* when using figures, and write *114* Street.) It is permissible, also, to use figures for any street number if such a word as *East* or *North,* either written out or abbreviated, appears between the number of the house and the number of the street.

Years and days of the month are always expressed in figures. It is wrong, however, to use a figure to indicate the name of the month. (Correct: August 11, 1974. Wrong: 8/11/74.) The form *11 August 1974* is not widely used in the United States except in the armed services.

THE SUBSTANCE OF LETTERS

Learning how to use correct form is no more than a starting point in the study of letter writing. The real opportunity to profit by such a study comes when attention is turned to the substance of letters.

To be sure, some letters deal with routine situations and call for no special qualities beyond the completeness, clearness, and accuracy that are essential to any type of writing. Yet even a routine letter is a personal contact; and in letters that are not matters of mere routine, the personal element becomes extremely important. Thus it is well worth while to learn something about the attitudes and techniques that give letters the best chance of obtaining satisfactory results.

At the outset, you should understand the difference between a letter that merely is not bad and one that is really good. A letter may not be bad if it is correct in form, is free from serious errors in English or content, and does the obvious job without giving offense. A letter that is really good, however, is one *that exploits to the maximum, for the bene-fit of the writer or his organization, the contact that the letter represents.*

Standards of Appearance, Correctness, and Accuracy

You should make your letters neat for much the same reason that you would try to present a neat appearance in a personal contact. The appearance of a letter influences its reception, for the reader forms some sort of impression before he even starts to read. A messy-looking letter can create an impression of carelessness that lessens the reader's confidence in the accuracy of its contents. Rightly or wrongly, the reader may subconsciously assume that your standards in other matters are comparable to your standards in your letters. If you have an opportunity to look at letters sent by successful American businesses, you will find that they invariably are neat and attractive.

Correctness involves both letter form and English. Correct form should become so habitual that it is taken for granted. And bad English in a letter, like bad English anywhere, is so damaging to prestige that it is not tolerated by modern American industry. The standards of correctness, however, should be those of liberal up-to-date authorities. These authorities base their judgment on the way language is actually used by literate people; thus many forms that were branded incorrect by textbooks years ago are now entirely acceptable except on extremely formal occasions.

Accuracy should include both the accurate reading of the letter you are answering, and accuracy in the information your own letter contains. A surprising amount of confusion, waste of time, and annoyance is caused by writers' failure to notice what is really said in the letters they attempt to answer. Frequently, an additional exchange of letters is the result. And as for making sure that their own letters are accurate—that is a matter that demands more care than most writers realize. Such matters as dates, amounts, the numbers of orders or statements, and file numbers demand methodical checking. Errors such as a reference to *Monday, May 18,* when May 18 falls on a Tuesday provide another example of the need for close attention to detail. For that matter, errors in ordinary words are almost as easy to make as errors in figures and can create confusion and sometimes ill feeling. If you write, "We are *not* making an exhaustive study of the cause of the trouble" when you intend to write "We are *now* making an exhaustive study," you convict yourself of a fault you have not committed. And again, if you write "*Neither* the manager *nor* the field agent would have been seriously concerned about this matter" when you intend to write "*Either* the manager *or* the field agent," you seriously misrepresent the facts and may lose a great deal of good will.

Language in Business Letters

The first point to realize about the language to be used in business letters is that there is no special language for business letters. However, many writers whose conversation is perfectly natural fall into an unnatural jargon as soon as they start a business letter. They use long, flowery expressions such as *I beg leave to inform you.* They use wordy phrases such as *at an early date* for *soon* and *at the present time* for *now.* They make excessive use of the passive voice, writing, for example, *Your letter has been received by us* and *It would be greatly appreciated by us* instead of *We have received your letter* and *We should appreciate it.* They use pomposities such as *It would be in accordance with our desires if . . .* for the simpler *We should like to have you. . . .*

A good way to avoid unnatural language is to ask yourself, "Would I feel ill at ease if I talked this way in a personal conversation with the man to whom I am writing?" It would be going too far to say that a letter should exactly duplicate the language of ordinary conversation, but it is certainly true that language which would sound pompous or unnatural in a conversation is bad style in a letter also.

Additional examples of undesirable business-letter jargon are:

beg to advise, state, inform, acknowledge, etc.

your letter of recent date (Refer to the date specifically; or if the date is not known and the situation does not justify looking it up, merely use *your recent letter.*)

same (as in "Fill out the questionnaire and return *same* to us.")

party (with the meaning of *person.* This use of *party* is appropriate only in legal documents.)

favor us with

kindly for *please*

enclosed please find a copy of (Use *enclosed is a copy* or *a copy is enclosed.*)

We wish to *take this opportunity* (unless the phrase has some meaning. If you were mentioning something incidentally when writing upon another subject, the phrase would be appropriate, but most of the time it is entirely devoid of meaning.)

Sentences in Business Letters

Sentences in business letters tend to be shorter than sentences in most writing. This results from the fact that many letters must be dictated, or written rapidly and mailed with little if any revision. Under these circumstances, long sentences are likely to become jumbled and to slow up the process of writing. You should not form the habit, however, of making all sentences so short that you write in the "primer style."

Paragraphs in Business Letters

Because most business letters are not themselves very long, their paragraphs tend to be short. The first and last paragraphs are often single sentences and are seldom more than two sentences long. The other paragraphs characteristically run from two to four or five sentences. Unless a letter is longer than average, it is unlikely to contain any paragraph of more than 100 words. Single-sentence paragraphs are permissible if they do not become too numerous, but you should not fall into the habit of beginning a new paragraph with every sentence. Unless most of the paragraphs are longer than one sentence, there is no value in paragraphing at all except the psychological value of making the letter look easier to read. As a matter of fact, most long or medium-length letters that contain too many one-sentence paragraphs are found on examination to be poorly organized.

Though paragraphs in business letters are often too short to call for the use and full-scale development of topic sentences, paragraph unity is nevertheless important. The opening of a paragraph may cause a reader to form an instant impression about the topic that the paragraph concerns, and it is bad writing to mislead him by presenting information on other topics. In letters that are longer than average and are therefore likely to contain longer paragraphs, attention to paragraph unity is especially desirable.

Beginnings of Letters

The first paragraph of a letter, being mainly a contact paragraph, should usually be short—often consisting of a single sentence and rarely of more than two sentences. Nevertheless, it may need to (1) indicate the subject and purpose of the letter, (2) refer to previous correspondence if such correspondence needs to be borne in mind, and (3) establish a satisfactory tone. Obviously it is not always necessary to refer to previous correspondence, but the first and third of these functions always call for attention.

To say that the opening of a letter must always indicate the writer's subject and purpose seems to be dwelling on the obvious, yet this need is often overlooked. For example, a person inviting the reader to address some organization might tell about the organization for two or three paragraphs before coming to the point that concerns the reader—the fact that he is being invited to deliver an address. Therefore it pays to remember: Make both the subject and the purpose clear at once so that the reader can see how *he* is concerned with what he is reading.

When you answer a letter you have received, you should immediately identify it by mentioning its subject and if necessary its date. Also, you should say enough to demonstrate that you have read it accurately. Such key facts as dates and amounts of money may well be restated, so that if any error has been made in the other person's letter it will be discovered at once. Though such errors are not frequent, they can occur often enough to justify the precaution. The manner in which facts are restated should not be crude and obvious. It is poor writing to say, "I have received your letter of June 15. In this letter you say that you would like to have me make reservations for you for July 17, 18, and 19." Rather, let the facts seem to be mentioned incidentally, as for example, "Thanks for letting me know, in your letter of June 15, about your need of reservations for July 17, 18, and 19." Ostensibly, the sentence was written to say "Thank you." Actually, its purpose was to repeat the dates.

Mention of your own previous letters on a subject is advisable because the reader may have forgotten them and because your present letter may not reach the same person who read the earlier letters.

The establishment of tone is especially important. This subject will be covered extensively later in the chapter. Here, suffice it to say that the reader's original impression influences his reaction to all that follows. If the tone of the beginning is offensive, it will be doubly hard for the material that follows to get a proper reception. Hence the opening should usually be friendly in tone, and should never, regardless of the circumstances, be harsh or sarcastic.

At the very beginning you should try to get in step with the reader. Even if it will be necessary, later, to include material that takes issue with him, the opening should be an attempt to suggest that you are not an opponent who is trying to defeat him in an argument, but a person who has approached the subject with an open mind. Even if a letter deals with a situation so serious that a display of friendliness would clearly be insincere, it is always possible to be completely courteous in the opening and to avoid any display of anger.

Some might feel that it is a nuisance to have to make the opening of a letter friendly. Yet when you adopt a friendly or at least a courteous tone at the beginning, you are merely opening your letter as a tactful person usually opens personal conversation.

Endings of Letters

The best method of ending a letter depends on whether you are attempting to secure some specific action. If not, the ending should be a final brief paragraph showing friendliness and good will. Here again, there is a parallel between a letter and a personal conversation. One usually tries to end even a business conversation in a friendly manner.

A typical letter not attempting to secure action might be one giving some information that had been requested. You might well conclude such a letter by expressing hope that the information will be useful and that the reader's project will be successful. On the other hand, if you had been unable to provide the information asked for, it would still be possible to express a hope that the reader would find it elsewhere and that you could be of some service to him in the future.

If the purpose of a letter is to secure action, the action desired should be suggested by the ending. The suggestion should be as specific as the circumstances permit. Sometimes the request for action may be in the form of a command: "Sign up today." Sometimes, tact would make it advisable that the command be veiled: "To make sure that you receive the accommodations you desire, make your reservations today." Sometimes it must be toned down still more: "We should appreciate it if you could let us know by April 1 the dates that you prefer." The degree of urgency must always be a matter of judgment. If you need immediate action when requesting a favor, it is especially important to be tactful. In such a case your letter is less likely to arouse a negative reaction if it tells why immediate action is important—for example, "If you accept this invitation, we are anxious to give your address plenty of advance publicity. Therefore we hope you will let us know your decision at your earliest convenience."

An ending aimed at securing action need not entirely neglect the element of good will. A skillful writer can usually phrase his request for action so that good will is apparent.

Whether a letter does or does not aim at action, one fundamental point about endings should never be overlooked: The ending of a letter should never dwell upon anything that you would not want the reader to continue thinking about. A reader tends to retain in his mind whatever he has read last. Suppose a letter ends by saying, "In closing, let me express once more my regret that, as originally written, my report did not contain all the information you needed." The reader's mind is thus focused on the fact that the original report was unsatisfactory. On the other hand, the letter might end, "In the version I am now sending, I am sure you will find the full information that you will need as you proceed with the project." The reader's mind is thus directed to the merits of the new version rather than to the faults of the old. Although there are many occasions when a letter must express regret, the end should, if possible, be optimistic.

Need of Planning before Writing

Unless a letter is to be extremely short, you should plan it before writing it. Your plan may be nothing more elaborate than rough notes, but it should at least indicate the points to be covered and the order in which you will cover them. As you make your notes, it will pay you to make sure that you have at hand all the facts that must be included, so that you will not have to stop writing or dictating while you look up some bit of information.

The time used in making notes will be short and will be more than compensated for later because you will be able to forge ahead with assurance when you write. Also, planning in advance will usually improve the quality of your letter. That is, it will ensure your putting related material in the same part of the letter, will result in better arrangement than you could achieve while struggling with questions of phraseology, and will permit you to concentrate on how to express each idea rather than being constantly preoccupied with the question of what you are going to say next.

To be sure, as you gain experience and deal with situations that have become familiar, you will eventually be able to base many a letter on a plan you have used on similar occasions in the past. Even so, it will usually be advisable to have in mind or on paper a checklist of points to be covered; and it will also be advisable, before you write even a routine letter, to pause long enough to make sure that a routine letter will really cover the situation. Nothing is more annoying to your reader than to receive a letter that ignores some of the relevant facts.

Arrangement of Material in Letters

In writing a letter, as in writing anything else, you will need to be governed by two considerations as you plan the arrangement of the material—the subject matter and the reader. Whatever may be the message of a letter, there are two basic methods of arranging the material. These might be called "the order of saying *Yes*" and "the order of saying *No*." In any specific situation, one or the other should definitely be preferable.

When you are saying "yes" to a request, you should say it at once —sometimes even in the opening sentence. This is appropriate because a reader who learns that his request has been granted is in a more receptive state of mind to hear anything else you wish to tell him. Perhaps he must be told that granting such requests cannot be made a regular policy, or that careful consideration was necessary before consent could be granted. Perhaps you need to explain the circumstances that gave rise to the whole incident. All such material should be held back, however, until the affirmative answer has been given.

Some writers neglect this principle. They tell all the reasons that might justify a refusal before informing the reader that his request is granted. Thus they make the reader antagonistic, or, at best, secure only superficial attention to this material while the reader skims down the page looking for the definite answer. Moreover, by delaying the consent too long, they seem to give it grudgingly, and thus lose the good will that the consent might otherwise have gained. In view of all this, when you intend to say "yes," you should say it at once and say it graciously, then offer any explanations that are necessary, and finally, bring the reader's attention back to the fact that his request has been granted.

When saying "no," however, you should not blurt out your refusal at the beginning, but should hold it back until you have given the reasons for the decision. Thus the decision seems to be the outgrowth of the facts. If the "no" comes before the reasons, the reasons seem to have been scraped together to justify an answer that had been decided upon at the outset. The proper method is to try to make the reader himself see that a negative answer is indicated by giving him the evidence and leading him to that answer gradually. When you make your decision known, you should do so as tactfully as possible. Although no letter should be long-winded, the very fact that you have taken pains to write a full explanation suggests that you have not rejected the request peremptorily.

Naturally, it is not possible to generate so much good will with a refusal as with a consent, but the arrangement of material recommended above at least makes the best of the situation. A refusal that is tactful will create much less resentment than one that is tactless.

Not every letter, to be sure, forces you to grant or refuse a request, but the basic principle can be applied in almost every situation. If your letter bears good news, let the reader know it at once; if the news is bad, break it gently; if it is partly good and partly bad, first tell the good news, then tell the bad news, and next direct the reader's attention toward whatever aspects of the situation are most cheerful.

As you apply these principles, it is a good idea to decide very early how you want a letter to end. You need not go to the extreme of writing the last sentence of a letter first; but it is only common sense to decide what idea you want that last sentence to express. If you do so, you can move toward that ending more smoothly. Also, you can avoid the misfortune of using, in the middle of a letter, the idea that would be the best possible ending—therefore having either to use an ending that is not as good as it might be or to repeat ideas that you have already expressed.

Negative, Neutral, and Positive Approach

When we say that a writer's approach to a letter is negative we mean that by the power of suggestion he creates in the reader's mind an undesirable impression about his (the writer's) attitude or about the way in which a situation should be interpreted. When we say that the approach is neutral, we mean that there is nothing in the writer's manner that creates either a desirable or an undesirable impression. When we refer to the approach as positive, we indicate that the power of suggestion operates in a favorable manner.

The difference between the negative, the neutral, and the positive approach can be seen in the following sentences, all of which convey the same basic fact.

1. I was surprised to discover in your letter of June 26 that you claim I have made an excessive charge for investigating your lighting needs.
2. I have received your letter of June 26 concerning my charge for investigating your lighting needs.
3. I have been glad to do as you requested in your letter of June 26, and look into the question of whether I have made the correct charge for investigating your lighting needs.

The first sentence sounds indignant; it suggests that the writer's purpose is to prove the reader's letter was not reasonable; it shows that he regards the reader as an opponent in an argument. The second is completely noncommittal as to the writer's attitude. The third, without implying that a concession is to be granted, makes it appear that the writer welcomes the opportunity to correct a mistake if one has been made or to explain how the charge was arrived at if it is correct. A positive approach is always best when it is possible, and in any event the negative approach can always be avoided.

One cause of the negative suggestion is the use of words with a negative flavor. For example, "We have received your letter in which *you claim* that the last shipment was improperly packed," implies that the writer questions whether the statements of the reader are accurate. (Even if such were the case, it would be undesirable to make that fact apparent at the outset.) Other characteristic negative terms are "you neglect," "you fail," "your complaint," and "your error."

The negative tone may also be a matter of basic outlook—a tendency to see the dark side of a situation rather than the bright side. One writer may write, "We regret to say that unfortunately we cannot comply with your request until . . . ." Another and better writer will say, "We shall be glad to comply with your request as soon as . . . ." The difference is that one sounds like a refusal and the other sounds like a consent. One stresses the fact that immediate compliance is not forthcoming; the other stresses the fact that eventual compliance is possible. Again, one writer may write, "My experience is limited, consisting of nine months as assistant to . . . ." Another writes, "My qualifications have been increased by nine months' experience as assistant to . . . ." The first suggests that the reader consider nine months as very slight experience. The second encourages the reader to think of how much a person may know at the end of nine months that he did not know before.

Finally, certain kinds of material may in their very nature be negative. Consider the sentence, "Now that it's fall and you're getting warm clothes for the children, hadn't you better make sure that your automobile chains are in good condition?" This is negative because it makes the reader think of other expenses, which may be so hard to meet that he will economize by making his old chains serve. Another example of negative material is this ending to a letter: "I hope that our decision will not interfere with the pleasant relationship that has always existed between our companies." The writer showed by this ending that he himself thought of the decision as one that *might* interfere with the pleasant relationship. He might better have written, "I believe that the reasons I have offered to explain our decision show it to be appropriate, and I shall look forward to a continuance of our pleasant relationship."

Tone

A discussion of tone is really an extension of what has been said about the negative, the neutral, and the positive approach, for it concerns the impression created in the mind of the reader about the writer's personal attitude. Just as in a personal conversation, this impression can either increase or decrease the likelihood of satisfactory results.

The tone of letters can be infinitely varied. A letter can sound friendly, cheerful, courteous, serious, formal, informal, cordial, hostile, ingratiating, blunt, sarcastic, indignant, flippant, whining, evasive—in fact it can suggest any attitude that can be suggested by a person's manner in a face-to-face conversation. It is possible, of course, for a letter to have a purely impersonal tone. The best that can be said for such a tone, even on a routine occasion, is that it does no damage. But on any occasion that is not entirely routine, there is often a danger that it will create a damaging impression of coolness and indifference.

Obviously some of the tones named above—sarcasm, bluntness, hostility—are undesirable. A warning against them is nevertheless called for because many a person who would agree that they are undesirable will, under the stress of emotion, write letters that possess them.

As for letters that sound apologetic or ingratiating, their tone is often the result of carrying such qualities as reasonableness and friendliness too far. No quality is so good that it cannot be overdone.

What tone, then, should you try for? So far as it is possible to generalize we might say: Your letters should almost always sound friendly; and on the few occasions when friendliness would obviously be insincere, they should as an absolute minimum be entirely courteous. They should usually be cheerful. And the tone of modern letters is strongly in the direction of informality except when informality would suggest lack of suitable respect.

None of this means that a letter need be unbusinesslike or that it cannot be intensely serious when seriousness is called for, any more than a personal conversation would need to lack seriousness because one's manner is friendly.

How is such a tone achieved? Skillful letter writers, especially in letters that go outside their own companies, form a mental picture of the image they want people to hold of their companies, and they bear this image in mind as they write. Eventually it becomes practically automatic for them to adopt a suitable manner when they write letters. This is comparable to the way in which a person in military service automatically adopts a manner suitable to the occasion when he talks in the performance of his duties. It is comparable to a salesman's habit of talking to a prospect in a certain manner, regardless of his own state of mind at the moment. Unless you are fortunate enough to possess instincts that will always guide you to letter tone that is appropriate, you should keep reminding yourself to adopt a desirable tone until it finally becomes second nature for you to do so. If you fail to do this, you are likely to form habits that produce a contrary result.

The "You Attitude"

Almost every book on letter writing speaks of the "You Attitude." This phrase means merely that the writer attempts in every way possible to show that he has thought about the reader's interest as well as his own. A good example is one letter written by a manufacturer refusing to accept an order from an individual consumer. This manufacturer might have offered as his reason the nuisance it would be to handle such small orders. Instead, he stressed the advantage to the purchaser of being able to see the various sizes and colors and of obtaining products without delay, as he could do if he were served by a retail merchant.

This "You Attitude" can be achieved even in letters that would seem to offer little opportunity for it. For example, a letter written to collect back payments on a refrigerator purchased on the installment plan would have as its actual purpose the securing of the payment; yet the writer might indicate that he was writing because he did not want the reader to lose the refrigerator, as would happen if he did not make the payment; or he might stress how desirable it would be to the reader to protect his credit rating.

Good Will in Business Letters

Whatever its message may be, a letter should present that message in a manner that generates or retains as much good will as possible. The way to create or retain good will, so far as you can do so by means of writing techniques, is mainly to express good will. It is hard for a reader to remain unfriendly in the face of consistent friendliness on your part. It bothers him more to say "no" to someone who has been pleasant than to someone who has been indifferent or antagonistic. Showing good will pays off by making the reader want to see things as you see them.

Also, making good will permeate your letters is a sensible precaution for it can prevent inadvertently offensive remarks from doing damage. If a statement in a conversation is taken in an unpleasant way that you did not intend, you can notice the effect at once and correct your slip. Moreover, once said the remark passes by at once. But when a letter gives offense, you are not present to observe its effect and have little opportunity to set the matter right. Also, what you have said is there to be read repeatedly. Consequently, it is necessary, when you are writing, to take extra precautions to avoid giving offense unintentionally. There is no better way to do this than to make good will so evident that the reader will realize an unfortunate slip could not have been meant as it sounded.

The time when it is hardest to express good will is when you must answer an unpleasant letter from an angry writer. It is entirely natural to feel an impulse to answer such a letter sharply, but the impulse should be resisted. In refusing to lose your temper because he has lost his you are not necessarily giving evidence of weakness. And one point not to be overlooked is this: Having blown off steam, he may no longer feel as belligerent as before. He may even feel a little ashamed at having lost his self-control. If you answer in kind, you relieve him of his worry about whether he has gone too far; but if you are courteous as well as firm in your manner—if you refrain from entering a contest in recrimination— you have a good chance of making him regret his remarks and wish to show that he is not quarrelsome and unpleasant. This may go far toward making him easier to deal with.

If you are tempted to regard anything said above as soft or emotional, remember that American industry is spending millions of dollars annually in the belief that good will is profitable. Expressing good will in your letters will increase the chance that they will be effective both in their immediate results and in their long-range contribution to a favorable image of you or your company. And just as it is worth while to include material for the sake of good will, it is often sensible to omit remarks that, though justified by the facts, are not really essential and would sacrifice good will. It is important to remember that you cannot assume every reader to be a fair-minded, unprejudiced judge. Therefore your manner and material should be determined not by what you would be *justified* in saying but by what will produce the result you desire. The satisfaction of "telling off" a reader who has been unreasonable is a luxury you cannot afford. Constant attention to good will is by no means sentimentalism. We may hope, of course, that it will usually be a sincere expression of your actual feelings; but it is definitely called for by intelligent self-interest.

EXERCISES

Exercise 1

Change the following addresses, salutations, and complimentary closes so that they conform to the best modern usage. Do not change anything that is already correct. You may assume that where abbreviations appear in company names, the forms used are correct official forms.

(1)

Mr. John O. Clark, president
Cascade Paper Company
Box 687
Cascade, Washington 98498

Gentlemen:

Very truly yours,

(2)

E. A. Thompson, Inc.
1224 Grove Ave.
Millwood, Arkansas 72189

Dear Mr. Thompson:

Very truly yours,

(3)

Dr. Doyle M. Hobbs, M.D.
Walters Dental-Medical Bldg.
Springfield, Ohio 44903

My Dear Dr. Jackson

Sincerely yours,

(4)

Henry R. Norton
4223 13th St.
Middleton, Vermont 05982

Dear Sir:

Very Truly Yours,

(5)

Hon. Walter P. Woodcock
The House of Representatives
Washington, D.C. 20515

Dear Hon. Woodcock:

Sincerely yours,

(6)

Collins Pre-Mix Company
Tenth and Franklin Streets
Caulfield, Mississippi 38925

Attention of Mr. Henry Jackson

Dear Mr. Jackson,

Very truly yours,

(7)

Mr. Richard I. Dormer
Upland Chemicals, Inc.
Box 71
Cranston, Texas 78201

Dear Mr. Dormer;

Sincerely yours,

(8)

Prof. Winston V. Upjohn
Dept. of Chem. Engineering
Kiley Tech. Institute
Kiley, Penn. 15695

Dear Prof. Upjohn

Very truly yours,

(9)

Ames and Coope, Incorporated
2363 Bristol Way
Longview, Mass. 01033

Dear Sirs:

Very truly yours,

(10)

Victor R. Milbrook
Principal of the High School
Westbank, Nebraska 68569

Dear Principal Milbrook:

                              Sincerely Yours,

Exercise 2

Rearrange the material in each of the following items so that it is an inside address in acceptable form, with an attention line if one is called for. Add a suitable salutation after each address.

1. Frederick B. Neumann, who works for the Atlas Filter Company, 1114 Jefferson Avenue, Toledo, Ohio. His position is that of Applications Engineer. Zip code is 43609.
2. The Zenith Manufacturing Company, located in Minneapolis, Minnesota at the street address 18256 Central Avenue. Zip code is 53414.
3. The Excelsior Steel Company of Gary, Indiana. The address is box number 794. Zip code is 46444.
4. Two men, Dr. Albert A. Boyd and Dr. William O. Jenkins, who are on the faculty of Midland University which is located in Des Moines, Iowa. Zip code is 50343. Both are members of the Department of Physics.
5. Kenneth Zellerbach, a doctor of medicine at the Kramer Clinic, 789 Arbor Way in Kansas City, Missouri. Zip code is 64195.
6. Kellerman Plastics, Inc. with attention of the Sales Manager requested. The company is located in Atlanta, Georgia at 6636 Riverside Drive. Zip code is 30364.
7. The director of educational relations of the Cambridge Electronics Company, located in Philadelphia, Pennsylvania at 10824 Commercial Circle. Zip code is 19127.
8. The Denver chapter of the American Association of University Women. The address is Box 232 in Denver Colorado. Zip code is 80375.
9. Elliot B. Eaves, who is purchasing agent for Harbor Forests, Inc. at 366 2nd Street, Tacoma Washington. Zip code is 98468.
10. Arthur M. Baker, who is an associate professor of Economics at Southern Technical College of Houston, Texas. He holds a degree as Doctor of Philosophy. Zip code is 77018.

Exercise 3

Each of the following sentences occurred, as indicated, at or near the beginning or the ending of a letter. Most of them are faulty in some respect, either because they contain unsuitable ideas for the positions that they occupy or because their tone, ideas, and language would be unsuitable regardless of their position. Write an improved version of each sentence that needs improvement. If it is necessary to interpret a situation more definitely or to supply additional facts, feel free to do so.

Beginnings

1. In your letter of June 16 you have agreed to consult with us about locations available for a marina on the Wheeler Dam Reservoir, but you have neglected to set a date when you would do so.
2. I realize that a man in your position has heavy demands on his time, but I hope the facts below will persuade you to join one more committee.
3. Though questionnaires are sometimes regarded as a nuisance, I hope that the one enclosed will not be consigned to your wastebasket.
4. I have received your complaint about the nauseating odor that you claim is caused by this company's feed lot.
5. I hope that you will not take offense, but I feel justified in objecting to the uncontrolled growth of weeds in the vacant lot that you own, which is next to my residence.
6. We are sorry to inform you that for the reasons stated below we must refuse to repair your floor-cleaning machine without charge.
7. We have received your letter saying that the paint on your building is peeling, but feel that you are mistaken when you say that the paint we used must have been of poor quality.
8. We are sorry to learn that your houseboat on the St. Joe River was damaged on October 19, and have carefully checked our records to see whether, as you believe, the damage was caused by one of our tug boats.

Endings

1. We believe that our installation is now in condition to give you good service, and are sorry that our first attempt to make the necessary repairs was unsuccessful.
2. In closing I wish once again to express our regret that the oil spill caused you so much inconvenience.
3. We believe that the air filter gauges we have installed will take care of your filter problems and trust that the past difficulties will not prevent you from using our equipment in the future.
4. Since the valves in your heating system deteriorated because you used hot water valves in a system that heats by steam, we must decline to replace them at our own expense.
5. If you cannot make the reset valves in our air-conditioning system function properly without further delay, we will look elsewhere when we install air-conditioning equipment in the new wing we shall soon construct.
6. We hope that your preventive maintenance service succeeds in preventing further trouble so that we will feel justified in renewing our contract with you when the present one expires in June.

Exercise 4

Study the following letters to discover the respects in which the principles explained in this chapter have not been complied with. Try also to think of ideas that might, if they had been added, increase the letters' effectiveness. Criticize or rewrite letters designated by the instructor.

(1)

Dear Mr. Smith:

This is to let you know that Mr. Howard Jones, who is to speak to your chapter at a meeting open to the public on November 25, is now filling a different position in the national organization. He is now acting president rather than vice president, having taken the place of Harvey Swank, who resigned the presidency because of ill health. Mr. Jones' new title should be used in newspaper stories and other publicity.

Mr. Jones is not yet sure whether he will drive or fly to your city. His decision may be affected by the weather. If he drives, he will want to drive directly to his motel or hotel. In order that I may make reservations for him, I would appreciate your suggesting one that is suitable.

Could you also give me the name and the telephone number of the person whom he should contact after arrival if he comes by car and therefore will not need to accept your offer to meet him at the airport.

I am enclosing a sheet giving some biographical facts about Mr. Jones that should be useful when you publicize his address. I should also mention that he has decided not to show any slides, so our previous request to have a projector and operator available can be disregarded.

Finally, could you let me know whether there are any other events on the conference schedule that it would be desirable for Mr. Jones to attend, so that he can arrange his travel plans accordingly.

Thanking you for your attention to these details, I remain

Very truly yours,

Administrative Assistant

(2)

Dear Mr. Peterson:

I have received your letter in which you request that we change our schedule for sanding icy streets so that Washington Street will be sanded earlier.

I am unable to comply with your request. I realize that Washington Street is somewhat steep, but the schedule we follow is the result of careful study, taking into consideration both the grades and the traffic count. All the streets that we now sand before Washington are at least as steep and carry more traffic.

You might be interested in knowing that yours is only one of a great many requests involving a great many streets. Everyone seems to feel that the particular streets he travels should be among the first to be sanded. If we neglected other streets to give Washington special consideration, the people who use them would deluge us with complaints.

Let me further point out that Monroe Street, which like Washington runs east and west, is sanded extremely early because it is a through street. By driving only three blocks extra you could use it when the sand trucks have not yet reached Washington.

I trust that you will see that our decision in this matter is the only one possible.

Very truly yours,

Superintendent of Streets

(3)

Dear Sir:

I am writing to complain about the manner in which the county road crews have for a long time consistently neglected the condition of Derby Road between Highway 93 and the bridge over Cottonwood Creek.

This stretch of road is so full of chuck holes that it looks like it had been bombed by an airplane, but the road crews ignore its existence. Not long ago they went over the entire part of Derby Road that lies east of the highway. A little earlier they had patched the black top on the portion that lies west of the bridge. I am unable to understand why the half-mile stretch between the two should go without attention when it is rougher than either.

As a taxpayer who helps to foot the bill for the upkeep of all the roads in the county, I see no reason to tolerate the neglect of a stretch which I and others who live upon it must use every time we leave home. I trust that you will put an end to this discrimination.

Yours truly,

(4)

I have received your letter in which you complain about what you claim is neglect by county road crews of a stretch of Derby Road west of Highway 93.

There is one fact, however, that you ignore. The stretch of bad road lies within a tract that was annexed to River City over a year ago.

I agree that the condition of this stretch of road is bad and that the county road crews have done nothing to make it better, but we have our hands full caring for our own roads, and cannot be expected to do the city's work.

Since the city is now responsible for the stretch of road you complain about, I recommend that you address your complaints to the city government.

Very truly yours,

Superintendent of
County Roads

15

SPECIAL TYPES
OF LETTERS

Not every letter can be classified according to type, and the inclusion of this chapter is not intended to imply that such is the case. Rather, its purpose is merely to show how the general principles apply in a number of characteristic situations. No effort is made to include all the types that are generally recognized. Credit letters, for example, are omitted because they are unlikely to be written by anyone in a position that calls for technical writing. Sales letters are omitted for the same reason, though some of the sales letter techniques are explained in the discussion of letters urging action. The types included justify the attention they receive not only because they illustrate the application of general principles but also because you may some day occupy a position where any of them may have to be written.

INVITATIONS

Though invitations of a social nature are beyond the scope of this book, there are many times when invitations are essentially business letters. For example, anyone who is active in a professional organization may need to write an invitation to attend or participate in a meeting.

In writing an invitation, make the purpose of your letter clear very early. Don't leave your reader wondering, as he reads, why you are telling *him* whatever it is you are saying. Also, in or immediately after the opening, include something to catch his interest. One good way to do this is to refer to something he himself has done or something you have read or heard about him.

Try to persuade your reader that the expenditure of time involved in the acceptance of your invitation will be justified. The importance of the occasion, the number expected to attend, facts about the organization, names of others who will be present—any such information can be useful in this connection.

Make a subtle appeal to self-interest if you can plausibly do so. Many a person accepts an invitation, partly at least, because he hopes to gain good will or publicity for the organization he is associated with. Take care, however, not to be crude and obvious in appealing to self-interest. Your tone should make it clear that though you hope your reader will somehow benefit from accepting your invitation, you consider that basically he will be rendering a service.

Make sure that you include all the facts your reader will need in making his decision—especially the date, the time, and the place of meeting. Make it clear what you expect or hope he will do—deliver a major address, participate in a panel discussion, attend social events—in general, make him feel confident that he has the complete picture. If he is to deliver a major address, give him some idea about how long he will be expected to talk, and either suggest a subject or let him know that you will leave the choice of a subject to him. (One possibility is to mention something you would be glad to hear him talk about, but indicate your willingness to let him use either that subject or one of his own choosing.) If the occasion calls for any comment on financial matters, say what is necessary.

When all these points, or as many of them as seem appropriate, have been covered, end an invitation by suggesting action, that is, by referring to your hope to receive a letter of acceptance. The main difficulty in writing the ending is the question of how to induce the reader to write soon without seeming to rush him unreasonably. (In this connection, be considerate. Don't delay so long in sending your invitation that you must ask your reader to answer in haste.) Perhaps the mildest way to request early action is to use the phrase, "at your earliest convenience," which is not jargon, because it is the simplest way of expressing the idea. If it is necessary to ask for action by a specific date, mention the reason —for example, the need of printing a program or beginning publicity.

Make your ending pleasant, in fact make the letter pleasant, by showing a desire to be hospitable and to make the occasion enjoyable for the person you are inviting. An offer to make hotel reservations or to meet him at the airport is one way of showing hospitality and consideration. In any event, your ending can easily be written so as to be complimentary as well as being a request for action.

Though all the preceding suggestions are worth considering, you should not try to follow so many of them in one letter that the letter grows unusually long. Usually, a letter from two-thirds to three-quarters of a page is long enough. If the reader accepts and needs further information, you can supply it later. In the original invitation it is better to select the ideas that best suit the occasion rather than to try to use every device you can think of.

The following letter illustrates the application of these principles.

Dear Mr. Kiley:

Like other snowmobile owners in Centerton I have been greatly interested in the work that the Mountainview Snowmobile Club has done under your leadership to improve the opportunity of snowmobilers to enjoy their hobby.

Several snowmobilers in Centerton have formed a committee to organize a similar club in our area. As a first step we have called a meeting at the Community Center, to be held at 8:00 p.m. on December 3. We would like to extend to you a cordial invitation to attend this meeting and speak to us about the way you organized your highly successful club and about its accomplishments since it was organized.

We believe that if we direct our efforts wisely, we can develop a successful organization. The group I mentioned, which has asked me to serve as its chairman, has been sounding out snowmobile owners and their response has been enthusiastic. The local snowmobile dealers are helping us to arouse interest, and I am confident that at least fifty people will attend the meeting—probably more if we are able to announce that you will be there.

Since the existence of a strong organization here would help your own organization to accomplish its purposes on a statewide level, I hope that you will feel it worth while to accept this invitation.

If you can do so, I also hope that you can spend enough time here that some of us on the committee I am speaking for will have a chance to extend our personal hospitality. We shall be very much pleased if we receive word of your acceptance in the near future, so that we can spread the word among potential members.

Sincerely,

LETTERS GIVING INSTRUCTIONS

Letters giving instructions, as we will discuss them here, are those primarily intended to tell the reader that he is to do certain things and to explain, if necessary, how to do them.

The first requirement, if those who receive instructions are to follow them properly, is absolute clarity. As a Prussian general once remarked, "Anything that can be misunderstood will be misunderstood." True, it may take extra time to make instructions unmistakably clear; but the extra time devoted to clarity is usually less than the amount of time it takes to repair the damage caused by having things done wrongly.

Clearness is partly, of course, merely a matter of writing. Everything about clearness said elsewhere in the book, especially in the discussion of style, applies to clearness in writing instructions. Completeness is also an element in clearness. In giving instructions you will find the time well spent if you think the subject through, visualize every contingency, and cope with it if you think it might arise and cause uncertainty. And in dealing not with contingencies but with regular essentials, you will have to use good judgment in deciding which details would be useful and which can be omitted.

Usually instructions will be clearer if at least a minimum amount of information is provided about the situation that makes them necessary. A reader who is aware of reasons and purposes will usually grasp instructions faster, remember them better, and follow them more intelligently than would be likely if he were expected to follow them blindly. But the actual orders should stand out sharply and distinctly rather than merging with the explanatory matter.

A letter of instructions can often be improved by the use of mechanical devices such as numbering, headings, and extra indentation. These devices were discussed in Chapter 5, in the section dealing with instructions. To be sure, a letter is seldom long enough to permit the use of very many mechanical devices, and also they may take more time to work out than it is practical to spend on a letter to a single individual. But they should not be overlooked in a letter that goes to a number of people.

Finally, tone can be important in a letter of instructions as in other letters. Instructions are not always given exclusively to subordinates. For example, instructions on such matters as safety procedures, accounting methods, or use of cars from a company car pool might be issued to people in other lines of authority. And even in giving instructions to subordinates, a pleasant and courteous tone increases the chance of securing willing cooperation.

Following is a letter giving instructions.

Dear Mr. Kennedy:

We are glad to know that you have decided to add an index to your book about the early history of your home town of Pavelle, which we are publishing for you.

For your assistance we are enclosing a specimen page of a good index. If you use it as an example, you can avoid any uncertainty about what is customary in regard to such matters as capitalization, punctuation, indentation, and wording.

We think it will also be helpful if we explain how experienced index makers go about their work to avoid confusion and unnecessary effort. Basically, it calls for the use of 3- by 5-inch cards, each of which bears a single entry. In using them, proceed as follows.

1. When you receive the page proofs, look over each page; and when you find an item that a reader may wish to look up, record it near the upper, left-hand corner of a card, with the page number following it. If the item is extensive enough that the reader might be helped by reference to separate parts of it or to related items, make a subordinate card for each such item. Each subordinate card should show the main item where it would appear on the main card and the subordinate item below it but more deeply indented.

2. When you make out a card for a main entry, pick out the key word—the word that best identifies the actual subject. Start the entry with this key word, put a comma after it, and then add what is necessary. For example:

Wild Rivers, Areas Classified as

3. If information on some point might be looked up under either of two words, make a card for each. For example you might have a main-entry card for Camp Grounds followed by cards naming streams or other locations where camp grounds are available; and you might also have a main card for each of several streams, followed in each case by subordinate cards—one referring to camp grounds on the stream and others to other information about it.

4. When you have made out cards for everything you believe a reader is likely to look up, arrange them so that the main cards are in alphabetical order, followed in each case by the subordinate cards. Then alphabetize the subordinate cards in each set on the basis of the first important word in each—which means, for example, ignoring a preposition even if it is the first word in a subentry.

5. Cross out the main entry on each subordinate entry card— lightly enough that it is still legible. (The reason for showing the main entry on a subordinate card is to identify it if it gets out of place.)

6. When you have made and arranged your cards in this manner, you need only copy them onto your ordinary manuscript paper and your index will be complete. We advise, however, that you look over the copy you have made so as to eliminate needless duplication by means of cross references.

If you use this method of working with cards you will find that it facilitates alphabetizing and prevents confusion. And if you write the entry on each card exactly as it should appear in the finished index, you can copy the cards without being concerned about anything except copying correctly.

We hope that these instructions plus the enclosed page will make it easier for you to produce your index so that your book can be published as soon as possible.

Very truly yours,

INQUIRIES AND ANSWERS TO INQUIRIES

Letters of Inquiry

The task of writing letters of inquiry may fall to the lot of a person in almost any position. When an inquiry is sent to a person or organization that has a responsibility for supplying information—for example, a chamber of commerce or a governmental bureau—it poses no special problems. The same is true when the receiver may have a chance to profit by supplying information because he increases his chance of selling goods or services. But when it goes to a recipient who is under no obligation to answer and who does not profit by doing so except in good will, it must be written with more care. Most organizations, to be sure, answer legitimate inquiries as they answer other correspondence. But whether they supply the desired information depends in part on whether the writer succeeds in making them really want to be helpful and in making the answer as easy as possible to write. Our discussion will concern mainly inquiries to be sent when the circumstances call for tact and care.

The resemblance of one such inquiry to another is great enough that the following instructions may be applied in most occasions.

1. Open your letter by making it clear that your purpose is to obtain information, and identify the subject that the information concerns. Don't leave your reader wondering why he should be concerned with the situation that you are dealing with.
2. Tell your reader why the information is needed. Though your explanation should not be long-winded, it is neither courteous nor reasonable to ask your reader to use his time for your benefit unless you let him see that your purpose is neither trivial nor contrary to his own interests.
3. Unless it is obvious, let him know why he or his organization has been chosen as a source of information. As you do this, you will often be able to say something complimentary that will increase your reader's desire to be obliging.
4. Ask the questions that you hope your reader will answer. These

questions, of course, are the actual reason for your letter's existence and will be discussed below in more detail.

5. Offer to return the favor by supplying similar information to your reader, or perhaps by supplying information on the results of your investigation if you are writing to a number of other people. Of course no such offer should be made if it would be an empty gesture, but there are many times when a mutual exchange of information is natural among individuals or organizations engaged in similar activity.

6. End with a tactful suggestion of action combined with the assurance that you will appreciate whatever help you receive. In expressing your appreciation it is best to avoid saying "Thanks in advance," either directly or by implication, unless your personal relationship with the reader justifies your taking his assistance for granted. There is a real though subtle difference between saying "I shall appreciate whatever information you send me" and saying, "Thank you in advance for answering these questions" or "Thank you for this information."

The action requested should be made as easy as possible to perform. For example, you might encourage your reader to jot down his answers at the bottom or in the margin of your letter, and enclose a stamped and self-addressed envelope. If you do this and if your reader can answer your questions without looking up the information, he may answer you at once rather than waiting until some time when he is writing letters.

The actual questions, as mentioned above, are the heart of the letter. Unless your needs are so simple that you are asking only one or two questions, both courtesy and self-interest dictate that your questions be worked out carefully. They should be well organized, with those dealing with a single aspect of the subject placed consecutively. Each question should be as brief as clarity permits. The answers called for should be brief facts or estimates rather than general discussions, for a discussion is likely to be vague and to be filled with hedging and qualifications.

If there are several questions, it is advisable to number them. When there is room, it may be desirable to make them stand out as a single block of material by widening the side margins. The numbers make it easy for the reader to check up on precisely what facts he must have in order to answer. Also, they make it impossible for him to overlook any questions by accident. And finally, they cause him to arrange his answers in the exact order of the questions, which is a great help if you are tabulating the answers from many persons.

Though an inquiry should request information mainly by asking specific questions, it is usually advisable, also, to invite the reader to add his own comments or other information that he might consider relevant. Especially, he should be invited to supply any additional information that affects the interpretation or the significance of the answers he has supplied.

Closely related to an inquiry is a questionnaire, which of course would be accompanied by a covering letter. In fact, a questionnaire is merely the result of removing the questions from the letter, placing them on a separate page, and providing room for answers. Unfortunately, though a questionnaire must be used when questions are numerous, many people who answer straight inquiries disregard questionnaires. If a questionnaire seems necessary, the letter that accompanies it should merely be a brief, normal letter of inquiry without the questions. The page bearing the questions should contain enough information, however, to be answerable if separated from the letter.

A letter of inquiry follows.

Dear Professor Saunders:

I have heard about your system of teaching students in your technical writing classes how to write reports by having them help faculty members in the professional schools produce reports on their research projects. Since I teach technical writing, I am very much interested. I would be really grateful if you would help me by supplying answers, based on your experience, to the following questions:

1. Have you received satisfactory cooperation from the faculty members of the professional schools?

2. Has the benefit to the students depended to any great extent on the faculty member's performance rather than on their own efforts?

3. How large a percentage of the student's total work in the course does his faculty-related work amount to?

4. Does the system provide an opportunity for the student to demonstrate his own judgment as to what the report should include and what it should omit?

5. Does the system test the student's sense of responsibility for the accuracy of data?

6. Does the system force the student to do his own thinking— for example, to interpret the significance of data?

7. Since two to four students work on the same project, and join in the production of the same report, is it possible to pass judgment on their individual contributions?

8. In the same vein as number 7, have you found a way to tell whether, in each student team, some single student who is a natural leader or unusually competent does the lion's share of the work though the others get equal credit?

As I have indicated above, I can see that the system might solve some of the problems of teaching the course, but I am uncertain about some of its aspects. If you will give me the benefit of your experience by answering my questions, I will really appreciate your assistance. I shall of course be pleased if you add any further comments that you think would be of interest.

Very truly yours,

Answers to Inquiries

The following suggestions apply when the answer to an inquiry is sent to a reader who is not a potential customer for goods or services. The answer to a potential customer is essentially a sales letter and is therefore beyond the scope of this discussion.

When your answer to an inquiry contains most or all of the information requested, the opening should make it clear that the reader's request is being complied with, and of course it should cover all the points that must be covered in any opening. Then should come the information requested, as far as it can be provided. If part of the information cannot be provided, a statement to that effect comes next, accompanied by a suitable expression of regret and whatever explanations are called for. The ending is usually an expression of good will. For example, it can be a statement of your hope that the reader's project will be successful or an offer to provide additional information if such an offer is suitable.

When your answer to an inquiry does not provide the information requested, its contents should be arranged in the pattern suggested for saying "No" under *Arrangement of Material in Letters,* Chapter 14. Rather than blurting out the fact that the request is being refused, you should first attempt to show that you would like to comply with it. Then should come a brief statement of the reasons that you cannot do so, leading to a tactful but not evasive statement that the information requested cannot be supplied. In stating reasons you should not be unduly apologetic but should avoid bluntness.

After the actual refusal you should include at least one or two sentences so that the reader's final impression will be your good will rather than your refusal. You can create such an impression by saying that you hope he can find the information elsewhere, that his project will be successful, or that you can be of service to him in the future. Many times, when you cannot answer his exact questions, you may be able to send printed material or suggest other sources of information. A well-written refusal can at least prevent the reader from feeling rebuffed.

The following letters are examples of how the answer to an inquiry may provide information or may say that the information requested cannot be provided.

Gentlemen:

We have received your letter asking about the Mohawk garbage compressor and are glad to provide the following information in answer to your questions about how it has worked out in the Excelsior Apartments.

1. The Excelsior Apartments consists of about two hundred units housing about 450 to 500 residents.

2. The garbage to be disposed of averages about 1000 pounds per day. We estimate its original bulk to be great enough to fill forty garbage cans.

3. The compressor handles all kinds of garbage—cans, bottles, food waste, and paper.

4. The compressor reduces the garbage to about one sixth of its original bulk.

5. The operation is automatic. When a certain amount of garbage reaches the container, the compressor is turned on by an electric eye. Very little attention is necessary except for removing the compressed blocks.

6. The system has been trouble-free since we installed it 18 months ago.

7. We estimate that the saving in labor resulting from our use of this system will pay for it within two and one half to three years.

Since you ask us to add any information that we think might be helpful, I will mention that our building was designed with the use of a compressor in mind. The tenants on each floor can dump their garbage into chutes that carry it directly into the compressor, located in the basement. Thus it is handled only once, and then in the form of compressed blocks. The system would be less efficient if the tenants had to put their garbage into cans and we had to move these cans to the basement, dump the garbage, and then return the cans to their former locations.

To be sure, the compressor has a door through which garbage can be inserted, but only a trivial fraction of our waste material needs to be handled in this way. I do not know whether the apartment you are planning will provide chutes such as those that we use, and I believe that the advisability of your using a compressor like ours may well depend on the answer to that question.

I hope that this information is helpful and will be glad to answer any further questions that may occur to you.

Sincerely yours,

Dear Mr. Jones:

We have received your letter asking about our cloud-seeding operation during the early part of the year on the Warren River drainage—its extensiveness and our appraisal of the results.

As you may know, this operation was carried out under a contract with Central Light and Power Company. Consequently we do not feel free to release information about it to anyone except our client.

In the hope that it will be helpful, however, I am enclosing our brochure that contains a substantial amount of information about our methods and the results that we have obtained on other occasions.

Central Light and Power is of course free to release any information we developed for them. If you send them the same set of questions that you sent to us, it is possible that they may provide you with answers.

In any event, since we are naturally interested in seeing the public made aware of the potential benefits of weather modification, I hope that you can find enough material so that you can write the article you are working on.

Sincerely yours,

COMPLAINTS AND ADJUSTMENTS

Complaints

A letter of complaint is one of the types most likely to be badly written. It is likely to be written by a person who is irritated—perhaps justifiably—by both the facts to be complained about and the necessity of writing the letter; and with the reader at a distance, the temptation to write a harsh, angry letter may be hard to resist. Many a person who would be moderate if he were face to face with someone throws restraint aside and is unpleasant in a letter.

The purpose of a letter of complaint is not to relieve your emotions. Rather, it is to affect the attitude and perhaps the actions of the reader. Thus, even if harshness would be justified, you defeat your own purpose by being harsh, angering the reader, and rendering him incapable of passing fair judgment on the merits of your case. All in all, you are unlikely to gain any desirable result by a display of irritation that you could not gain without it and yet retain good will.

A letter of complaint should, at the least, be completely courteous, and it is often more effective if the tone is friendly. You should address the reader as if you consider him a fair and reasonable person. Thus your letter will seem intended to make sure he becomes clearly aware of certain facts and realizes their significance—facts that you would presumably wish to know if the case were reversed.

This does not mean that you need to be weak and apologetic, nor that you need to minimize the seriousness of a serious occasion. But, it is possible to be direct and intensely serious without being quarrelsome or sarcastic. Even when a situation is so serious as to call for a clear warning that drastic action may become necessary, the warning can be phrased so that it does not sound like a threat. In keeping with the "You Attitude," it can be expressed in such a way as to show your desire to avoid an action that presumably would be to his disadvantage.

Another attitude to guard against is self-righteousness. Though it may be desirable to present yourself as fair and reasonable, you are not likely to secure the reaction you desire if you seem convinced that you are perfect and your reader is highly reprehensible.

One may hope that in avoiding personal unpleasantness in a complaint you are not falsifying your feelings; but whatever your feelings may be, your letter should not be written to express them. It is written to produce a desired effect on the reader, and a letter along the lines recommended is most likely to secure the effect desired.

When a complaint must contain a request that a reader take some specific action—for example, grant an adjustment—you should assume that he accepts his moral and legal responsibilities, tell him the facts on which your request is based, and ask him to do what you think the situation calls for. The tone of such a letter is improved by an expression of regret that the request is necessary. The conclusion should of course be the specific request for action.

Following is a letter of complaint.

Dear Mr. Zimmerman:

For the last three years your garage has been doing the service work on our trucks and company cars. During most of this period we have been pleased with the results.

For the last two or three months, however, there have been a good many occasions when the service received has not been as good as before. The latest incident occurred last week, when the cap was left off after brake fluid had been added to one of our trucks.

Earlier incidents included such slip-ups as failure to replace the cap on a gasoline tank, leaving tires underinflated, and a good many other and similar cases of carelessness. Also, there have been occasions when work has been done that we had not authorized —for example, replacing spark plugs during a tune-up when the ones in the car had not been in service long enough to need replacement, and installing a full tail pipe behind a muffler when an extension was all that was needed.

We have brought these matters to the attention of the head of your service department when they have occurred, but have not noticed any improvement.

I would really regret the necessity of taking our work elsewhere, but I shall be forced to do so if we have further cause for concern. I am bringing this matter to your personal attention in the hope that you can bring the service we receive up to its former high quality, so that we can continue to bring our work to you in the future.

Very truly yours,

Answers to Complaints and to Requests for Adjustments

Three kinds of letters are covered in the following discussion: answers to complaints that have not asked for any specific adjustment, answers that refuse adjustments, and answers that grant adjustments. All three have a good deal in common; in fact, everything that is said about the first applies to both the others.

Answers to Complaints Not Involving Adjustments. The key to writing a good answer to a complaint is the same as the key to success in writing anything else: Form a clear picture of the result you hope to accomplish, and write in a manner designed to accomplish that result. You will find it is easy to forget this, however, in answering a complaint. The complaints you will answer will not always be justified, and whether justified or not they will sometimes be unreasonable or unpleasant. Consequently, you will be tempted to say, and even be justified in saying, things that it would be better to leave unsaid. You may constantly have to remind yourself that you are not writing to express your emotions or to win an argument. You are writing in order to affect your reader in a certain manner, usually to make him feel less injured or antagonistic. The tone and content should be settled upon with this definite purpose in mind.

This general advice is outlined in the following suggestions:

1. Even if you are answering an angry, unreasonable letter, your tone should be pleasant, or at least completely courteous. This does not mean that your letter must be subservient, but only that it should be a calm, reasonable attempt to reach an understanding.

2. You should avoid the role of an opponent who is trying to defeat the reader in an argument. To accomplish this result it is well to mention very near the opening some point or principle on which you and your reader will agree.

3. Detailed restatement of the complaint refreshes the reader's memory of his original annoyance and revives his anger. Therefore you should limit your restatement of the complaint to the minimum that is necessary to show you have read it with care and understand it correctly.

4. If the complaint contains misstatements, do not repeat them for the express purpose of proving that they are false or erroneous. Of course you must state the facts as you know them to be, but it is better to

refrain from calling attention to the fact that what you are saying con-
tradicts what the reader said. In doing this you give him a chance to
back down with a minimum of embarrassment.

5. Expressions of regret should be proportionate to the seriousness of
the occasion. They should be neither so perfunctory as to suggest
indifference nor so excessive as to magnify a small matter into a large
one.

6. Explanations should not be too long and involved. To have them so
creates a risk of seeming unduly apologetic or unduly argumentative.
If the tone creates a satisfactory impression, a simple explanation of
what went wrong will usually be sufficient.

7. The ending should not revert to the cause of the complaint. It should
be based on the assumption that the matter at hand has been cared
for and that all will be well in the future.

Some of the preceding suggestions picture your reader as a person
whose complaint has been unpleasant or immoderate. This emphasis is
justified by the fact that unpleasant complaints are the hardest to answer.
Your problem will of course be simpler when you answer a complaint
that is free from a quarrelsome tone. The opening should make it clear
that you are glad your reader let you know he was disturbed. It should
show him that you have been glad to look into the case open-mindedly
and are anxious to set matters right if you have been at fault. (There is
no reason, however, that on every occasion blame must be fastened upon
someone.) Include whatever explanations seem relevant and express an
appropriate degree of regret. If his complaint called for action, say what
you have done or propose to do. End as recommended in the suggestions above.

The main consideration here is to convince your reader that you
are really concerned about whatever it is that has disturbed him. It is
possible, of course, to magnify a small matter into a large one; but even
a moderate complaint should not be answered in such a perfunctory
manner that you appear indifferent.

The following letter is an answer to a complaint.

Dear Mr. Parr:

Thank you for your letter of May 10 letting me know of your
concern about the manner in which our trucks are driven in your
neighborhood. We are sorry that they must use the residential area
where you live to reach the construction work on the new levee.

We certainly do not want to have our trucks driven in any but a
safe and legal manner. Consequently I have brought your letter
to the attention of all of our drivers and made it entirely clear to
them that they are to drive properly in the future.

I think it likely that most of them have been doing so already, and would be glad to receive information that would help me to single out the exception or exceptions. Each of our trucks can be identified either by its license number or by the number painted on its door. If the offenses continue, and if you can supply me with one of these numbers on the truck that is at fault, and tell me when and where the offense occurred, I will take the matter up with the individual driver concerned.

I believe, however, that what I have already said to all drivers will remedy the situation and that you will have no further cause for concern.

Sincerely yours,

Letters Refusing to Make Adjustments. Included in Chapter 14, on business correspondence in general, was a discussion of the order of saying "no." What was said in that discussion applies, as might be expected, to a letter refusing to make an adjustment.

In brief, the order of saying "no" calls for holding back the actual statement of a refusal until the facts and reasons for refusing have been presented. This arrangement conveys the impression that the request was received by a person who was willing to consider it with an open mind and grant it if the facts showed that an adjustment was due. Having the actual statement of refusal precede the statement of reasons conveys the impression that the writer decided to refuse and then looked for reasons to justify his decision. This, plus the fact that being refused does not put the reader into a receptive state of mind, makes it undesirable for the refusal proper to be placed near the opening.

The opening of a refusal to make an adjustment should of course acknowledge the letter you are answering and should include enough facts to show that you understood it correctly. It should convey the impression that your first reaction was, "Let's look into this matter and see whether we should grant an adjustment." Even if you know before you start writing, as of course you will, that you are not going to comply with the request, you can open your letter by saying that you are glad your reader felt free to bring the case to your attention, or that you are sorry he has encountered trouble. Often it will be useful if you state the principle on which you base your decisions on matters of the kind in question. For example, you might mention your willingness to make repairs without charge if the product you had sold developed trouble within the guarantee period and had been used for the purpose intended.

Next should come the facts that you consider relevant and perhaps your reasoning as you considered these facts. In refusing to grant a request that is based on incomplete or inaccurate statements, present the facts that justify your decision but do not go out of your way to prove that the reader is guilty of misstatements. Nothing is gained by making your reader feel embarrassed. If you show up the weakness of his case too devastatingly, he may dislike dealing with you in the future, or may become angry at you in order to escape a low opinion of himself. Likewise, in trying to demonstrate that you are not to blame for something that has gone wrong, do not make a point of putting the blame on the reader. Most of the time, the less said about blame, the better.

Having stated the reasons for your decision, you may logically go on to state the decision itself. Be tactful as you do so. Avoid the word *refuse* if possible. It would be better, for example, to avoid such phraseology as, "We must therefore refuse to grant your request for a reduction of our charge from $275 to $200" and say instead, "We believe that in view of these facts, our charge of $275 is correct." The effort to be tactful should not be carried so far, however, that it makes your answer unclear or evasive.

When your decision has been stated it is usually desirable to add a few lines turning attention to the future. You can say that you look forward to a continuance of pleasant and mutually rewarding relations, that you would like to be of service in the future, or whatever else the specific circumstances permit you to say. Though any idea or attitude can be carried too far, the general principle of ending a letter on a cheerful, forward-looking note is sound.

It is possible, of course, that a letter constructed along these lines may not have the desired effect; but after all, letters cannot achieve miracles. Sometimes no amount of skill in saying "no" can prevent the reader from feeling resentment. But when refusal is necessary, the method explained should hold resentment to a minimum.

The letter that follows is an example of a well-written refusal to make an adjustment.

Gentlemen:
   I have received your letter telling us that the sidewalk we laid for you, leading from your parking lot to the rear door of your building, is not holding up as well as you had expected.
   We make it our policy to stand behind our work, so I sent one of our people to inspect the walk that you refer to. He agrees that it is in poor condition. He also learned, however, that loaded trucks often use it to get closer to the building to unload.
   As you requested at the time, we designed and laid an installation suitable for any amount of pedestrian traffic. We could, of course, have provided a driveway, but its cost would have been substantially higher. Neither the preparation of the subgrade nor the thickness and reinforcement of the concrete for a sidewalk make it strong enough to be used by trucks.
   We feel that the deterioration of your sidewalk has occurred not because our work was defective but because it was used for a pur-

pose for which it was not designed. Under the circumstances, I believe you will understand why we cannot agree to repair it at our own expense. We would be glad, however, to do a repair job—or to build you a driveway if trucks must use it, at the lowest cost that will permit us to do a satisfactory job.

If you would like to have us submit an estimate, let me know and we will do so immediately.

Sincerely yours,

Letters Granting Adjustments. Sometimes you may need to grant an adjustment on an occasion that is important enough to call for more than a routine consent. In this event, the letter differs from a refusal in one important respect: The action being taken should be indicated extremely early in the letter. Thus the letter will contain: (1) an acknowledgment of the complaint (though the term *complaint* would not be used); (2) a statement that the request for an adjustment is being granted (these two parts may be combined); (3) whatever expressions of regret are called for unless regret has already been expressed in the opening; (4) whatever explanation seems necessary; (5) an attempt to turn the reader's thoughts away from the occasion for complaint and focus them toward the future.

This order is psychologically sound. The reader will listen to explanations more understandingly after he has learned that the request has been granted.

When as a matter of policy you must grant an adjustment to which you do not feel the claimant is entitled, you should grant it graciously. Some writers will offer all the reasons that might justify a refusal before finally saying that an adjustment has been granted. By adopting this grudging, argumentative method, they sacrifice the good will that they are making the adjustment in order to retain. A letter will not have the desired effect if it shows the reader that though his request is being granted, he is considered unreasonable to have made it.

The letter that follows is an answer to a complaint in which an adjustment was granted.

Dear Mr. Smith:

I was sorry to learn from your letter of March 10 that the tile drain we installed for you has not been functioning properly, and agree that it is our responsibility to take care of the trouble.

When I received your letter I sent one of our foremen out to see what was the matter. By means of a probe he discovered that the tile was 15 inches under the surface of the ground. Apparently, it froze during the cold spell last month; and when the weather changed and the heavy rain began, it did not thaw out and drain away the water as fast as it should have done.

It is highly unusual for the ground to freeze so deep in this area, and I think that a recurrence of the trouble is very unlikely. However, we stand behind our guarantee, and if you feel it necessary, as soon as the weather permits, I will send out a crew to dig up the tile and bury it deeper. Though this means that the work we did last summer will have to be done over again, which adds to our expense, there will be no charge for this service.

I am really sorry that the drain tile did not solve your problem when the rains came; but since I have never known of the ground in this area freezing to a depth of two feet, I believe that there will be no trouble in the future.

Very truly yours,

LETTERS URGING ACTION

Many letters written for the purpose of urging action do not fit into any of the types discussed elsewhere. There are letters intended, for example, to urge the reader to contribute to some fund, to join a program for smoke or erosion control, or to join some organization. The writing of such letters may be demanded of men in any business or profession.

Letters of this type resemble sales letters closely enough that they can often be based on the sales letter pattern. This pattern involves four steps: (1) attracting attention and securing interest; (2) arousing desire; (3) convincing the reader that the act he is urged to perform will satisfy that desire; (4) securing action. The key words are *interest, desire, conviction,* and *action.* Though the lines that mark off one step from another need not be sharply drawn, the four functions must always be performed in any letter that attempts, under one cover, to persuade the reader either to make a purchase or to perform an action of some other kind.

To obtain *interest* it is best to get the letter under way with a minimum of the preliminaries that usually occur at the opening. The reader should immediately encounter something that he himself is concerned with rather than material that concerns the writer. If the opening can suggest that the reader's own welfare is involved, or the welfare of his organization, so much the better.

One of the poorest openings encountered recently was in a letter by a social fraternity attempting to raise funds. It read: "It seems as though every time we write to the alumni we are asking for something." This showed that the writer himself was uncertain of the appropriateness of his request. Moreover, it put the reader on guard and warned him to steel himself to refuse. The writer should have aroused enthusiasm about the fraternity and its project before letting the reader realize that he was about to be asked for money. In other words, he should have adopted the "You Attitude."

In attempting to arouse *desire,* remember one fundamental point. You do not attempt to *create* desire. Rather, you try to think of some desire that the reader already feels, fan it into intensity, and then associate the action aimed at with its satisfaction. Technical writers can profit by seeing how advertisers use this principle. For example, a person may not desire an air-cooling system; but he does desire comfort, so his desire for comfort is exploited in a letter selling an air-cooler. Similarly, a man may not desire membership in an organization, but be eager for professional advancement; or a company may be induced to join a smoke-control program not so much to provide smoke control as to obtain good will. The starting point in arousing desire is to ask yourself, "What are some of the desires that the reader already feels?" and then, "How can I connect the idea I am trying to sell with one of those desires?"

This is not to urge you to attempt anything specious or far-fetched. The connection you attempt to show should really exist. In fact, if the connection between the action you try to obtain and the desires the reader feels is not genuine, the letter has little chance of success.

In arousing desire you do not emphasize every idea you can think of. You should make one single "appeal" dominate. Other possible appeals may be mentioned but should be kept subordinate. If a second letter is sent, a different appeal may be emphasized; but in any single letter, one incentive is emphasized and the others are kept in the background—visible, perhaps, but not conspicuous.

To secure *conviction,* simply use whatever evidence is best suited to the occasion. No generally appropriate methods can be suggested, as have been suggested for attracting interest and arousing desire. Whatever evidence you offer, you should carefully avoid an argumentative tone. Try to make it appear that you and your reader are both following certain facts to their logical conclusion.

In the effort to secure *action,* three points must be borne in mind: (1) The action suggested should be complete and specific. Rather than "Let us know your decision," you should write, for example, "Check the enclosed card to indicate your decision." (2) The action should be made as easy as possible. For example, it might be helpful to supply an envelope or card that is already stamped and addressed, or perhaps you can supply blanks to fill in rather than forcing the reader to write a complete letter. (3) There should be an effort to make the action immediate, even if delay would do no harm. Once the reader starts postponing action, he is likely to do so indefinitely, for his impulse to act is stronger at the moment he finishes reading the letter than it will be later, when its contents have faded from his memory. Therefore it is well to offer some plausible reason he should act at once, preferably some benefit to him if he does so.

Certain final cautions are needed in regard to a letter aimed at action. First, the opening paragraph should be brief. This will make it look easy to read and thus encourage the receiver to start reading. If your action letter is a form letter, the hardest problem, quite often, is to induce the reader to read it at all. Second, the length of the letter needs careful consideration. If it is too long, a reader may not complete it—in fact may not even start it; but if it is too short, he finishes reading it before it has time to make any impression on him. Finally, you should not discuss any expense that you are asking the reader to incur until you have aroused his enthusiasm about the project concerned. Talking about expenses too early arouses the reader's defenses and dampens his enthusiasm.

An example of a letter urging action is in the section that covers form letters.

FORM LETTERS

Everyone is familiar with form letters, yet they pose some problems in writing that you may not realize when reading them.

A form letter is reproduced in quantity, so it is less impressive than a personally written letter, for the reader immediately realizes that the message is of no more concern to him than to many other people. This is a disadvantage it is frequently necessary to accept, however, when the same message must in fact be sent to many people and when the time and expense of individual letters would be prohibitive. Moreover, the disadvantage can be somewhat reduced by obtaining a good reproduction job on a good grade of paper and by sending the letter as first-class mail in a sealed envelope. Sometimes, also, individual addresses and salutations can be inserted by typewriter, so that the effect is less impersonal. As any of these measures is considered, its expense must of course be taken into consideration.

Unless individual inside addresses and salutations are inserted with a typewriter, they are the parts of a form letter that pose the first problem. Sometimes it is possible to improvise two or three lines to replace the inside address, as for example:

To the Officers and Advisors of
Student Chapters of
The American Society of Civil Engineering

Gentlemen:

More frequently, however, the inside address is omitted, and except for the heading, the letter begins with the salutation. This can be *Dear Sir* or *Gentlemen* unless the letter reaches both men and women. Sometimes, instead, you may be able to use more specific salutations such as *Dear Purchasing Agent, Dear Subscriber, Dear Member, Dear Motor User,* or *Dear Friend.* These variations should be carefully chosen, however, for they are in danger of seeming too coy. Still another device is to replace the salutation with such a line as *To All Users of Drafting Instruments* or *To All Users of Power Saws.*

The only other question of form arises in connection with the signature. Sometimes, in form letters, an effort is made to make the letter seem more personal by reproducing the writer's handwritten signature. Often, however, the primary signature (that is, the first line below the complimentary close) is typed, usually capitalized throughout.

There is no reason that the body of your form letter really needs to differ from the body of individual letters, either in form or content. However, since a form letter is not likely to be dictated and sent without revision, you can often spend more time planning its layout. For example, you can emphasize a paragraph by widening the margins, or you can use the device of having the first paragraph end with three double-spaced periods in the middle of a sentence, and thus tease the reader into the second paragraph. Sometimes these eye-catching devices help to offset the disadvantage a form letter suffers in comparison with an individual letter, but they should not be so extensive that they make a letter look cluttered.

The soundest way, however, to compensate for these disadvantages is to take more care with a form letter than it is possible to take with most individual letters. Having more opportunity to revise, and being free to write a form letter when you can do it justice, you can often produce a much better letter than you can dictate as part of the day's routine.

One more caution is needed: The message of a form letter must be so expressed that it is appropriate for *every* person who will receive it. For example, if some of the receivers are in universities and others in business, or if some have been members of an organization and others are not yet members, the phraseology must be appropriate for either group. And if the letter is to be sent at different times of the year as need arises, it must avoid seasonal allusions.

Following is a form letter urging action.

Dear Sir:
    As you know, the southern entrance to our city of Lakeland is unattractive because the highway runs parallel to the levee that was constructed twenty years ago as protection against high water that occurs during the spring runoff of the Canterbury River. Since you are a member of one of the service clubs that have always been interested in civic improvement, we are writing to solicit your help in getting rid of this unsightly feature.

The Army Corps of Engineers has scheduled a meeting in room 116 of the courthouse, to begin at 8:00 p.m. on Thursday, April 16. Their purpose is to determine whether our community would be seriously in favor of having the height of the levee reduced, having its present slope made more gradual, and having it landscaped where it runs beside the highway.

The Corps now has the authority to do this work as a pilot program like the one that has already led to a similar undertaking at Longchamps on the Weaver River.

Reducing the height of the levee would not create a danger of flooding. The two dams that have been built upstream since the levee was constructed have ample storage capacity to hold back excessive runoffs.

Whether this project is placed high on the Corps' list of priorities depends a great deal on whether the citizens of Lakeland show a keen desire to have it undertaken. If we let this opportunity slip by, there is little likelihood that we will have another like it for many years.

Feeling sure that you would like to have a conspicuous and unsightly stretch of rocks and dirt replaced by a parklike stretch of green lawn, shrubs, and trees, we urge you to attend the meeting and speak out in favor of this new project.

Sincerely yours,
John R. Doe, President
Lakeland Civic Improvement Society

ASSIGNMENTS

In writing the letters called for in the following assignments, make up headings, addresses, and signatures. When such an assumption is reasonable, assume that the letterhead occupies the top two inches of the page and use only a dateline as a heading. If you type your letter, place it properly on the page.

In developing the details of a situation, you may need to assume that the date is different from the date on which you actually write the letter. You may also need to make up additional details. However, you should *not* make up facts that would change the essential nature of the problem. In particular, do not yield to the temptation to imitate the specimens that appeared earlier in the chapter. Rather, try to develop your own skills.

1. Write a letter assuming the existence of the following circumstances: You are responsible for obtaining some well-known person to serve as a judge in a contest or to speak at a meeting. The con-

test might be something such as a competition in speech, a science fair, a competition to create a landscape design, a music contest, or the choice of school newspapers or annuals to receive awards at a high school journalism contest. The meeting could be a retirement ceremony, a banquet where trophies are to be awarded, or the launching of a drive to build a retirement home.

2. Write a form letter to be sent to numerous people inviting them to attend an event of the kind described in Assignment 1. (If this is to be an invitation to a banquet, do not offer complimentary tickets.) Though this is a form letter, it is to have an individual name and address on each copy, so provide what is necessary. If study assignments have not already covered form letters, you should read that section before writing the assignment.

3. Assume that you have received the letter called for in Assignment 1 but are unable to accept the invitation. Write an answer. If it seems appropriate, offer to have someone else in your organization come in your place. (How definite such an offer should be would depend on the role that you have been asked to play.)

4. Write a letter of instructions to be sent to a substantial number of people. Those you are writing to might be serving in several areas as sanitary inspectors, for example, in such establishments as restaurants or migrant labor quarters; they might be safety inspectors responsible for conditions in nursing homes or rest homes; they might be proctors in dormitories, or faculty responsible for counseling students during registration. The letter might be either a first set of instructions, an effort to improve performance, or a change in procedure. If you choose a situation in which past performance has sometimes been poor, take pains not to antagonize those whose performance has been satisfactory.

5. Write a letter of instructions to be sent to a single person. A possible subject would be an investigation requested by a district office about the qualifications of someone who wished to obtain a dealership or franchise for a water-softening service in residences. It could be assumed that the instructions are sent to a company dealer in a nearby town.

6. Write a letter of inquiry addressed to one person under one of the circumstances described below: (a) You have been approached to see whether you would be interested in a dealership such as the one referred to in Assignment 5, and you write to a dealer in a nearby town asking for information that will help you come to a decision. (b) You plan a long trip abroad and write to someone who has used the services of a certain travel agency to see whether they have been satisfactory. (c) You are responsible for securing lodgings for an athletic team or other group in a distant town, and you write to someone in a similar position in another school to see whether a certain hotel his team used was satisfactory. Though you may ask for general comment in whatever letter you write, include several specific questions.

7. Write a letter answering an inquiry and providing most or all of the information requested. This may be an answer to an inquiry called

for in Assignment 6 or to some other imaginary inquiry that you have the information to answer.

8. Write a letter answering an inquiry—in this case an answer that contains little if any of the information requested. The specific situation may be similar to one suggested in Assignment 7.

9. Write a letter asking for an adjustment based on the situation that follows or on a similar case of your own choice: You are the manager of a motel or hotel where an athletic team from a school in another town spent one night or more. After the team had departed, it was discovered that four towels and one blanket were missing, that the shade of a floor lamp was ruined, and that the arm of an easy chair was loosened. Write to an official of the school in question asking reimbursement for the damage and the return of or reimbursement for the missing articles. (You may estimate the amounts.)

10. Write a letter based on the following case or on one that is similar. (The letter is to indicate that action will be taken unless an undesirable situation is remedied.) You manage a hotel. A women's organization has written to ask that you permit them, without charge, to use one of your two large dining rooms for special groups for a used book sale, the proceeds of which go to some good cause. You have done this during each of the two preceding years, but it was understood that they would clear the room and rearrange the tables and chairs when the sale was over. They failed to make good on this commitment last year, and you had to use your own employees to do the job in a rush because a lunch meeting was to be held there the next day. Answer the letter, giving permission to use the room but tactfully letting them know that future use will depend on whether the clean-up job is done as agreed upon.

11. Write a complaint—one that does not ask for an adjustment—based on the following case: You own a home in an attractive residential neighborhood. A crew has recently worked in your street, cutting out parts of trees that were too close to power lines. You feel that they have gone to needless extremes and have damaged the appearance of your home and others on the same block. Write to the power company letting them know your feelings—with reasonable restraint —and trying to prevent such damage when the trees are cut back again.

12. Write an answer to the letter called for in Assignment 11. Do what you can for the sake of good will. Assume that your company hires a contractor to do such jobs, so the work was not done by one of your own crews, but promise to bring the case to the contractor's attention and insist that he do no more cutting than is necessary. You might stress the necessity of keeping branches away from the power lines for the sake of safety and to prevent outages.

13. You are connected with a company that does a mail order business. A man writes to say that he purchased a tent or some other article ranging in cost from $75 to $100, for which he is paying in monthly installments. A friend of his ordered the same article "a very short time later" and was charged $15 less. You find the facts to be as follows: The writer ordered the article early in August, while the

camping season was still at its peak. On September 10 you issued a catalog offering reduced prices on summer-seasonal merchandise because the season was practically over. You cannot make it a policy to sell articles at clearance prices before the date when clearance is necessary because of the change of seasons. Write a letter declining to make a price adjustment.

14. You are a general contractor operating on a small scale. A home owner writes to tell you that water still leaks into his basement though one of your men supposedly plugged the leaks. He also says that your statement charges him for three and one half hours of labor, but that the man was on the job an hour less than that amount. The writer feels that you should do the job over again without charge.

Facts not in his letter are as follows: You had told the man that you could not guarantee an inside job; but outside repairs were prevented at the time by the weather and, besides, would cost substantially more. You were given the go-ahead under such circumstances. Also, the workman who did the job on an emergency basis had to leave another job to do it, had to go to a store for materials, and had to drive back the five miles, afterwards, to the job that he had left. This accounts for the apparent overcharge for time. Write a letter explaining these facts, but saying that in an effort to be fair, you will reduce the labor charge to the amount your customer says it should be. You will not do the job over without charging him. Try to sell the customer on the idea of waiting until the weather is good and then having the leak patched from the outside, which will cost more but can be guaranteed.

15. Write a form letter 250 to 300 words long in which you try to convince the readers to perform some action such as promoting a project desired by your school, supporting or opposing a zoning change, changing the season for hunting—anything of current interest in your school, your home community, or your future profession. As an alternative you might consider soliciting pledges to support some program or activity.

The date may need to be different from the date on which you actually write. Those who receive the letter should have something in common, such as being farmers, parents, sportsmen, people interested in public health or environment. Supply an inside address that would be suitable for all who receive the letter. (Unless the sort of group to which the letter is sent is obvious from the letter itself, add a note giving this information to the instructor.)

16

LETTERS

CONCERNING EMPLOYMENT

For a person about to graduate from college, the problem of securing employment varies from year to year depending on the current job market. Even in poor years, however, large corporations and governmental units send representatives to colleges and universities for the purpose of recruitment to such an extent that a letter of application is not always the first contact between the employer and the prospective employee. On the contrary, when a graduating senior writes a letter in connection with employment, he often has been interviewed already, has received an application blank, and may have filled it out and submitted it. As a result of the interview a company representative will have formed and recorded his impression about the applicant's appearance and personality to supplement the facts about qualifications which he obtained from the applicant and from other sources.

Under such circumstances, the traditional instructions about how to write a letter of application are not entirely applicable. Rather than starting from scratch and writing a full-scale application, the applicant will base his letters on the way he sizes up the situation as a result of the interview and on any leads he may have picked up regarding what he should emphasize among the many things he may feel justified in saying about himself.

Still, the full-scale letter of application is not a matter to be neglected. There are still numerous occasions when it may be written by students seeking their first employment after graduation, and it is as important as ever to the person who already has a position but wants to apply for something better in a different company. Therefore this chapter will first cover the traditional letter of application, because even those who do not need to write the full letter may need to use parts of the material it contains. Next, attention will be centered on letters written after an applicant has been interviewed. And finally, some concrete suggestions that apply to letters written under either circumstance will be presented.

THE CONVENTIONAL LETTER OF APPLICATION

The Opening

When you write a conventional letter of application for a specific position, start with a short paragraph indicating the position you are applying for and telling, perhaps, where you learned that the position was open. If necessary, identify yourself—for example, as a graduating senior or as the occupant of whatever job you hold. Make it clear that your letter is definitely an application, if such is the case, rather than just an expression of interest that includes some facts about yourself to justify a request for information. In general, let your opening paragraph merely establish contact and provide the information that will enable your reader to judge your qualifications without having unanswered questions in the back of his mind.

Try to put something that will arouse your reader's interest either in the first paragraph or in the opening of the second paragraph if you can think of anything that will serve such a purpose. This might be, for example, a reference to some extraordinary qualification or to an unusual combination of qualifications. Not everyone is fortunate enough to be able to include such a bid for interest, but if you can somehow, at the outset, set yourself apart from the run-of-the-mill applicants, what follows is more likely to receive attention.

The Central Section

The central section should follow smoothly after the opening. Its contents will often fall naturally into sections on education, experience, personal information, and references, but there is no reason to regard this pattern as mandatory, nor to arrange the sections in this particular order. The best arrangement is one in which the strongest points are placed first. It is easier to convince a reader that you are a strong candidate at the beginning than it would be later, when the first sections have already created the impression that your qualifications are about on a par with those of other applicants.

Education. As a minimum, name the higher educational institutions that you have attended and the degrees you have earned or are about to earn. If you have an outstanding academic record, mention it. Do not hesitate to name your extracurricular activities, for they help to round out the impression about the kind of person you are.

The amount of detail about education should be governed by the amount of other information you expect to present, the strength of your scholarship, and the extent to which your academic training is directly related to the work you would do on the job. Facts about education that one applicant should mention—for example, courses taken—might well be omitted by an applicant whose other qualifications are stronger.

Experience. Nothing is better evidence of your fitness for a job than successful experience in a job of similar nature. If you have had such experience, mention it early. Do not feel, however, that experience in a job of any kind is unimportant. In almost any kind of work, you have had a chance to demonstrate good or bad personal qualities, and it is a well-known fact that personal qualities are often a decisive factor in an employee's success or failure. In general, the picture of your experience should be complete. If anything is omitted, there will be noticeable gaps of time in your record, and these will make a reader wonder whether you have something to conceal.

Personal Information. As a minimum tell your age, place of birth, weight, height, health, and marital status. It might be desirable also to mention freedom from physical handicaps. Reference to leisure interests will sometimes help the reader to think of you as a person. Additional facts about your background help to make the picture complete and are especially likely to be valuable if your background is in any way related to the field you are seeking to enter. Of course the value of personal information, like the value of any information, is relative. Whether certain facts are worth including may depend on how much else you have to say, and it might be helpful for one applicant to include information that another could omit.

References. You should usually name from two to four references, preferably former employers or instructors. If you are just completing your work in college, include at least one reference from the institution that is awarding your degree. The number of references should of course depend on how many are necessary to confirm what you have said in your letter. Courtesy and self-interest make it advisable to secure the consent of anyone whom you name as a reference. And by all means make sure that you have the name (spelling included) and position of each reference absolutely correct. Unless it is obvious, make sure that the reader knows why a person whom you name as a reference is qualified to judge your fitness for the job you are seeking.

Though the logical place to list references is the latter part of the letter, do not hesitate to mention an especially important reference early—even in the opening paragraph. The name of an important person or of a person whom the reader knows personally is effective in arousing interest.

The Ending

Because a list of references does not make interesting reading and because the references are usually named near the end of the letter, you will improve your letter if you can say something to give it a lift at the end. A single-sentence summary of your qualifications might serve the purpose. Another possibility would be a comment on the work you have done to qualify yourself for such a job and the effort you will consequently put forth to make good. Whatever you may say to strengthen your ending must be the outgrowth, of course, of material that your letter has included, and it is therefore impossible to offer a formula for your guidance.

The major function of the ending is to suggest action. This action may be the offer of a job, the invitation to come for an interview, or a letter providing information or telling you what to do next. You will need to appraise the situation, decide what action is the next likely development, and phrase your suggestion accordingly.

Try to phrase your effort to secure action in such a manner that it avoids negative suggestion. When you write, "*If* my qualifications are of interest, I shall be glad to come for an interview," you imply uncertainty whether he will be interested. Though you should not appear to take too much for granted, you might avoid this implication by writing something such as, "I shall appreciate your consideration of my qualifications and shall be glad to come for an interview on any date that you suggest." Such an offer gets rid of the *if* without seeming overconfident.

Unless your potential employer is hard-pressed for help, it is best to refrain from trying to hasten his action. If it is essential, however, that you secure an early response and if you are tactful, you might be able to speed matters along without giving offense. No resentment would be aroused, for example, by such a sentence as, "Since I shall be leaving town for one month on June 10, I should appreciate it if I might have an interview before that date." Most employers are normally considerate and are not likely to resent a reasonable request tactfully made. But pressure must be applied with care or your effort to obtain early action may reduce the chances that action will be favorable.

The Use of a Resumé

When you have not filled out an application blank, you will probably find it advisable to use a resumé. The resumé is a complete picture of the facts, in tabular form. It should bear your name and address (future as well as present address if a change is imminent), should be dated, and should indicate the position you are applying for. It should contain the routine personal information that would otherwise be placed in the letter, and should give full information on education, experience, and references. More details can be included in the resumé than it is usually advisable to place in the letter itself.

There is no standard form for the resumé, but it should be a tabulation rather than a discussion in sentences. The major headings are likely to be *Personal Information, Education, Experience,* and *References,* under which may be used any subheadings needed. To decide on the system of headings, jot down as rough notes all the facts to be covered, and then work out a plan of headings to cover them. Arrange the material on the page so it has a pleasant appearance, and make it as neat and accurate as the letter itself.

A clear understanding of the relationship between the resumé and the letter is important. The function of the resumé is to free the letter from certain routine information that is uninteresting but necessary for completeness. In the resumé you provide the full record. In the accompanying letter you focus the reader's attention on the particular facts that are most important, and bring out their full significance.

A resumé cannot perform all the functions of a letter; rather, it performs the functions of an application blank. However completely the resumé may present the facts, it remains for the letter to create a personal interest, for it is only in the letter that you seem to be talking to the reader personally.

LETTERS FOLLOWING INTERVIEWS

As mentioned earlier, for a student who is about to earn a degree in a large or middle-sized institution, the routine of obtaining a position may begin with an interview on the campus and filling out of an application blank. If he writes a letter at all, it will be only to supplement the considerable information already in the hands of the potential employer, or to cope with some later development. The person who writes under these circumstances may be able to apply some of the preceding instructions but his letter will not stand independently as the major instrument for obtaining employment.

Many authorities feel that after an applicant has been interviewed he can help himself by writing a letter even if he must use his imagination to create a pretext for doing so. One possibility is a letter ostensibly written just to say "Thank you for the interview," but expanded by the addition of anything else that the writer thinks would be helpful. The conditions that can exist after interviews are so varied that it would be impossible to generalize on the question of whether an applicant's chances would be helped by a letter written on mere pretext. An applicant has to use his own judgment about whether to write a follow-up letter. The decision should certainly be based in part on the atmosphere that developed during the interview.

Assuming that you face this problem and do decide to write a follow-up letter (either because of a concrete development or on a pretext you have decided would be plausible), what kind of material can you use? Some of the possibilities are as follows:

> You can answer or supplement your earlier answers to questions that were not entirely settled.
> You can describe the growth of your interest in some matter that you had not thought about intensively before the interview—for example a kind of work that is different from what you had had in mind.
> You can dwell upon facts about the company that increase its appeal to you.
> You can say that you have been giving additional thought to what qualifications you can offer in some particular respect, and supplement what you said earlier on that point.
> You can say that you have been thinking a great deal about some remark made by the interviewer, and review some aspect of the case in the light of this added consideration.

Admittedly it takes imagination to translate these general ideas into definite, concrete terms. There is no guarantee that every applicant can produce a letter that will help his prospects. But it is probably true that most job seekers, after an interview, think of things they wish they had said, or wonder whether they made themselves entirely clear on certain points, or think of questions they wish they had asked. The exercise of ingenuity will often make it possible to use some such afterthought as a point of departure for a letter.

And what can such a letter do for you beyond supplying concrete facts not supplied before? The answer to this question depends largely on your adroitness as a writer. Assuming that you write it skillfully, such a letter can help you to gain credit for initiative, originality, imagination, and aggressiveness. It can convey the impression of a lively interest in the company, beyond the fact that you need a job.

Also, it can demonstrate your literacy and your skill in writing—a skill that improves your potential value in any large organization. Just as there are some people who write badly but make a good personal impression, there are others who can present themselves better when they write than when they talk. If you are one of these, your letter may help to counterbalance your conversational shortcomings. And even if you are fortunate enough to make a good impression in a conversation, you will strengthen that impression by showing that you write well in addition to talking well.

Of course if you are weak in writing, there is no reason to feel that you can help your prospects by writing when it is not essential to do so. The best short-range policy under these circumstances is to keep every letter as simple as possible and at least to avoid the pitfalls that are pointed out in the remainder of this chapter. The only suitable long-range policy is to improve your skill in writing rather than reconciling yourself to a weakness in one of the most basic skills in science and industry. Poor writing abilities are a handicap that will pursue you during your entire working life.

GENERAL SUGGESTIONS

Neatness

Neatness is a must. Before the person who receives your letter can begin to read it, he forms an impression about you on the basis of its appearance which affects his reception of its contents. If sloppiness and carelessness are qualities of your letter, he associates you, at least subconsciously, with those qualities, and you are handicapped from the outset.

Further, a failure to make your letter neat is not very flattering to those to whom you send it. You force your reader to conclude either that you regard him and the position as unimportant or else that you are satisfied to turn in an inferior performance on an important occasion. You can hardly expect him to believe you will do your utmost to serve the interests of his company if you are careless even when working directly in your own interest.

In view of these facts you should no more send a slovenly looking letter than you should appear for an interview with uncombed hair and a dirty shirt.

The "You Attitude"

Don't let your desire for a position make you overlook the fact that a potential employer is less interested in your anxiety for the job than in the question of whether you can become a valuable employee. Your tastes, ambitions, and desires are irrelevant unless they will cause you to do better work. Sometimes, to be sure, they will improve your work. The fact that a job is in line with your long-range ambitions might make you try harder to do it well. Your liking for a certain region might reduce the likelihood that you would resign because of a desire to live elsewhere. Still, you are wasting space if you talk about such matters without relating them to the way you will function as an employee.

In this connection, before you apply for one specific position you will do well to carefully consider the demands of the job as well as your own qualifications, so that when you write your application you can make it clear to the reader that you have done so. The effort to create such an impression must be made carefully; you do not wish to appear to be telling the reader what he wants. But if you do the job well, you can suggest that before you decided to apply, you carefully considered the question, "Is this the right kind of job for a person with my qualifications?" Every one of us knows things about himself that no one else knows. Your potential employer realizes that try as he will, he cannot discover all of your limitations. If you can convince him that you yourself have coolly analyzed your strong points and your weak points and are applying for the job that your abilities qualify you to perform, he will take your application far more seriously.

Objectivity

Keep your letter objective. You will gain nothing from making unsubstantiated claims that you are self-reliant, dependable, tactful, and the like. Possibly you are, but such claims are just as likely to appear in the letters of applicants who have no justification for making them. Anyone who has had much experience in hiring employees knows this, and nothing will lose his interest faster than claims that are not supported by objective evidence.

Tone

A satisfactory tone is usually the result of striking the right balance between extremes. You should aim to appear self-confident but not conceited; obliging but not ingratiating; earnest but not pompous.

The tone of your letter will result in part from your choice of language. Many a person whose manner in a conversation is entirely normal becomes stiff and pompous when he writes a letter of application. Phrases such as *earnest desire, intense determination,* and *it is my wish to* are unlikely, however, to increase your chance of obtaining a position.

On the contrary, they suggest that you are so eager to be impressive that you have been unable to express yourself in a natural manner. Everything said about the language of business letters in Chapter 14 applies with special emphasis to letters of application; and if it is not fresh in your mind, you would do well to review the section dealing with that point.

Excessive Use of "I"

Since you are telling about yourself when you write a letter of application, you will probably find it difficult to avoid using "I" at the beginning of every sentence, and thus making your letter monotonous. It is better, however, to use *I* liberally than to try so hard to avoid it that your sentences become involved and unnatural, or that you sacrifice parallel structure where it might be appropriate. The best way to solve this dilemma is to write the first draft of your letter without worrying about how often *I* appears and then, as you revise it, change a sentence here and there until the use of *I* is no longer conspicuous.

Discussion of Salary Expected

Much of the time you will know whether a position pays an acceptable salary before you apply for it; and even when this is not the case, you are usually in a better strategic position if you let the employer be the first to bring up the subject. If you feel it necessary, however, to mention the salary you would expect, the place to do so is near the end of the letter. Your first job is to convince your reader that your services would be of value, for until he is convinced of this, he is not interested in what it will cost to secure them.

Making Sure All Questions Are Answered

Before you consider your letter complete, make sure that it will not leave any unanswered questions lurking in the reader's mind and making him uneasy. For example, make it easy for him to see that there are no substantial periods of time left unaccounted for in your history. Also, unless it is apparent that the position you apply for is an advancement, let him know why you prefer it to the one that you already hold. If you are not working, let him know the reason unless it is obvious, as it will be of course if you are just finishing school.

Failure to make the picture complete by the inclusion of such information may delay a decision about whether to hire you until there has been another and needless exchange of letters, or may even result in the job's going to someone else who seems to be qualified and who has provided all the desired information.

Seeing Things as the Reader Sees Them

One final suggestion is in order. When you are keenly interested in a job and sincerely convinced that you could do it well, you will be strongly tempted to make statements about yourself that sound less impressive to someone else than to you. You will probably feel that these statements are justified, and you may be right in thinking so. The reader, however, does not see things through your eyes. It is therefore a good idea to put yourself in his position. Imagine yourself receiving the letter—from someone whom you do not know any more about than the reader knows about you. Ask yourself, "If I were the reader, would these statements impress me as I hope they will impress the man I am writing to?" If you are capable of using your imagination in this manner, you may eliminate some lines that sounded good to you when you wrote them, but by doing so you will produce a letter that is not diluted by ineffective material and will therefore be more likely to carry conviction.

SPECIMEN LETTERS OF APPLICATION

Following are two letters of application, each accompanied by a data sheet. These letters are in line with the principles described in the preceding discussion; but they are specimens, not models. If a letter of application is assigned, do not borrow phraseology from the specimens.

There is no reason, however, that you should not adapt the specimen data sheets to your needs, for a data sheet is merely a tabulation of facts and makes no pretense of being a personal communication.

1149 Park Hill Road, Apt. C
Corvallis, Oregon 97330
April 10, 197_

Personnel Director
Environmental Protection Agency
National Field Investigations Center
Denver Federal Center
Denver, Colorado 80225

Dear Sir:

I would like to apply for a position as Field Investigator with your agency. I shall be receiving my Bachelor's degree in Forestry from Oregon State University on June 10, and shall be available at any time after that date.

As indicated in the attached resume, I have worked two summers as a Fire Guard. Between fires, I was assigned to a variety of duties, but the one most applicable to the kinds of tasks I would undertake with your agency was the Camp Ground Patrol duty. For this duty, I was required to check the two camp grounds in the Big Creek district to make sure campers were obeying regulations and were not causing needless damage. This duty involved contact with the public, and frequently required a great deal of diplomacy in convincing campers to obey the rules. I wore no uniform and had no real power except to report to the ranger, so I had to use persuasion. Throughout the period I had this duty, I had to resort to calling the ranger only once. It seems to me that the skills I used would be similar, though of course not on the same scale, as the skills I would need as a member of your agency in discussing possible violations with individuals being investigated.

With this in mind, besides the regular courses leading to a Bachelor of Science in Forestry, I have taken courses in speech, persuasion, group dynamics, and psychology. They have taught me a great deal about people, and as a result they have made my contacts with the public more enjoyable for me, as well as, I think, more successful.

I have always enjoyed the outdoors, and have for a long time planned to find work in an area where I could help so that others could enjoy it as I have. I feel that a position with your organization would help me achieve this goal, and I feel I could contribute to your agency's goals as well. If you feel that I might fit your requirements, I would be available for an interview at your convenience any time but the first week in June, when I take my final exams. Thank you very much for your consideration.

Sincerely,

John Q. Doe

John Q. Doe

FIGURE 17. Well Written Application Letter. The writer has not done the exact kind of work applied for but makes good use of other material.

John Q. Doe

School address: Permanent home address:
    1149 Park Hill Road, Apt. C   Box 67, Route 4
    Corvallis, Oregon 97330   John Day, Oregon 97845
Phone: 503 754-1266   503 693-7659

Date of Birth: July 10, 1955   Place of Birth: Bend, Oregon

Education:
    John Day Consolidated High School   B plus average
        Lettered three years in basketball and track; wrote for school
        paper; was on debate team.
    Oregon State University   B average
        B.S. in Forestry, expected June 10, 1975
        Intramural basketball, four years; active member of the
        Outdoor Program, four years; campaigned for student
        government; was a volunteer Big Brother; member of Forestry
        Club four years, secretary one year.

Work experience:
    Ranch work: part time and summers on my grandfather's ranch
        while in high school and first summer in college.
        I did everything there is to do on a ranch including set fences,
        drive a truck, plow, nurse sick animals, paint outbuildings,
        and help out generally. The last summer I was overseer of one
        of the haying crews. I think this experience has helped me
        become more versatile in meeting unusual situations and has
        helped me become self-reliant.
    Fireguard: summers of 1973, 1974, Big Creek, Montana.
        As a Fire Guard I worked on six two-man fires and two crew-
        sized fires. Between fires I strung telephone wire, planted
        trees, cut slash (debris left from logging), picked pine cones
        for seed, and had Camp Ground Patrol duty. The second summer
        I was head of a ten-man planting crew.

References:
        William B. Gardner   Thomas H. Jackson, Ph.D.
        Chief Ranger   Forestry Department
        Big Creek Ranger District   Oregon State University
        Columbia Falls, Montana 59912   Corvallis, Oregon 97331

        George C. Douglass   Mark Baker
        John Day High School   Assistant Professor
        John Day, Oregon 97845   Forestry Department
                             Oregon State University
                             Corvallis, Oregon 97331

FIGURE 18. Specimen Resumé (to be sent with letter shown in Figure 17).

389

739 Circle Drive
Pullman, Washington 99163
April 10, 1975

Mr. Eugene M. Johnson
The Asphalt Institute
College Park, Maryland 20740

Dear Mr. Johnson:

During a recent visit to Pullman Mr. Harvey Darrow, who was associated with The Asphalt Institute for many years before his retirement, told me that you usually have openings at this time of the year for two or three chemical engineers. In June I will receive an M.S. degree in Chemical Engineering from Washington State University, and I would like to apply for one of these openings.

I believe I could do satisfactory work for you because both my education and my work experience have been in line with your operations as Mr. Darrow explained them.

My graduate study has been in a field that involved work with asphalt. I participated, for example, in studies at the Riedesel Testing Facility-- studies including both the design and testing of various asphalt mixes to determine their suitability for highway surfacing and other uses.

Before I returned to the university to earn an advanced degree, I was employed for two years by the Superior Paving Company of Richland, Washington and supervised the laying and inspection of highway surfacing. Then for a third year I was a laboratory assistant in the asphalt laboratory of the Corland Oil Company in Martinez, California.

I have talked at length with Mr. Darrow about your work and have read several of your reports describing one or another of your projects. The more I learned, the more interested I became. I would welcome an opportunity to join your staff because your research work is the kind I have been preparing to do and because the atmosphere in which it is carried out, as described by Mr. Darrow, would make my work enjoyable.

To make more information about me immediately available, I am enclosing a resume, but I would be glad to fill out an application blank if you will send one.

I would greatly appreciate any word you might send me about my prospects.

Sincerely yours,

John R. Doe

John R. Doe

FIGURE 19.   Well Written Application.

QUALIFICATIONS OF JOHN R. DOE
Applicant for a position as a chemical engineer

Address   739 Circle Drive
          Pullman, Washington 99163   Date April 10, 1975

PERSONAL INFORMATION

Age - 27   Health - Excellent
Place of birth - Bend, Oregon   Marital status - married, two
Weight - 165   children
Height - 5 feet 10 inches   Leisure interests - golf, chess

EDUCATION

Cheney High School, Cheney, Washington. Graduated in 1966
Eastern Washington State College - September, 1966 to June, 1968
Washington State University - September, 1968 to June 1970
                             September, 1973 until present
  Degrees: B.S. in Chemical Engineering - June, 1970
            M.S. in Chemical Engineering - will receive in June, 1975
  Academic record: undergraduate 2.9, graduate 3.1
  Activities: member of American Society of Chemical Engineers (student
               chapter)

EXPERIENCE

Experience related to professional objective

    July, 1972 to September, 1973 - Laboratory assistant in asphalt labora-
    tory of Corland Oil Company, Martinez, California
    June, 1970 to July, 1972 - Superior Paving Company, Richland, Washing-
    ton. Duties: to inspect and supervise laying of asphalt surfacing
    on roads and airport runways.
    September, 1969 to June, 1970 - Assistant (part time) Materials-Testing
    Laboratory, Washington State University

Experience - General

    Summers of 1967, 1968, and 1969 - Truck driver, A & S Produce Company,
    Kent, Washington

REFERENCES

Duane V. Kellogg   Martin A. Driscoll
Professor of Chemical Engineering   Associate Professor of Chemical
Washington State University   Engineering
Pullman, Washington 99163   Washington State University
                                    Pullman, Washington 99163

Donald R. Newsome, Manager   Frederick S. Whitcomb
Superior Paving Company   Director, Asphalt Laboratory
1860 Front Street   Corland Oil Company
Richland, Washington 99352   Box 895
                                    Martinez, California 94553

FIGURE 20.   Resumé (to be sent with letter shown in Figure 19).

ASSIGNMENTS

Assignment 1

Write a letter of application setting forth your own qualifications for a position that you might plausibly apply for. Let it be assumed that you definitely know that an opening exists and that the reader knows little if anything about you until he reads your letter. In stating your qualifications limit yourself to facts except that if the instructor permits, you may assume that you are about to complete or have just completed the course you are taking. Obviously, this assumption would include your having gained any experience you are likely to gain before graduation. Do not make unrealistic offers, such as an offer to travel farther for an interview than you would really be willing to travel under the circumstances.

Assignment 2

Assume that you had been scheduled for an interview with some potential employer's representatives who were on your campus or in the town where you are located. For some reason, such as an attack of influenza, you were forced to cancel the interview. You have learned that in the near future a representative of this employer will be holding interviews in some other city perhaps fifty or one hundred miles away. Write to him asking him if he will interview you there.

You may assume, if you like, that you have already submitted an application blank or, if you prefer, that you are sending it with your letter. If other assumptions would make the assignment more plausible in your particular case they are also permissible. The basic situation, however, should remain unchanged. In this letter you are attempting to demonstrate your interest in the company concerned and seize the opportunity to say enough about yourself to arouse interest.

Assignment 3

You have been negotiating with a potential employer and have reason to believe that the offer of a position is likely. Another employer with whom you have also been in contact makes you a definite offer. You feel that a position with the first employer mentioned would be preferable, but the one you are offered is good enough that you do not wish to lose your chance for it.

Write a letter to the employer who made the offer. You should probably tell him that you have been negotiating for another position but the major purpose of your letter, at least ostensibly, should be to ask some questions you would like to have answered before you make a definite decision.

Assignment 4

Under the same conditions described in the preceding assignment, write a letter to the potential employer who has not offered you a job, telling him about the offer that you have received. Without letting your letter sound like a threat, let him know that you will soon have to accept or decline the offer you have received. Try to make your reader understand that though a position with his company would be your first choice, the other position is good enough that you are reluctant to turn it down unless you can feel confident of something better. Try in a tactful manner to obtain a definite decision as soon as possible, or at least to obtain definite information about when a decision will be made.

Assignment 5

Assume that you have been interviewed by the representative of a potential employer four or five days ago. Write a follow-up letter of the kind described in the chapter.

PART IV

HANDBOOK OF FUNDAMENTALS

PARAGRAPHS

General Comment

The following discussion is concerned mainly with the ordinary, run-of-the-mill paragraphs that comprise the bulk of most writing, technical or otherwise. It also contains brief comments on paragraphs of isolated statement. To be sure, there are other kinds of paragraphs: introductory, transitional and concluding paragraphs, and paragraphs of dialogue; but the first three of these four kinds do not give rise to much trouble and the fourth is rarely seen in technical writing.

The function of a paragraph is to group together sentences that concern the same topic and combine to form a thought unit. Some of the points to bear in mind in producing paragraphs that perform this function are as follows:

1. Effective paragraphs do not result from merely beginning a new paragraph whenever the one you are writing has grown long enough, nor even from merely beginning a new paragraph whenever you take up a new topic. Your paragraphs will be effective only if you know, each time you begin one, what point you intend it to make or at least what ground it will cover. That is, as you begin a paragraph you should have a definite idea about where you will be at its end.

2. A paragraph is likely to be more effective if its basic idea is expressed in a topic sentence. A topic sentence is usually most useful near the beginning of the paragraph, but it may be placed at the end; occasionally, for the sake of emphasis, it may be placed at the beginning and restated at the end in different words. Wherever it appears, it should dominate the paragraph, and the rest of the paragraph should serve mainly to develop it.

3. "Developing" a topic sentence means providing the additional facts and ideas that are needed to make it clear and acceptable, and possibly to indicate its importance. If a sentence does not need such development, then either the idea it conveys is not big enough to form the basis of a paragraph or else—much less frequently—it should stand alone as a single-sentence paragraph. If an idea cannot be developed in a paragraph of reasonable length, then it should be broken up into ideas of paragraph size.

4. If the topic sentence is merely implied and the reader must grasp the point of the paragraph from the details themselves, the first sen-

397

tence should never fail to create an accurate impression about the topic and the basic idea of the paragraph. A reader who is skimming rapidly should not be misled about the contents of a paragraph when he looks at its first sentence.

5. Sometimes it is necessary, for the sake of fairness and completeness, to include material that runs counter to the main point of the paragraph. This material should not be placed at the end, for the end should strengthen, not weaken, the impression the paragraph is supposed to make. If some facts in a paragraph amount to concessions, the best pattern is likely to be: (1) a topic sentence making the main point clear, (2) the facts that must be conceded—introduced in a manner that shows them to be concessions, and (3) the material that establishes the main point of the paragraph despite the concessions.

6. If it is necessary to use transitional material to bridge the gap between two paragraphs, such material should open the second of the two rather than close the first. Such placement avoids weakening the emphasis of the first, and helps to introduce the topic of the second. Also, since most writers do place transitional material at the beginning of the new paragraph, that is where the reader will expect it to be.

Length of Paragraphs

One need only look at contemporary writing to see that paragraphs vary widely in length. The special kinds that were mentioned at the beginning of this discussion are often extremely short. The same can be said of the paragraphs in popularized treatments of technical subjects, and of paragraphs in publications with columns so narrow that long paragraphs would appear difficult to read.

Nevertheless, normal paragraphs serve their purpose best when they contain from three or four to seven or eight sentences—which would be from 75 to perhaps as much as 200 words. Experience has shown that paragraphs of such length usually present the material to the reader in thought units that he can grasp most effectively at one time. Occasionally it may be desirable to have a paragraph consist of only one or two sentences, but when a substantial piece of writing consists mainly of one-sentence paragraphs, the reader receives no help in determining which sentences are related to the same topic.

Paragraphs of Isolated Statement

Paragraphs of isolated statement are statements that do not need to be expanded upon but do not merge into the development of a single larger idea. For example, in presenting a series of conclusions each consisting of a single sentence, it is often desirable to treat each item as a separate paragraph. The same treatment is often suitable for some instruction material and for various other kinds of material, including much of the material in this Handbook.

Writing paragraphs of isolated statement obviously does not call for any special technique. They should be recognized as a distinct and acceptable type, but their existence should not be regarded as justifying the excessive use of one-sentence paragraphs when the subject matter calls for related rather than independent statements.

CORRECTNESS IN STANDARD USAGE

The following discussion has only one purpose—to help you to avoid the errors in usage that occur most frequently. Consequently its contents are limited to the information that will best serve that purpose. A definition of grammatical terms appears later in the Handbook.

There are times, of course, when literate people disagree about whether some form is right or wrong. Many a usage that was once considered incorrect is now acceptable to liberal authorities. The point of view in this book is liberal, but in many cases both sides are presented when a form is disputed. This has been done on the grounds that if there are still many literate people who consider something to be an error, those who use this book are entitled to be warned. In this connection it is interesting to note that many a liberal authority concedes—in fact takes pains to point out—the acceptability of forms that he avoids in his own work.

Case of Nouns and Pronouns

Nouns and pronouns may be nominative, possessive, or objective in case. Nouns are identical in nominative and objective cases, but change form when they become possessive. Pronouns vary in form as shown by the following list:

Nominative	Possessive	Objective
I	my, mine	me
you	your, yours	you
he	his	him
she	her, hers	her
it	its	it
we	our, ours	us
they	their, theirs	them
who	whose	whom

All pronouns except those listed above and others derived from them, for example *whoever,* which becomes *whomever* in the objective case, are identical in nominative and objective cases and form the possessive, if at all, just as nouns form it.

The possessive forms *your, its,* and *whose* should not be confused with *you're* (you are), *it's* (it is), and *who's* (who is).

The three cases are used in accordance with the following rules:

1. **The nominative case is used for the subject of a verb.**

> VIOLATION: We shall hire *whomever* applies.
> CORRECT: . . . *whoever* applies. (The word in question is the subject of *applies* rather than the object of *hire.* The object of *hire* is the entire clause *whoever applies.*)
> VIOLATION: *Us* draftsmen need better light.
> CORRECT: *We* draftsmen . . . (*We* is correct, for it is in apposition with *draftsmen* and thus is a subject of *need.*)
> VIOLATION: He is older than *me.*
> CORRECT: He is older than *I.* (*I* is correct for it is the subject of the implied verb *am.*)

2. **The nominative case is used for a subjective complement.**

> It is *I.*
> It was *we* workers who objected.
>
> VIOLATION: It was *him.* It was *me* whom you saw.
> CORRECT: It was *he.* It was *I* whom you saw.

Note. In spite of the rule, such forms as "It was *me*" and "This is *him*" are widely used in conversation even by literate people, because the forms that are formally correct sometimes sound stilted. It is still best, however, to follow the rule in any piece of writing where formal style is called for; and even in conversation many discriminating people try to phrase their remarks so that they do not need to choose between stiffness and violation of the rule.

3. **The possessive case is used to indicate possession.**

> VIOLATIONS: A violation of this rule almost always results from the fact that the possessive and some other form are pronounced exactly alike. (Examples: *it's* and *its, who's* and *whose, ladies* and *lady's. It's* is correct only for *it is,* and *who's* for *who is. Ladies* is plural, not possessive.) Thus the errors do not indicate ignorance of the need for the possessive case, but rather ignorance of (or carelessness about) how the possessive case is formed. Information on forms that are correct for the possessive case appears in the list of pronouns at the opening of this discussion and in the discussion of the apostrophe in the section on punctuation.

4. **The possessive case is used for a noun or pronoun that precedes and modifies a gerund.**

VIOLATION: They objected to *him* altering the records.
CORRECT: They objected to *his* altering the records.

Note. When a pronoun has no possessive form, for example *this* or *that,* Rule 4 must of course be disregarded. It is also disregarded if the presence of other words between the noun or pronoun and the gerund modified would make the possessive sound unnatural.

CORRECT: There is no record of *this* being done before.
CORRECT: The chance of *anyone* in the vicinity wanting to buy the property is slight.

5. If a noun or pronoun stands for something inanimate, the possessive case should be avoided, if possible.

UNDESIRABLE: The *furnace's* grates; the *street's* surface.
PREFERRED: The grates *of the furnace;* the surface *of the street.*

Note. There are many exceptions to this rule, especially when the noun in question involves time, as "an *hour's* work," or "a *day's* pay." Exceptions are also permissible to avoid awkward constructions.

6. The objective case is used for the object of a verb, verbal, or preposition.

Direct object of verb: The letter encouraged *him.*
Indirect object of verb: The letter gave *him* useful information.
Object of verbal: Whom do you intend to hire? (*Whom* is the object of the infinitive phrase *to hire.*)
Object of verbal: Answering *him* was difficult. (*Him* is the object of the gerund *answering.*)
Object of preposition: Three of *us* were chosen. (*Us* is the object of the preposition *of.*)

COMMON VIOLATIONS: They offered my partner and *I* a good contract. (*Me* would be correct, for the term is used as the indirect object of *offered.* This type of error results from excessive concern about avoiding *me* as a subject, as in "John and *me* are going." Uncertainty as to what form to use after *and* can be settled quickly by eliminating *and* along with the term that precedes it. "They offered *I* a contract" would obviously be wrong.)
All of *we* draftsmen were questioned. (*Us* would be correct, for *us* along with *draftsmen,* is the object of the preposition *of.* The subject of the sentence is neither *we* nor *draftsmen,* but *all.*)
He is a man *who,* in spite of his youth, we can trust. (The parenthetical phrase *in spite of his youth* obscures the fact that the objective *whom* rather than the nominative *who* is needed, for the term is the object of *trust.*)

Who do you give the answer to? (*Whom* would be correct, for the term is used as the object of *to*.)

Note.   In conversation and in informal writing there is a strong tendency to permit *who* at the beginning of a sentence despite the rules. In formal writing, however, it is still desirable to follow the rule.

7. The objective case is used for either the subject or object of an infinitive.

> CORRECT: They expect *him* to do well. (*Him* is the subject of the infinitive *to do*. The object of *expect* is not *him*, but the entire infinitive phrase, *him to do well*.
> CORRECT: We knew *him* to be competent.
> CORRECT: We knew it to be *him*.

Agreement of Verb and Its Subject

A verb must agree with its subject in number, person, and gender. Practically all of the violations of this rule have to do with disagreement in number. Person and gender will therefore not be discussed here. Rules involving agreement in number are as follows:

1. When the subject of a verb is compound and its parts are joined by *and,* it ordinarily takes a plural verb.

> The time and the place *are* uncertain.

2. When the subject of a verb is compound in form but singular in meaning, it takes a singular verb.

> Blue and gold *is* a pleasing color combination.

3. When the subject of a verb is compound and its parts are joined by *or,* it takes a singular verb if its parts are singular and a plural verb if its parts are plural. If its parts differ in number, it agrees with the part that is closer.

> CORRECT: Either the oak or the elm *is* to be removed.
> CORRECT: Either the chairs or the tables *are* to be refinished Monday.
> CORRECT: Either the cow or the pigs *are* to be slaughtered.

4. The word *number,* numerical quantities, and fractions take either singular or plural verbs according to their meanings.

> The number of complaints *has* been increasing.
> A number of changes *have* been made.

Three days *is* a long wait. (*Three days* refers to a single period of time, even though *days* is plural in form.)
Three days *have* passed since your letter was mailed. (Each day is conceived as a unit rather than as part of a single period of time.)

5. A collective noun takes a singular verb when the group it refers to is regarded as a unit, but takes a plural verb when the statement concerns the members of the group as individuals.

The crowd *was* breaking up.
The crowd *were* going to their homes.
The committee *has* adjourned.
The committee *have* taken their seats.

Note. When you feel that following the rule would result in an awkward sentence, revise the sentence. For example: "The *people* in the crowd (rather than *the crowd*) were going to *their* homes."

6. Some nouns that are plural in form take singular verbs when singular in meaning—especially nouns ending in *ics*.

CORRECT: The *news is* encouraging. *Mathematics is* difficult. *Tactics wins* battles.
BUT: Our *tactics are* proving successful.

Note. For an excellent discussion of this subject, see the entry under *ics* in *Webster's New Collegiate Dictionary*.

7. Intervening words should not be allowed to interfere with the agreement of a verb with its subject. Many of the errors in agreement are caused by intervening words.

CORRECT: Good management plus favorable economic conditions *was* responsible for the improvement. (The phrase introduced by *plus* is not part of the subject.)

8. A verb agrees with its subject even though a subjective complement that follows is different in number.

The border *was* shrubs of several species.

Note. When you feel that following the rule would result in an awkward-sounding sentence, revise the sentence. For example: "The border contained shrubs of several species."

Agreement of Pronouns with Antecedents

The Rules. The basic rule is that a pronoun and its antecedent should agree in person, number, and gender. When a pronoun has a compound antecedent the following additional rules apply:

1. If the antecedent consists of two or more nouns connected by *and,* the pronoun should be plural.

    The table and the chair showed *their* age.

2. If the antecedent consists of two nouns connected by *or,* the pronoun should be singular if both nouns are singular, and plural if both nouns are plural. If the nouns differ in number, the pronoun should agree with the nearer.

    You may pay the bill if the manager or the auditor will give *his* permission.
    You may use either bolts or screws if *they* are large enough.
    Either the bolts or the clamp must have been loose in *its* place.
    Either the clamp or the bolts must have been loose in *their* places.

3. If the antecedent is two or more nouns connected by *nor,* the pronoun should be singular if both nouns are singular, but *plural if either noun is plural.*

    Neither the manager nor the auditor gave *his* consent.
    Neither the bolts nor the screws had fallen from *their* places.
    Neither the employer nor the employees were willing to change *their* attitudes.
    Neither the employees nor the employer were willing to change *their* attitudes.

The Errors.   When a pronoun does not agree with its antecedent, the disagreement, almost always, is in number. The error does not usually result from ignorance of the rule, but from uncertainty about whether the antecedent itself should be treated as singular or plural. Numerous though the errors are, a surprisingly large proportion of them result from one of three conditions. Any person who can master these three conditions will thenceforth have little trouble with the reference of pronouns: (1) when the antecedent is a *collective* noun; (2) when the antecedent is *a noun of common gender;* (3) when the antecedent is not a noun but *an indefinite pronoun.*

1. *Reference of a pronoun to a collective noun:* A collective noun is sometimes treated as singular and sometimes as plural. It is treated as singular when the statement where it appears applies to the group as a group, and plural when the statement applies to the members of the group as individuals.

    CORRECT: The board of directors makes *its* report.
    CORRECT: The board of directors took *their* seats.

    The forms that are wrong or at least undesirable occur when a writer first treats a collective noun as singular, then as plural—usually making it the subject of a singular verb and the antecedent of a plural pronoun. The best way to avoid such inconsistencies is to change the construction of the sentence.

DOUBTFUL: The orchestra *waits* with *their* instruments ready.
IMPROVED: The orchestra waits with instruments ready.
DOUBTFUL: The team *was* photographed wearing *their* new uniforms.
IMPROVED: The team was photographed wearing new uniforms.
IMPROVED: The members of the team were photographed wearing *their* new uniforms.

2. *Reference of a pronoun to a noun with common gender:* Writers sometimes make the error of using a plural pronoun because of uncertainty about whether to use a pronoun of the masculine or the feminine gender in referring to a noun such as *spectator, student,* or *employee.* The correct form to use, in case of doubt, is the masculine.

WRONG: A person should be able to rise above *their* environment.
CORRECT BUT STILTED: A person should be able to rise above *his* or *her* environment.
CORRECT: A person should be able to rise above *his* environment.

3. *Reference of a pronoun to an indefinite pronoun:* Sometimes the antecedent of a pronoun is not a noun, but an indefinite pronoun. Most of the indefinite pronouns are listed below. Each word on this list is treated as singular, even though some of them are plural in meaning. Hence any pronoun referring to a word on this list should be singular in number.

Indefinite Pronouns

one	everyone	somebody
anyone	anybody	nobody
someone	everybody	each
no one		

All the words on the list are common in gender. Hence, in an effort to achieve agreement in gender, writers use the plural *their,* which can be common gender, in order to avoid *he* or *she.* This tendency is intensified by the fact that some of the words, for example *everyone,* are actually plural in meaning. Nevertheless, all the words listed are treated as singular (no one would think of writing "Everyone *are* here"), and except as noted below are not to be referred to by the plural *their.* Unless the feminine singular is known to be appropriate in the individual case, the masculine singular (*he, his,* or *him*) should be used. *His or her* is of course correct but is so stilted as to be undesirable.

DOUBTFUL: No one should neglect *their* responsibilities.
CORRECT: No one should neglect *his* responsibilities.
DOUBTFUL: Everyone should pay *their* bills promptly.
CORRECT: Everyone should pay *his* bills promptly.
WRONG: Each of those hired will be required to supply *their* own tools.
CORRECT: Each of those hired will be required to supply *his* own tools.

Exceptions. The rules just explained apply to writing that is formal in style, and many people who are careful with their language follow them even when writing or speaking informally. There is a contemporary tendency, however, to relax the rules. Some of the more liberal modern authorities would regard it as correct to use *their* when the indefinite pronoun referred to is plural in meaning. At least one president and more than one senator, speaking on television, have been heard to use *their* when the antecedent was *everyone*.

Agreement of a Demonstrative Adjective with its Object

A demonstrative adjective (*this, these, that,* and *those*) should agree with the word it modifies.

WRONG: *These* (or *those*) kind of brakes may give trouble.
CORRECT: *This* (or *that*) kind . . .

Note. In the error illustrated, the demonstrative adjective agrees with a closely related plural noun when it should agree with the singular noun that it actually modifies.

Adjectives and Adverbs—General Comment

1. An adjective may modify a noun or pronoun directly (in which use it is called an attributive adjective) or as a predicate adjective. Used as a predicate adjective, it applies to a noun or pronoun by assistance of a verb.

   The *old* building should be abandoned. (Attributive modifier. *Old* modifies *building*.)
   The building is *old*. (Predicate adjective. *Old* relates to *building* by assistance of the verb *is*.)

2. An adverb is used to modify a verb, an adjective, or another adverb.

   The snow had melted *rapidly*, so what might have been a *very* difficult trip was made *quite easily*. (*Rapidly* modifies the verb *melted; very* modifies the adjective *difficult; quite* modifies the adverb *easily*—which itself modifies the verb *was made*.)

3. Use of an adjective instead of an adverb to modify a verb is usually a serious error.

> WRONG: The car runs *good*. (The adjective *good* should not modify the verb *runs*.)
>
> CORRECT: The car runs *well*. (*Well* is correctly used as an adverb. In some meanings, *well* is also an adjective; but *good* is never correct as an adverb except when the phrase *as good as* appears in such a sentence as "He was *as good as* elected.")
>
> WRONG: I can do it *easy*.
>
> CORRECT: I can do it *easily*. (*Easily* is the adverbial form of easy.)
>
> WRONG: I can do it *easier* than he can.
>
> CORRECT: I can do it *more easily* than he can.
>
> WRONG: Among all those who protested, he talked the *angriest*.
>
> CORRECT: Among all those who protested, he talked *most angrily*.

Note. The error of using an adjective to modify a verb is especially prevalent, as shown in the last two examples of errors, in the comparative and superlative degrees. (For discussion of degree, see Rule 7 below.)

4. Use of an adjective to modify another adjective or an adverb (as contrasted with an adverb modifying a verb, discussed in Rule 3) is not formally correct, but sometimes may pass as colloquial rather than being considered a serious error.

> WRONG: The car was running *considerable* better. (*Considerable* is an adjective and should not modify the adjective *better*.)
>
> WRONG: The next day, I felt *some* better. (*Some* is an adjective, not an adverb. It should be replaced by the adverb *somewhat*.)
>
> WRONG: It was *real* nice of you.
>
> CORRECT: . . . *very* nice of you.
>
> WRONG: It was made of *pretty* good material.
>
> CORRECT: . . . *rather*, or *quite*, or *fairly* good material.
>
> WRONG: He was late *most* every morning.
>
> CORRECT: . . . *almost* every morning.

Note. *Real, pretty,* and *most* might be regarded as colloquialisms above, but it would be better, except on an extremely informal occasion, to replace them by words that are recognized as adverbs. There is no way, except by familiarity with the specific word, to determine what adjectives are considered merely colloquial rather than actually wrong when used as adverbs.

5. Adverbs are rarely misused for adjectives as direct modifiers, but use of an adverb when a predicate adjective is needed is an extremely common error. A predicate adjective follows the verb but does not modify it. Rather, it relates to the noun or pronoun used before the verb as a subject. This use should not be confused with the use of an adverb to modify the verb. A predicate adjective characteristically follows a "linking" verb such as *be, become, seem, appear, smell, feel, look, sound, taste.*

> CORRECT: The glue is *sticky.* (Predicate adjective. *Sticky* refers to the noun *glue* rather than modifying the verb *is.*)
>
> CORRECT: The man looked *angry.* (Predicate adjective. *Angry* describes the man rather than changing or adding to the meaning of the verb *looked.*)
>
> CORRECT: The man looked *angrily* at the intruder. (Adverb modifying verb. *Looked,* in this sentence, has a different meaning than in the example above. It is not used as a linking verb, but indicates an action. *Angrily* tells the manner in which the action was performed.)
>
> WRONG: The dinner smells *well.* (*Well,* except in the sense of healthy, is an adverb, not an adjective. An adjective is needed here, for *smell* does not indicate an action and *well* tells a quality of the dinner rather than modifying *smells.*)
>
> CORRECT: The dinner smells *good.* (Predicate adjective.)
>
> CORRECT: The dog smells *eagerly* at the bone. (Note the different meaning of *smells,* and note that *eagerly,* an adverb of manner, modifies *smells.*)
>
> CORRECT: The clerk felt *bad* about his error. (Predicate adjective.)
>
> CORRECT: The motor was running *badly.* (Adverb modifying verb.)

Note. Those who make this type of error do so because they have been corrected for such errors as "The motor is running *bad,*" or "He reads *good.*" In an effort to avoid using an adjective when an adverb is needed, they are reluctant to use an adjective after a verb, even when an adjective is correct. One use of *badly,* however, calls for special comment: There is a strong tendency at present to accept "The clerk felt *badly* about his error" as correct despite the rule. *Bad,* of course, is also correct; in fact, it is preferable.

6. When the question of whether to use an adjective or an adverb has been answered, it may still be necessary to determine whether the particular word one wishes to use is an adjective, is an adverb, or may be used as either. Often, the adverbial form of a word may be distinguished from the adjective form because of its *ly* ending, as in *strong, strongly; clever, cleverly; sure, surely.* The *ly* ending, however, is not always a means of identification, for there are some adjectives that end in *ly* (*friendly, manly*) and many adverbs that do not end in *ly.* In case of doubt, it is advisable to consult a dictionary to determine the part of speech of the word in question.

7. Most adjectives and adverbs may be positive, comparative, or superlative in degree. The change in degree is achieved by more than one method, as illustrated in the following examples:

Adjectives

Positive	Comparative	Superlative
easy	easier	easiest
large	larger	largest
curious	more curious	most curious
good	better	best
bad	worse	worst

Adverbs

Positive	Comparative	Superlative
slow or slowly	slower or more slowly	slowest
easily	more easily	most easily
quickly	more quickly	most quickly
well	better	best
badly	worse	worst

Note. There are some who point out that terms such as *perfect* and *unique* cannot logically be used in the comparative or superlative degrees. *More perfect* is a contradiction in terms, as are *more unique, more empty,* and so on. It would be logical, however, to use such expressions as *more nearly* perfect, unique, full, or empty. The case, as stated above, is logical, but English is not always a logical language. The Constitution of the United States refers to a *more perfect union.*

The comparative rather than the superlative degree must be used to indicate comparison of two objects. The superlative degree is used when three or more objects are compared.

> WRONG: His was the *best* drawing of the two drawings that were submitted.
> CORRECT: His was the *better* drawing of the two that were submitted.
> CORRECT: His was the *best* of the three drawings that were submitted.

UNITY AND COHERENCE IN THE SENTENCE

Unity of Thought

In this discussion we are concerned only with the declarative sentence —the sentence that makes a statement. Such a sentence may lack unity either because it fails to include all the material needed or because it contains extraneous material. The first of these two weaknesses may be prevented by the avoidance of primer style. The second may be guarded against by being alert to the question of whether all the material in the sentence really merges to make a single statement. Consider, for example, the sentence, "Mack trucks, which are now being produced in three foreign countries, would be the best kind for us to buy for hauling logs and lumber." There is no clear reason for placing the fact about foreign production in a sentence with the purpose this one seems to have. There are many times, of course, when certain facts may be presented as either one sentence or more than one, but this does not justify inserting into a sentence facts that are not related to its purpose.

Sentence unity also demands that we do not end one sentence and begin another at an illogical place. Consider these sentences:

> The Ellsworth machine has ample power and those who have used it say it is easy to operate, but it is extremely noisy. Also, it is somewhat bulky.

The first of these two sentences names two good features and one bad. The second mentions another bad feature. If we assume that two sentences are desirable, sentence unity would be improved by telling the good features in one sentence and the bad in the other. A general rule to follow is: For the sake of sentence unity, construct your sentences so that they draw together the facts that are most closely related.

Unity of Structure

When a sentence lacks unity of structure it may be faulty in either of two respects: (1) It may be a mere fragment because it lacks some element necessary for grammatical completeness. (2) It may contain the error variously called a "comma splice," "comma fault," or "comma error." (A "run-on sentence" is basically the same error as a comma splice and is covered under that heading.)

Fragment Treated as a Sentence. A declarative sentence, to be complete grammatically, must contain a subject and a finite verb. Thus a participial or gerund phrase, a prepositional phrase, or a dependent clause is not a complete sentence, regardless of its length or complexity. Actually, it should not be necessary to subject a sentence to grammatical analysis in order to determine whether it is complete. If it is read attentively, a fragment may be distinguished from a complete sentence by the simple fact that it does not actually make a statement.

The error may usually be corrected by either of two methods: joining the fragment onto a sentence with which it is naturally connected, or changing it so that it may stand by itself as a complete sentence. The following examples show how fragments may be joined to other sentences or rewritten so that they are structurally complete. Whether the corrections indicated would be the best corrections possible would depend on the context.

> FRAGMENT: The volume grew lower and lower. *Finally becoming so slight as to be inaudible.* (Participial phrase rather than sentence.)
>
> CORRECT: The volume grew lower and lower, finally becoming so slight as to be inaudible.
>
> FRAGMENT: He was employed by the government for six years. *First in Washington, and later in New York.* (Prepositional phrases.)
>
> CORRECT: He was employed by the government for six years, first in Washington and later in New York.
>
> FRAGMENT: The company was losing money steadily. *Although sales were as high as in previous years.* (Subordinate clause rather than independent clause.)
>
> CORRECT: The company was losing money steadily even though the sales were as high as in previous years.
>
> FRAGMENT: Some of the expenses were increasing. *Overhead, for example, and the cost of raw materials.* (The italicized expression has no verb; it is used in apposition to *some of the expenses* in the preceding sentence.)
>
> CORRECT: Some of the expenses were increasing. Overhead, for example, and the cost of raw materials were higher than before.

Note. Skilled authors do sometimes use incomplete sentences as a device to secure special effects. Dickens, for example, wrote: "Dogs, indistinguishable in the mire. Horses scarcely better; splashed to their very blinkers." But this incompleteness was no accident. Until you are skillful enough never to violate the rules of grammar accidentally, however, you should not violate them on purpose to secure a special literary effect.

The Comma Splice. When a sentence contains two independent clauses, the punctuation between them must be a semicolon unless they are joined by a coordinating conjunction. Use of a comma instead of a semicolon on such an occasion is the error variously called comma splice, comma blunder, or comma error. It is more than a mere mistake in punctuation; it is a serious error in sentence structure—an indication that the writer does not know when he has completed a statement.

The only coordinating conjunctions are *and, but, for, or, nor,* and sometimes *so* and *yet.* These should be distinguished from conjunctive adverbs such as *however, moreover, therefore, consequently, furthermore,* and many others. A handy rule of thumb for avoiding the comma splice is found in the fact that a conjunctive adverb may be buried within the clause that it introduces. Thus, if a connective word might come elsewhere than between the clauses it connects, a semicolon rather than a comma is necessary. For example:

> I have inspected it, *but* I shall be glad to inspect it again. (*But* could come only between the clauses.)
>
> I have inspected it; *however,* I shall be glad to inspect it again. (*However* might, if one desired, be placed after *glad* rather than at the beginning of its clause; thus it is a conjunctive adverb, and a semicolon must be placed between the clauses.)

There are various ways in which a comma blunder may be corrected. The simplest but not always the best is to replace the comma by a semicolon or a period. Another possibility is to change the conjunctive adverb to a coordinating conjunction. Often, however, it is better to change one of the independent clauses to some other form.

> COMMA SPLICE: We shall be forced to find a new route, the grades on the one suggested are too severe.
>
> CORRECT: We shall be forced to find a new route; the grades on the one suggested are too severe. (Correction by change in punctuation.)
>
> CORRECT: We shall be forced to find a new route, for the grades on the one suggested are too severe. (Correction by using a coordinating conjunction.)
>
> CORRECT: The grades on the route suggested are so severe that we shall be forced to find a new route. (Correction by change in construction.)
>
> CORRECT: The severity of the grades on the route suggested makes it necessary to find a new route. (Correction by change in construction.)

Coherence

If writing is to be clear, every sentence must be coherent; that is, it must hold together. The relationship of every part to every other part and to the entire sentence must be unmistakable. To achieve this result it is necessary to consider the order of the parts, their structure, and the connectives that indicate their relationship. Lack of coherence often results from failure to apply the following principles:

1. Similar parts of a sentence should be expressed in parallel form so far as their contents permit.

> NOT PARALLEL: Wood was used for some of the parts, but others were made of metal.
>
> PARALLEL: Wood was used for some of the parts, and metal for others.

NOT PARALLEL: Students in the night school learn auto mechanics, and are also taught drafting.

PARALLEL: Students in night school learn auto mechanics and drafting.

PARALLEL: Students in night school are taught auto mechanics and drafting.

(a) There should be no unnecessary change in the subject or voice. (Change in one frequently results in change in the other.)

NOT PARALLEL: While the mechanic was tuning the motor, the tires and battery were checked by his assistant

PARALLEL: While the mechanic was tuning the motor, his assistant checked the tires and battery.

NOT PARALLEL: Washing the equipment was one of his duties, and he was also expected to keep the supply bins filled.

PARALLEL: He was expected to wash the equipment and to keep the supply bins filled.

(b) The elements of a series should be parallel in form.

WRONG: A technician must learn the *use, upkeep,* and *how to repair* equipment.

CORRECT: A technician must learn the *use, upkeep,* and *repair* of equipment.

If *the* is used before any element in a series except the first (in which case it applies to all the elements), it must be used before every element.

WRONG: Information is needed about the length, the width, thickness, and weight of the sample.

CORRECT: Information is needed about the length, width, thickness, and weight of the sample.

If *a* or *an* rather than *the* is the suitable word to precede any element in a series, it should usually be used before *each element where it would be suitable.* This is especially necessary if some elements are plural.

We must bring nails, screws, *a* hammer, *a* chisel, and glue.

When the elements of a series cannot be made parallel in form without producing an awkward result, the phraseology should be changed so as to make it clear that a series is not intended.

WRONG: The chips are mixed with liquor and steam, fed into the apparatus, and the cooking process takes place as the mixture moves through the tubes.

CORRECT: After the chips are mixed with liquor and steam, they are fed into the apparatus, where the cooking process takes place as the mixture moves through the tubes.

(c) The elements that follow correlative conjunctions (*either* . . . *or, neither* . . . *nor,* and so on) should be parallel in form.

> WRONG: It can either be shipped by freight or by express.
> CORRECT: It can be shipped either by freight or by express.

2. Connectives should clearly and accurately indicate the thought relationship of the elements that they connect. The connectives that give most trouble are *as* and *while.*

(a) *As* is sometimes ambiguous because it might indicate either time or cause.

> AMBIGUOUS: *As* the mixture was heated, its color changed.
> CLEAR: When the mixture was heated, its color changed.
> CLEAR: Because the mixture was heated, its color changed.

(b) *While* has two legitimate meanings, "during the time that" and "although." (Unfortunately, it is also loosely used in place of *and.*) As a result, many a sentence in which it appears is ambiguous. Unless the context makes its meaning instantly clear, it should be used sparingly for *although* and not at all for *and.*

> AMBIGUOUS: *While* his breathing was becoming more rapid, his fever had fallen.
> CLEAR: *Although* his breathing had become more rapid, his fever was falling.
> CLEAR: At the same time that his breathing was becoming more rapid, his fever was falling.
> AMBIGUOUS: Their truck tires are good, *while* their passenger car tires are even better.
> IMPROVED: Their truck tires are good, *and* (or *but*) their passenger car tires are even better.

3. Reference of a pronoun to its antecedent should be clear and unmistakable.

(a) There should be only one word to which a pronoun may plausibly refer.

> AMBIGUOUS (BECAUSE OF INDIRECT DISCOURSE): The manager notified the foreman that *he* was being transferred. (Manager or foreman?)
> CLEAR: The foreman received notice of his transfer in a conference with the manager. (*His* has been related to *foreman* before the manager is mentioned.)
> CLEAR: The manager, as he mentioned to the foreman, was being transferred.

Note. It is often suggested in textbooks that unclearness caused by indirect discourse be eliminated by changing it to direct discourse. This is rarely appropriate, however, in technical writing.

AMBIGUOUS: Dairy workers should not be allowed to care for cows when *they* are ill.

CLEAR: When dairy workers are ill, they should not be allowed to care for cows.

CLEAR: Dairy workers should not be allowed to care for cows that are ill.

(b) Reference of a pronoun to a "weak" antecedent should be avoided.

WEAK REFERENCE: The engineers must give special consideration to the smokestacks. *They* must be sturdy enough to withstand earthquakes. (Engineers or stacks?)

IMPROVED: The smokestacks must be sturdy enough to withstand earthquakes, so the engineers must give them special consideration.

(c) A pronoun should not be used when its antecedent must be inferred by the reader because it has been expressed in some other form than a noun.

VAGUE REFERENCE: An employee who is injured should receive first aid even though *it* may not be serious. (*It* has as an antecedent the idea of *injury;* but the idea is not present as a noun and must be inferred from the verb *injured.*)

CORRECT: An employee who is injured should receive first aid even though *his injury* may not be serious.

(d) Vague use of *it* or *they* when a specific noun could be used should be avoided. (This rule would not prevent the use of *it* as in "*It* is raining," or "*It* is apparent that . . . .")

VAGUE REFERENCE: *It* says in the report that the method is efficient.

EXPLICIT: *The report* says that the method is efficient.

VAGUE REFERENCE: In restricted areas, *they* require that visitors be identified.

EXPLICIT: *The company* requires that in restricted areas, visitors be identified.

EXPLICIT: In restricted areas, visitors must be identified.

(e) Use of *which* or *this* to refer to the general idea of a preceding clause or phrase rather than to a specific word is condemned by some textbooks but indulged in by many reputable writers. When this use can be avoided easily, it should be avoided.

VAGUE REFERENCE: He has been studying every night, *which* should increase his value to the company.

IMPROVED: He has been studying every night, so his value to the company should be increasing.

IMPROVED: The study he has been doing every night should increase his value to the company.

(f) Use of *you* for *one* is often inappropriate. Unless you are using an informal style, do not write *you* unless you mean the reader. If *one* sounds stiff, change the sentence.

INFORMAL: The factory is interesting, but *you* are not allowed to enter without permission.

FORMALLY CORRECT: The factory is interesting, but *one* is not allowed to enter without permission.

CORRECT: The factory is interesting, but *visitors* are not allowed to enter without permission.

(g) Use of the same pronoun to refer to two different antecedents should be avoided.

CONFUSING: Although *it* is less crowded on the second floor, *it* is not strong enough to support the heavy machinery. (The first *it* is indefinite in meaning; the second refers to *floor*.)

CORRECT: Although the second floor is less crowded, it is not strong enough to support the heavy machinery.

4. Modifiers (words, phrases, or clauses used to change or limit the meaning of other words) should be as close to what they modify as it is possible to place them without awkwardness. No possible doubt should exist as to what is being modified.

(a) Certain adverbs are especially likely to be put in the wrong position. These adverbs are *even, hardly, nearly, scarcely,* and especially *only* and *not.*

MISPLACED MODIFIER: We *only* tried three times.

CORRECT: We tried only three times.

MISPLACED MODIFIER: Every small business can*not* grow large. (This is literally a statement that *no* small business can grow large.)

CORRECT: Not every small business can grow large.

MISPLACED MODIFIER: The boat *almost* seemed ready to sink.

CORRECT: The boat seemed almost ready to sink.

MISPLACED MODIFIER: It *hardly* felt as if it were heavy enough.

CORRECT: It felt as if it were hardly heavy enough.

(b) Phrases and clauses, as well as words, should be placed as close as possible to what they modify.

MISPLACED MODIFIER: It made a good impression on the inspectors carrying a full load.

CORRECT: Carrying a full load, it made a good impression on the inspectors.

MISPLACED MODIFIER: After it was repaired, it was given a trial run by a testing laboratory that lasted one week.

CORRECTION BY CHANGE OF STRUCTURE: After it was repaired, it was given a one-week trial run by a testing laboratory.

MISPLACED MODIFIER: The factory reopened after a one-month shutdown on June 1.

CLEAR: After a one-month shutdown, the factory reopened on June 1.

MISPLACED MODIFIER: It was given an overhaul by a crew of expert mechanics that was long overdue.

CORRECT: It was given an overhaul that was long overdue by a crew of expert mechanics.

(c) "Squinting" constructions should be avoided. (Squinting constructions are those in which a modifier is placed between two objects, either of which it might modify.)

SQUINTING: The smoke jumpers who had been flown in *immediately* put out the fire.

CLEAR: The smoke jumpers, who had immediately been flown in, put out the fire.

CLEAR: The smoke jumpers who had been flown in put out the fire immediately.

(d) Though split infinitives are no longer regarded as a serious error, they are frequently awkward. Therefore one should not *unnecessarily* split infinitives in an effort to place a modifier nearer to its object.

AWKWARD: They decided *to quickly complete* the repairs.

IMPROVED: They decided to complete the repairs quickly.

(e) In attempting to place modifiers near their objects, one should not needlessly interrupt the smooth flow of thought through subject, verb, and complement.

AWKWARD: The crew extinguished, after a long and difficult struggle, the fire in the slashings.

IMPROVED ARRANGEMENT: After a long and difficult struggle, the crew extinguished the fire in the slashings.

5. Comparisons should be logical and complete.

(a) Objects that are compared should be similar in nature.

ILLOGICAL: The power of the diesel is greater than the steam engine.

LOGICAL: The power of the diesel is greater than that of the steam engine.

LOGICAL: The diesel engine is more powerful than the steam engine.

(b) Illogical use of *any* should be avoided.

ILLOGICAL: The first design was simpler than *any of the designs.*

LOGICAL: The first design was simpler than any of the later designs.

(c) If a comparison is not completely expressed, the words that are omitted must be clearly and unmistakably implied.

INCOMPLETE AND AMBIGUOUS: The manager trusts him *more than the superintendent.*

COMPLETE: The manager has more trust in him *than in the superintendent.*

COMPLETE: The manager trusts him *more than the superintendent does.*

INCOMPLETE: The highway is *as rough, if not rougher than* the side road.

COMPLETE BUT AWKWARD: The highway is *as rough as, if not rougher than,* the side road.

COMPLETE: The highway is as rough as the side road, if not rougher.

6. Use of mood and tense should reflect a consistent point of view.

INCONSISTENT USE OF TENSE: The production of grade "A" milk *required* the use of methods that *meet* official standards.

CONSISTENT: The production of grade "A" milk *requires* the use of methods that *meet* official standards.

INCONSISTENT IN MOOD: If you *would* give me an extension of time, I *shall* appreciate it.

CORRECT: If you *would give* me an extension of time, I *should* appreciate it.

INCONSISTENT IN MOOD: If the generated voltage *drops* below "E," current *would flow* through and help to rotate the armature.

CONSISTENT: If the generated voltage *drops* below "E," the current *will flow* (or *flows*) through and help to rotate the armature.

CONSISTENT: If the generated voltage *were* to drop below "E," the current *would flow* through and help to rotate the armature.

7. Dangling participles and gerunds should be avoided. A participle is a verb form used as an adjective. A gerund is a verb form used as a noun. (See Definition of Grammatical Terms.) When a participial or gerund phrase comes at the beginning of a sentence, it should be followed immediately by some term indicating who performed the action

indicated, so that the action will not be attributed to the wrong agent.

In respect to meaning, a dangling participle or gerund is undesirable because it is likely to be wrongly interpreted when first read. In respect to grammar, it is wrong because it is a modifying element with nothing to modify—thus the term *dangling*—and has no grammatical connection with the remainder of the sentence.

> DANGLING PARTICIPLE: *Having thought the case over carefully,* my opinion was unlikely to be changed.
> CORRECT: Having thought the case over carefully, *I* was not likely to change my opinion.
> CORRECT: *Since I had thought the case over carefully,* I was not likely to change my opinion.
> DANGLING PARTICIPLE: It can be built with either one or two doors, *depending on the wishes of the buyer.*
> CORRECT: It can be built with either one or two doors, *whichever the buyer wishes.*
> DANGLING GERUND: *By rotating the crystal,* the light is directed so that only the desired wavelength is reflected back to the absorption cells.
> CORRECT: By rotating the crystal, *one* directs the light so that only the desired wavelength is reflected back to the absorption cells.

8. Dangling infinitives should be avoided.

> DANGLING INFINITIVE: *To conduct the test properly,* the motor must run at a constant speed. (It sounds as if the motor were conducting the test.)
> CORRECT: To conduct the test properly, *one* must keep the motor running at a constant speed.
> CORRECT: *If the test is to be conducted properly,* the motor must run at a constant speed.

9. When elliptical clauses or phrases are used, a sentence should be constructed in a manner that does not permit the reader to form a false impression. (An elliptical clause, phrase, or sentence is one from which something is omitted, being implied by what is expressed.)

> MISLEADING ELLIPSIS: *While looking for a suitable location,* the truck broke down.
> CORRECT: While *we were* looking for a suitable location, the truck broke down.
> MISLEADING ELLIPSIS: *When working on the barn,* the problem of protecting the cattle hindered them.
> CORRECT: When working on the barn, *they were hindered by* the problem of protecting the cattle.

PUNCTUATION

The major function of punctuation is to make writing clearer and easier to read. The rules, except for those that are mere conventions, have come to be what they are because experience has shown us how punctuation can best contribute to clearness. Thus there is no fundamental clash between the idea of punctuating for the sake of clearness and punctuating in accordance with rules. The best way to assist clearness through punctuation is usually to follow the rules.

To be sure, the existence of rules has not caused usage in punctuation to be completely uniform. The rules are flexible enough to permit a writer, on many occasions, to exercise his judgment in deciding between equally correct alternatives. The lack of uniformity does not result from widespread disagreement about the rules themselves, nor from general disregard of the rules. Most of the rules are generally agreed upon by authorities, and the occasions when well-educated writers violate them are not frequent.

Three marks are especially important to anyone who has difficulty with punctuation: the period, the semicolon, and the comma. Even though familiarity with the other marks is desirable, you will find that if you can learn to use these three marks correctly—especially the comma—most of your problems with punctuation will be solved.

Finally, it should be emphasized that even though punctuation may be valuable, there is a limit to what it can accomplish. Punctuation is not a substitute for smooth, easy-flowing sentence structure. There are many times when a writer who is worried about how to punctuate a sentence should rewrite it rather than punctuating it, for difficulty in punctuation may well result from awkward writing.

The following rules are generally agreed upon for ordinary writing. They are not intended to cover footnotes, bibliographies, or the technical parts of letters. The punctuation of these special forms is illustrated where the forms themselves are discussed.

The Period

1. Periods are used at the ends of all sentences except those that are interrogative or exclamatory.

2. Periods are used after abbreviations.

a.m., a.d., Fig., i.e., R.F.D., U.S.

Exceptions. Many abbreviations made up of the first letters of words that comprise the names of organizations are written without periods

(NATO, UNESCO, TVA, NAM, UNNRA). Periods are often omitted, also, after abbreviations that are peculiar to technical style.

3. To indicate the omission of words, three spaced periods are used. If the omission occurs at the end of a sentence, a fourth period is added to mark the end of the sentence.

> The survey . . . covered 174 of the 192 colleges now accredited in this field.

The Comma

The comma is especially important because it is the main device by which the grouping of words, phrases, and clauses is indicated. Consequently it is used, and unfortunately misused, more than all the other marks combined.

Its use, however, is not haphazard. Competent writers almost always use commas for one of two purposes: to set off some element of the sentence from what precedes, what follows, or both, or else to separate two elements as they might be separated by a pause or a rising inflection of the voice if one were speaking. The few uses that fall into neither of these categories are the arbitrary use of commas on certain specific occasions, to be pointed out below, and the insertion of a comma when its presence is necessary for the sake of clearness.

The rules that follow, though numbered consecutively throughout, are grouped in accordance with the uses referred to above.

Commas Used to Set Off

1. An appositive or a term of direct address is set off by commas.

> The original factory, an old stone structure, is still standing.
> Your answer, Mr. Smith, is entirely satisfactory.

Note. No comma is used when a noun and its appositive are so closely related as to join in expressing a single idea.

> The invasion was led by my brother John.

2. An adverbial clause preceding its principal clause, or an adverbial phrase at the beginning of a clause, is usually set off by a comma.

> After the achievement tests had been completed, the results were tabulated.
> On all floors except the second and the fourth, the fire hazards have been removed.

Note. If an adverbial clause or phrase is extremely short, and if omission of the comma could not cause confusion, the comma may be omitted.

> When he arrived he was admitted immediately.
> During July the plant will be closed.

3. Independent elements, participial phrases, gerund phrases, and other such constructions at the beginning of a sentence are set off by commas.

> No, the shipment has not yet arrived.
> Worried by the complaints, we began an investigation.
> The contract having been broken, no payment was due.

4. A conjunctive adverb (*however, moreover, therefore,* etc.) is usually set off by commas when it comes *within* the clause to which it applies. When it comes at the beginning of a clause, it may or may not be followed by a comma but will always be preceded by a period or semicolon. (See Semicolon, Rule 2.)

> His objection, therefore, was ignored.
> I had heard the rumor before; consequently, I did not believe it.

5. Any mildly parenthetical element is enclosed in commas if it seems desirable to set it apart from the rest of the sentence. A writer is called upon to use his best judgment in applying this rule, for too many commas will make a sentence jerky and hard to read.

> The newer strains, to be sure, will survive the blight.
> The answer, when received, was unsatisfactory.
> The central section, for example, was undamaged.
> The frame, they insisted, was too tight.

6. A term such as *namely* or *that is,* used to introduce an example or a list, is usually set apart from that example or list by a comma. (The mark that precedes such an expression depends on the sentence structure.)

> The usual crops—that is, wheat, peas, and alfalfa—are in good condition.
> Three species of tree were observed, namely, pine, fir, and cedar.

7. Nonrestrictive clauses are set off by commas. Restrictive clauses, however, are not set off.

> The south side, which has been exposed to the sun, was badly faded.
> He moved to Arizona, where the climate was not so moist.

but

The road went to pieces where the permafrost had been disturbed.
All motorists who drive recklessly should be fined heavily.
I have never been there when the legislature was in session.

Note.   In the first two examples, the clauses introduced respectively by *which* and *where* merely add some additional facts. If they were omitted, the meaning of the remainder of the sentence would be unchanged. Hence they are nonrestrictive. The clauses introduced by *where, who,* and *when* in the other three examples are restrictive. Each is used to limit—to *restrict*—the meaning of the main statement, which would be radically changed if the clause in question were omitted.

Note.   Sometimes a sentence does not make sense unless a clause is interpreted in a single way—restrictive or nonrestrictive. When this is true, an error in punctuation merely increases the difficulty of reading. There are times, however, when restrictive and nonrestrictive interpretations are equally reasonable. When this is the case, an error in punctuation leads a reader to misunderstand the meaning. Note how the meaning of the two sentences that follow depends on punctuation:

The people from Troy, who had come early, obtained choice seats.
The people from Troy who had come early obtained choice seats.

8. A word or phrase that is placed in an abnormal position in a sentence should be set off by a comma or commas.

To a trained accountant, the problem would look easy.

9. A direct quotation is set off by a comma or commas.

"The tires are threadbare," he asserted, "and will blow out at any moment."
He asked me directly, "Will September delivery be acceptable?"

Exceptions.   A quotation that blends into the regular structure of the sentence is not set off by commas. A title in quotation marks is not set off by commas unless some other rule makes commas necessary.

The poet's prophecy about "airy navies grappling in the central blue" has become an unpleasant reality.
The rhythm of "The Raven" is very striking.

Commas Used to Separate

10. A comma is ordinarily used between two independent clauses that are joined by a coordinating conjunction. The coordinating conjunctions are *and, but, for, or,* and *nor.* (*Yet* and *so* may also be treated as coordinating conjunctions when this rule is applied.)

> The trees had been damaged by fire, and the wild life had been destroyed.
> The building is old, but it has been kept in good condition.

Note. When both clauses are extremely short and simple, the comma may be omitted.

> It was damaged but it still is usable.

Note. If a comma is used *within* one or both of two independent clauses, the comma between them is sometimes replaced by a semicolon. (See Semicolon, Rule 5.)

11. When a sentence contains a series, the elements in the series are normally separated by commas.

> Cattle, sheep, and hogs are now selling for higher prices.
> New deposits have been found in Canada, in Africa, in Central America, and in Alaska.

Note. If a comma is used *within* any element in a series, it is often better to use semicolons rather than commas *between* the elements.

Note. Opinions differ over whether to use a comma before a conjunction (*and* or *or*) that precedes the last item in a series. In technical and scientific periodicals and in material published by the United States Government, use of the comma is predominant. In journalistic and popular publications usage is divided. Sometimes a comma is essential for clearness because of *and* or *or* being used within one of the items. For example:

> The panels were painted red, green, yellow, and black and white.

Without the comma after *yellow,* it would be impossible to know whether *black* belonged with *yellow* or with *white.* In view of this, it seems advisable to regard the comma as normal punctuation rather than trying to check each series to see whether a comma is needed for clearness.

12. Two or more adjectives preceding a noun are ordinarily separated by commas. (The comma before the last adjective is omitted, however, if that adjective is so closely associated with the noun that the two merge into a single thought unit.) Likewise, a comma is used between adverbs that modify the same object.

> A big, powerful truck is needed.
> He has a modest, unassuming manner.
> It was housed in a large wooden structure.
> The watchman was a feeble old man.
> Slowly, relentlessly, the stream wore away the dike.

13. Commas are variously used to separate items in dates, places, and numbers as illustrated in the following examples:

In dates: Payment shall be made on September 15, 1956, at the main office of the company.

In places and addresses: San Francisco, California, is an important shipping point.
The company is located at 70 Fifth Avenue, New York 10011, New York.

To separate adjacent sets of figures: In 1950, 675 men were added to the payroll.

Between the digits of Numbers: 10,984. 234,617. 1,856,445. (The comma is omitted in a number with only four digits unless the number occurs in a column containing numbers in which commas are used.)

Commas Used Arbitrarily

14. A comma is used between the last and the first name of a person when the last name appears first, and also after the first name unless the sentence structure calls for some other mark.

Please insert the name Fitzgerald, Duane, in the proper place in the alphabetical list.

15. A comma is sometimes used to indicate the omission of one or more words.

July will be devoted to writing; August, to revision.

16. A comma may be used whenever it is necessary to force a pause for the sake of clearness.

Inside, the building was in better condition.
To begin with, his data are questionable.

The Semicolon

The semicolon is an intermediate mark, less emphatic than a period but more emphatic than a comma. There are many times when the rules would permit either a semicolon or a period, and a writer's choice must depend on his judgment about whether it would be better to continue the sentence or end it.

1. A semicolon is used sometimes between main clauses that are not joined by any connective.

Privately endowed schools must not be underestimated; they fill a genuine educational need.

2. A semicolon is used between main clauses connected by a conjunctive adverb rather than by a coordinating conjunction. (Conjunctive adverbs are such words as *also, accordingly, consequently, furthermore, hence, however, indeed, moreover, nevertheless, otherwise, still, then, therefore, thus.*)

> He had shown fine managerial ability; consequently his promotion was rapid.

3. A semicolon is used between main clauses when the second clause begins with an explanatory term such as *in fact, for example, that is.*

> All the costs have increased; for example, the cost of raw material has increased by 15 per cent.

Note. Rules 1, 2, and 3 might be summarized: A semicolon is normally the proper mark to use between independent clauses occuring in a single sentence except when the clauses are connected by a coordinating conjunction.

4. Even if independent clauses are connected by a coordinating conjunction, a semicolon may be used between them if it is desirable to set them apart more sharply than usual—for example, to set off one clause from two or more others to which it stands in contrast.

> Its paint was damaged, its lights were broken, and its fenders were a complete loss; but the motor was in good condition.

5. Coordinate elements of any type, clauses or otherwise, are often separated by semicolons if any of them contain commas.

> Some economies, perhaps, may be possible; but however hard we try to hold down expenses, a considerable increase will be unavoidable.

> The inspection was made by William Smith, representing the company; Walter Brosser, president of the union; and Boyd Anderson, inspector for the Bureau of Mines.

The Question Mark

1. A question mark is used at the end of a direct question.

> Is the price level rising or falling?

2. A question mark may be used to show that any expression is intended as a question, whether the form is interrogatory or not. (The use of a noninterrogative form to ask a question is seldom seen in any writing except reproduction of conversation.)

> You claim the records have been altered?

3. A question mark is replaced by a period at the end of a "courtesy question," which is actually a request though it may be interrogatory in form.

> Will you please send us this information as soon as possible.

4. A question mark, enclosed in parentheses, is used to express doubt.

> Thomas Hooker, a founder of Connecticut, lived from 1586 (?) to 1647.

The Exclamation Point

An exclamation point is used after a word, phrase, or sentence to indicate intense feeling or forceful utterance. There are few occasions, however, to use an exclamation point in technical writing.

> Ridiculous! The signature is forged!

The Colon

1. A colon is used before a long direct quotation that is being introduced formally.

> John Stuart Mill expressed his doubts as follows:

2. A colon is used to introduce a formal enumeration—especially after *follows* or *the following*.

> Bids were offered by the following contractors: Wilson and Taylor Construction Company, Toledo; Central Builders, Incorporated, Akron; Herman L. White and Company, Cleveland.

3. A colon is used between two phrases, clauses, or even sentences when the second is actually the equivalent of the first. In this use it conveys a meaning similar to that of *namely* or *that is*.

> The method that they used had one unique advantage: It could be used by personnel who had received only one week of special training.

The Dash

1. A dash can be used before introductory words such as *namely, in fact,* and *that is,* and before abbreviations such as *i.e.* and *viz.*, to introduce an enumeration. (See also Semicolon, Rule 3.)

> It is superior in three respects—namely, economy of operation, safety, and comfort.

2. A dash can be used to set off an informal enumeration or a list of examples that are separated by commas.

Some of the accessories—the heater, the fog lights, and the bumper —are really necessities.

3. A dash is used after a list that is followed by a summarizing expression.

Colds, influenza, sore throats—all the winter ailments were prevalent.

4. A pair of dashes may be used to set off interpolated material. In this use, dashes create sharper separation than commas but less sharp than parentheses.

Most of the additional cost—approximately 90 percent—was passed on to the consumers.

5. A dash may be used to indicate incomplete or interrupted thought. (This use of the dash would be unlikely in technical writing.)

The reasons for our decision—but no, I'll not bore you with them.

Note. The dash is a mark to use when clear-cut rules make it correct. It should not be used haphazardly, merely because one is uncertain what mark would be appropriate. Unless one is writing dialogue or writing very informally, the use of dashes to show interruption of thought indicates lack of smoothness and continuity.

Note. In typewritten material, a dash is indicated by a double hyphen, or by a single hyphen preceded and followed by a single space.

Quotation Marks

1. Quotation marks are used to enclose direct quotations.

"The modern automobile," he pointed out, "sells at about the same price per pound that is charged for beefsteak."

Note. If a quotation is more than one paragraph long, opening quotation marks are used at the beginning of each paragraph but closing quotation marks are not used until the end of the quotation.

Note. Quotation marks are not used when the material quoted is set in smaller type or when it has wider margins than those of the regular text.

Note. Quotation marks are not used to enclose widely known proverbs, such as *Honesty is the best policy,* or other well-known quotations that the reader will recognize as quotations without assistance.

2. Quotation marks are used to enclose titles of short poems, articles,

short stories—in general, the titles of writings that are not printed as independent publications.

> The statistics come from an article entitled "Science and Public Relations."

3. Quotation marks are sometimes used to enclose the names of ships, trains, airplanes, and the like, and to enclose words used as words. (Italics are used more frequently.)

> He had secured reservations on "The Portland Rose."
> In the fourth paragraph, the word "unique" is used incorrectly.

4. Quotation marks may be used to indicate that a word or phrase is used to convey the meaning it has acquired in some special field. For example, in writing addressed to readers in general, quotation marks might be placed around *pickle* as used in the metals industry, *flip flop* as used in connection with computers, or *mark sense* as used in connection with IBM cards.

> This use of quotation marks should be held to a minimum, for it is obtrusive if it occurs very often. In writing addressed to readers who are familiar with special vocabulary of a field, quotation marks are not needed around the terms in question. In writing addressed to readers who lack that familiarity, it is better to use standard English if it will convey the meaning efficiently. If frequent use of such terms is unavoidable, it is better to omit the quotation marks and insert a parenthetical explanation of each term the first time you use it.

5. A quotation within a quotation is indicated by single quotation marks.

> The instructions say, "You are to write 'Rejected' on the top of every imperfect copy."

Quotation Marks in Relation to Other Punctuation

1. A period or a comma ordinarily precedes closing quotation marks, even though it might logically belong outside.

> One of his poems, "Fuzzy Wuzzy," was especially popular.

2. A colon or a semicolon ordinarily follows closing quotation marks.

3. A question mark or exclamation point is placed inside the closing quotation marks if it applies to the quotation, but outside if it applies to the sentence as a whole.

> His exact words were, "Why was the gun concealed?"
> Had he ever read "The Third Ingredient"?

The Apostrophe

1. An apostrophe is used to form the possessive case of nouns and indefinite pronouns. (Indefinite pronouns include such words as *one, everybody, everyone, nobody,* etc.)

   (a) Singular or plural words that do not end with the sound of *s* form the possessive by adding *apostrophe* plus *s.*

the company's property	everybody's business
one's conscience	the men's wages

   (b) Singular words ending with the sound of *s* also form the possessive by adding *apostrophe* plus *s.*

Mr. Jones's desk	the horse's age
the boss's office	

Exception. If the form created by following this rule would be difficult to pronounce, it is permissible to add only the apostrophe, especially if the word in question is a person's name.

Moses' people (not Moses's people)	Mr. Jones' desk
Dickens' novels	

   (c) Plural words ending with the sound of *s* form the possessive by adding only an apostrophe.

the companies' policies	the workers' houses

2. An apostrophe is never used to form the possessive case of a personal or relative pronoun. (The possessive of *it* is *its,* not *it's.* The possessive of *who* is *whose,* not *who's. It's* means *it is; who's* means *who is.*)

3. The apostrophe to show the possessive case is often omitted in a formal title. In using any title, the form to follow is the form that has official sanction.

The Teachers Retirement Act	The Farmers Cooperative

4. A compound term such as *director of information* or *everybody else* is made possessive by adding the *apostrophe* plus *s* (or *apostrophe* only, if appropriate) to the last word.

Director of Information's statement	his father in law's tool chest

5. Joint possession is indicated by a change in the ending of only the last of two or more nouns.

   Fred and Henry's office (Joint possession.)
   Fred's and Henry's offices (Each has a separate office.)

6. An apostrophe is used to indicate the omission of one or more letters in a contraction.

can't	o'clock	they'll
isn't	it's (it is)	I'll
you're	who's (who is)	

7. An *apostrophe* plus *s* is used to form the plural of words used as words, letters as letters, figures as figures, etc.

The sentence contains too many *and*'s.
The *e*'s could not be distinguished from the *i*'s.
The *8*'s were blurred and looked like *3*'s.

Note. There is a growing tendency to omit this apostrophe when clearness would not be reduced. It would be clear, for example, to write Bs, Cs, 8s. Note, however, that omitting it before adding *s* to *A*, to *U*, or to *I* would result in *As*, *Us*, and *Is*.

Parentheses

Parentheses are used to set off a word, phrase, or clause that constitutes a definite interruption in continuity.

The store used "loss leaders" (articles priced below cost to attract customers) only on exceptional occasions.

Note. If material in parentheses is inserted into a sentence, no other punctuation precedes the opening of the parentheses, but the closing of the parentheses should be followed by whatever punctuation would have been used if the parenthetical material had not been inserted. If the parenthetical material is itself a sentence and is inserted at the end of a sentence, a period is used in the normal manner at the end of the sentence preceding the parentheses, and a period is placed inside the parentheses at the end of the parenthetical sentence.

Brackets

Brackets are used to mark off material that is inserted into a quotation but not quoted.

"This year [1949] the outlook is less favorable."
"It is definitely established that he [Mr. Schenley] signed the contract."

The Hyphen

A hyphen is used at the end of a line when lack of space makes it necessary to break a word and complete it on the line below. The break must come at the end of a syllable. It is best, when possible, to avoid breaking a word where only two letters would either precede or follow the hyphen. For hyphenation of compound words, see Chapter 4.

CAPITALIZATION

1. The first word of a sentence or of a line of poetry is capitalized.

> The air was smoky.
> "A voice by the cedar tree
> In the meadow, under the hill."

2. The first word of a quotation is ordinarily capitalized, but a capital letter is not used when the quotation merges into the structure of the enclosing sentence.

> His exact words were, "The request is refused."
> He objected to "the capitalization of unrealized anticipations."

3. Proper nouns, their derivatives, and common nouns used as proper nouns are capitalized.

(a) Capitalize personal and geographical names, names of races, nations, languages, and the like.

Henry A. Collins	American
English	France
Negro (but white; black)	Hudson River
King County	Lake Erie
The National Biscuit Company	Reed College

Note. Such words as lake and bay are always capitalized when they precede the word that indicates the specific bay or lake. Some magazines and most newspapers, however, do not capitalize them when they follow the word that indicates the specific name (for example, *Payette lake, Hudson bay*). Many other such words—for example *mountain, street,* and *railroad*—are similarly treated.

Note. When the terms discussed above are preceded by two proper nouns they are never capitalized.

> The Mississippi and Missouri rivers.

(b) Capitalize the names of organizations, businesses, governmental bodies, etc.

> Veterans of Foreign Wars
> The Builders Supply Company
> The Federal Communications Commission

(c) Capitalize the days of the week, months, holidays.

Monday   June   Memorial Day

(d) Capitalize titles of books, magazines, articles and other written materials. (The first word and each important word thereafter are capitalized. Such unimportant words as *a, an, the,* and most conjunctions or prepositions shorter than five letters are not capitalized. The word *The* is capitalized when it is the first word in the title of a book, but is not capitalized at the beginning of the title of a newspaper or magazine.)

Essentials of Microwaves   the Saturday Review
The World Almanac   Science and Public Relations

(e) Capitalize the title of a person if it precedes his name, but do not capitalize a title that follows a name unless it is a title of distinction.

Professor Page   Sergeant Keppler
Judge William E. White   Dr. Springer
President Johnson
Lyndon B. Johnson, President of the United States
Byron R. White, Associate Justice of the United States
   Supreme Court

but

Norman Page, professor of chemistry
William E. White, police judge
John Moore, president of the Rotary Club

Note. The illustrations above indicate capitalization in ordinary text. In the address of a letter or in most display printing, most titles would be capitalized.

(f) Capitalize words that are derived from the names of persons, places, or other proper names.

The crew lived in a Quonset hut.
The Plimsoll mark was visible.

Note. Eventually, such capitalization is usually discontinued, as in *sandwich* and *boycott*. Since the discontinuation is gradual, there will always, at any given time, be many terms in which usage differs in different publications. A dictionary will give guidance, but dictionaries necessarily lag behind technical books and magazines.

4. Nouns and pronouns referring to the Deity, and numerous other words with sacred significance are capitalized.

"Earth changes, but thy soul and God stand sure."
"So take and use Thy work."

5. The pronoun *I* and the interjection *O* are capitalized, but *oh* is not capitalized except at the beginning of a sentence or quotation.

6. In ordinary text, capitalization merely as a means of emphasis is poor form. Inexperienced writers tend to capitalize too frequently. The following precautions, especially, should be observed:

(a) North, south, east, and west are capitalized only when they are used as proper nouns, that is, the names of specific geographical regions. When these words indicate mere direction, they are not capitalized.

> He lived in the South.
> He lived south of the tracks.

(b) Such words as chemistry, psychology, and history are not capitalized except when they occur as part of a title—for example, the title of a specific university course.

> Being interested in physics, he decided to take The Physics of Sound.

ITALICS

Italics (indicated in typing or longhand by an underline) show that words are used in some special manner and hence should be set off from the rest of the material.

1. Italics are used for the titles of books, magazines, newspapers, bulletins, or other separately issued publications.

The Business Letter in Modern Form	*Basic Engineering Metallurgy*
Webster's Seventh New Collegiate Dictionary	the *Chicago Tribune*
	the *Farm Science Reporter*
	The American College Dictionary

Note. If the word *A, An,* or *The* is the first word of the title of a book, it is italicized; at the beginning of the title of a periodical, it is not italicized.

2. Italics are used for the names or titles of ships, trains, airplanes, works of art, etc. Some authorities, however, place such material in quotation marks.

> the *Lusitania*
> El Greco's *Cleansing of the Temple*
> the *Twentieth Century Limited*

3. Italics are used for words considered as words, phrases considered as phrases, and letters or numbers considered as letters or numbers. (Quotation Marks are sometimes used for the same purpose. See Rule 3 under *Quotation Marks*).

> The writer used *effect* incorrectly.
> The phrase *in case of* is overworked.
> The *i* looks like an *e*, and the 8 is so dim it looks like 3.

Note. A letter used to indicate shape, as in "V-shaped" or "I-beam," is not italicized.

4. Foreign words or phrases are italicized, whether written out or abbreviated. (This rule covers Latin scientific names). To determine whether some words are considered foreign or are accepted as English, it may be necessary to consult a dictionary.

> *coup d'état*   *ipso facto*   *de facto*

5. Italics are used to indicate special emphasis, but their use for this purpose should be limited to occasions when the emphasis is abnormal or when special emphasis is really necessary to make sure that the reader grasps the intended meaning.

> Do you mean to tell me that you met *the* Walter Huston?
> He definitely gave instructions that the deduction should *not* be made.

ABBREVIATIONS

The following rules for abbreviation apply to ordinary writing only. The additional abbreviations needed for technical style, and the use of abbreviations in footnotes, bibliography, and technical parts of letters, are discussed with those particular topics.

In ordinary writing, abbreviations should be used sparingly. Unless one of the following rules clearly calls for abbreviation, a word should be written out.

1. The abbreviations A.M., P.M., A.D., and B.C. are always used in place of the terms for which they stand.
2. The abbreviations *e.g.* (for example), *i.e.* (that is), and *viz.* (namely) may be used when informality is appropriate. At any other time, their English equivalents are preferable.
3. The abbreviation *etc.* is so extremely informal that it should usually be avoided in regular text that is offered as finished writing. It is acceptable, however, in a series of brief rules—as you see it used, for example, in Rule 2 of *Italics*, above.

4. In recent years many abbreviations such as NATO, TVA, and FBI have come to be used as if they were actual names. Their acceptance results from the length of the names they stand for. When the readers addressed are sure to understand their meanings, they are acceptable except on occasions so formal that nothing except full, official names would be appropriate.

5. Such abbreviations as *Jr., M.A., Ph.D.,* and *C.P.A.,* when following names, are preferred to the full words or phrases. Three abbreviations for titles—*Mr., Mrs.,* and *Dr.,* are used before names. In ordinary text, however, all other titles are in better form if written out whether they precede or follow a name. Examples of these titles are *Professor, Captain, President, Senator, Governor, Major.* Like these titles, terms of respect such as *reverend* and *the honorable* (these are not titles) should be written out.

6. Such words as *figure* and *number* preceding a number may be abbreviated: *Fig. 3; No. 7.*

7. Abbreviations are not used for the names of persons: *James, William, Robert,* not *Jas., Wm., Robt.*

8. Abbreviations are not used in ordinary text for the names of states or similar terms (except for *D.C.* and sometimes *N.Y.*); nor are they used in text for words such as *street, avenue, company, volume,* and *chapter.*

SPELLING

Among all the possible errors in English there are few that are so damaging as errors in spelling. Mistakes in spelling are noticed and found annoying by readers who pass over all but the most glaring errors. No person who expects to do technical professional work can afford to shrug his shoulders and say "I never could learn to spell."

The first necessity in overcoming weakness in spelling is the development of an adult attitude, a determination not to go through life handicapped by a juvenile weakness. Once such an attitude is acquired, improvement may not be easy, but it is hardly so formidable a task as might be anticipated. Even a weak speller, if he keeps a list of the words that he misses, is usually surprised at its shortness. The frequency of his errors does not ordinarily result from his misspelling a great number of words, but from his repeatedly misspelling certain words that he uses frequently. He can often eliminate most of his errors by learning to spell no more than 40 or 50 words.

If you are weak in spelling, you should be able to make steady improvement if you will carry out the following instructions:

1. Check your work carefully for the express purpose of correcting the errors that result from sheer carelessness rather than from ignorance.

2. Keep a list of every word you miss, and have that list in front of you when you write. Your individual list will be more valuable than a list taken from a book. You will soon come to recognize any word on the list when you start to write it, and can check the way you spell it without taking time to consult a dictionary.
3. Learn just a few rules—the ones that are violated most frequently.
4. Give special attention to certain words that are misspelled because of sloppy pronunciation, and to pairs of words that are confused because of their similarity.

The Rules

The following rules do not cover everything; they represent an attempt to eliminate as much as possible, so that the rules that are most helpful will not be buried in the mass of rules and exceptions that would be necessary for completeness.

1. *ie* and *ei:* When the pronunciation is long *e*, use *i* before *e* except after *c*, and use *e* before *i* after *c*. (When the pronunciation is not long *e*, the spelling is usually *ei;* but few words except those where the pronunciation is long *e* cause trouble.)

believe	receive	weigh
chief	conceit	neighbor
field	ceiling	foreign

Exceptions. either, neither, seize, leisure, weird, financier, species.

2. The effect of prefixes: When a prefix is placed before a word, the spelling of the word itself remains unchanged. (This rule will solve many troublesome questions about double letters.)

dis + appoint = disappoint	un + worried = unworried
dis + appear = disappear	un + noticed = unnoticed
dis + satisfied = dissatisfied	grand + daughter = granddaughter

3. Treatment of final silent *e* when a suffix is added:

(a) When a suffix beginning with a consonant is added to a word ending in silent *e*, the *e* is retained.

hope + ful = hopeful	shame + less = shameless
move + ment = movement	like + ly = likely

Exceptions. *duly, truly, argument, awful.* Also, dropping the *e* is preferred to retaining it in the spelling of such words as *judgment, acknowledgment,* and *abridgment.*

(b) When a suffix beginning with a vowel is added to a word ending in silent *e*, the *e* is dropped. (This covers the spelling of hundreds of words to which *ing* may be added.)

hope + ing = hoping	quote + able = quotable
move + ing = moving	promote + ion = promotion
locate + ion = location	change + ing = changing

Exception.   The letters *c* and *g* always have a hard pronunciation before *o* and *a* (*cat, coat, gave, go*). Hence a final silent *e* must sometimes be retained to keep the pronunciation of *c* or *g* soft.

enforce + able = enforceable	courage + ous = courageous
change + able = changeable	

Exceptions.   *Mileage, saleable,* and *useable* are exceptions to the rule for dropping the silent *e.* (*Milage* is an error. *Salable* and *saleable* are equally correct. *Useable* is acceptable but *usable* is preferred.) Also, there are a few words such as *singeing* or *dyeing* in which the *e* is retained to prevent confusion with other words (*singing, dying*).

4. Doubling a final consonant: When a suffix beginning with a vowel is added to a word ending in a consonant, the final consonant of the word is doubled under the following condition: (a) The word must end in only one consonant, preceded by only one vowel, and (b) the word must be accented on the last syllable. (This would include all one-syllable words.)

Final Consonant Doubled

occur + ed = occurred	stop + ed = stopped
refer + ed = referred	drag + ed = dragged
begin + ing = beginning	lag + ing = lagging

Exceptions.   gas + es = gases   gas + eous = gaseous

Final Consonant Not Doubled

rent + ing = renting	(Word ends in two consonants.)
read + ing = reading	(Two vowels precede the consonant.)
offer + ed = offered	(Accent not on last syllable.)
refer + ence = reference	(Accent shifted back to an earlier syllable.)

Note.   Though the rule just discussed seems somewhat complicated, its application may be greatly simplified, for the misspellings that may be corrected by applying the rule are almost exclusively the result of overlooking the portion of the rule referring to accent. Characteristic errors are *occured, begining, refered.* For purely functional purposes, the rule could be simplified to read: *If you are in doubt as to whether to double* *a final consonant when you add a suffix, notice where the accent lies. If it lies on the last syllable* (this obviously covers all words of one syllable) *double the final consonant. If it is not on the last syllable, do not double the final consonant.*

Note. The cases where this would not apply will not cause trouble. Even the weakest spellers would spell words such as *needed* or *spending* correctly without even considering the question of doubling a final consonant.

5. Adding a suffix to a word ending in *y:*

(a) When a suffix is added to a word that ends in *y* preceded by a *consonant*, the *y* is changed to *i* unless the result would be double *i*.

worry + ed = worried	mercy + ful = merciful
accompany + ment = accompaniment	hurry + ed = hurried
pity + able = pitiable	worry + ing = worrying

Exceptions. There are several exceptions, including words such as *shyness, citylike, secretaryship.* None of the exceptions is likely to be misspelled, or even to occasion doubt. Hence the rule may be applied, in case of doubt, without danger of causing an error in spelling.

(b) When a suffix is added to a word that ends in *y* preceded by a *vowel*, the *y* remains unchanged.

joy + ful = joyful	assay + ed = assayed
annoy + ance = annoyance	employ + ment = employment

Exception. *laid* (from lay + ed); *said* (from say + ed); *daily* (from day + ly).

Note. The most valuable fact to remember about errors resulting from addition of suffixes to words ending in *y* is that practically all these errors result from a single cause: failure to change *y* to *i* when it should be changed. The section of the rule expressed in (a) should receive major attention.

6. Words ending in *ede* or *eed:* Among the troublesome words ending in *ede* or *eed,* only three end in *eed,* namely *exceed, succeed,* and *proceed.* All the others end in *ede*—for example, *concede, recede, precede, supersede.* It is further worth noting, if one is doubtful whether *c* or *s* precedes the *ede,* that only one word *supersede,* ends in *sede.*

7. For rules that concern the apostrophe, consult the discussion of punctuation.

Pairs of Words That May Be Confused

Many misspellings result from confusion between the word desired and some other word that resembles it in spelling or pronunciation. Some of the words commonly confused are as follows:

accept, except	incidence, incidents
advice, advise	its, it's
affect, effect	lead, led
all ready, already	lightening, lightning
all together, altogether	loose, lose
born, borne	manufacturers, manufactures
brake, break	passed, past
breath, breathe	perform, preform
capital, capitol	personal, personnel
choose, chose	precedence, precedents
cite, sight, site	presence, presents
coarse, course	principal, principle
complement, compliment	prophecy, prophesy
dual, duel	stationary, stationery
dyeing, dying	their, there, they're
emigration, immigration	to, too, two
eminent, imminent	whose, who's
formally, formerly	your, you're
forth, fourth	

Words Misspelled Because of Pronunciation

Many words give trouble in spelling because they are mispronounced or pronounced carelessly. If the correct pronunciation of these words can be learned, the errors in spelling will be corrected automatically. Some of the most troublesome words of this sort are listed below:

accidentally (not accidently)	mathematics (not mathmatics)
arctic (not artic)	miniature (not minature)
athletics (not atheletics)	mischievous (not mischeevious)
boundary (not boundry)	performance (not preformance)
disastrous (not disasterous)	perseverance (not perserverance)
drought (not droughth)	prescription (not perscription)
foundry (not foundary)	quantity (not quanity)
height (not heighth)	similar (not similiar)
hindrance (not hinderance)	sophomore (not sophmore)
incidentally (not incidently)	temperament (not temperment)
irrelevant (not irrevelant)	temperature (not temperture)
laboratory (not labratory)	

Miscellaneous Spelling List

absence	desperate	led
accept	develop	leisure
accidentally	disappear	liable
accommodate	disappoint	lightning
accumulate	dormitories	loose
acquaint	dual	lose
across	duel	maintenance
address	embarrass	manufacturer
advice	emphasize	mathematics
aggravate	employees	mileage
allotted	environment	miniature
all right	equipped	missile
already	especially	nineteen
amateur	exaggerate	ninety
analyze	exceed	ninth
annual	excellent	noticeable
appearance	except	occasion
appropriate	exercise	occurred
argument	exhaust	occurrence
auxiliary	existence	omitted
beginning	familiar	optimistic
believe	feasible	parallel
beneficial	finally	particle
calendar	fluorescent	partner
ceiling	foreign	perform
changeable	foremost	permissible
coarse	forth	perseverance
committee	forty	personal
comparative	fourth	personnel
comparison	gases	plain
competent	gauge	plane
completely	government	practically
conscience	grievous	precede
conscious	guarantee	preferred
continuous	height	prejudice
controlled	hindrance	principal
convenience	hypocrisy	principle
course	immediately	procedure
criticism	inconvenience	proceed
criticize	independence	quantity
curriculum	indispensable	quiet
cylinder	interest	quite
deferred	its	receipt
definite	it's	receive
dependent	laboratory	recognizable
description	lead	recommend

repetition	similar	thorough
representative	sincerely	transferred
reservoir	soluble	unnecessary
schedule	sponsor	wherever
seize	studying	
separate	therefore	

GLOSSARY OF STANDARD USAGE

The following list includes many but far from all of the words and phrases that persistently trouble a great number of writers.

The term *colloquial* is used with the liberal meaning *informal* rather than being restricted to its original meaning of *conversational*.

Accept, Except.   *Accept* means *to receive willingly*. If you use *except* with that meaning, you are in error.

Affect, Effect.   *Affect*, as a verb, means *to influence*. *Effect* as a verb means *to achieve or accomplish*. *Affect* is never correct as a noun (except in one highly technical usage in psychology). *Effect* as a noun means *result* and is essentially the noun form suggested by *affect* as a verb.

> The weather *affected* their plans.
> The attorneys *effected* a settlement of the suit.
> The weather had no *effect* on their plans.

Already, All ready.   *Already* means *previously*. *All ready* means *entirely prepared*.

Among, Between.   *Among* is used in reference to more than two persons or objects. *Between* is used in reference to two only.

> The profits were divided *among* the employees.
> The profits were divided *between* the two partners.
> The profits were divided *between* the company and the employees.

Amount, Number.   *Amount* refers to quantity. *Number* refers to objects that can be counted.

> The *amount* of beef in storage was increasing.
> The *number* of cattle on the range was increasing.

And/or.   Useful though this expression seems, it is frowned on by most authorities. Never use it in ordinary text if you can avoid it.

> UNDESIRABLE: It had been colored by the use of ammonium carbonate, and/or ammonium chloride.
> BETTER: It had been colored by the use of ammonium carbonate, ammonium chloride, or both.

As. Often ambiguous when used to introduce a clause that might be introduced by *because* or *since*.

>AMBIGUOUS: *As* the time of departure was drawing near, he became very nervous.
>CLEAR: *Since* (or *because*) the time of departure was drawing near, he became very nervous.
>CLEAR: *When* the time of departure came he grew very nervous.

As per. A bit of commercial jargon undesirable, in normal use, as a substitute for *in accordance with*.

>UNDESIRABLE: The work was done as per instructions.
>BETTER: The work was done in accordance with the instructions.

Bad, Badly. See *Adjectives and Adverbs*.

Balance. The word *balance*, except in connection with bookkeeping or accounting, should not be used for *remainder*.

>WRONG: We will do the *balance* of the work next week.
>CORRECT: Our *balance* in the bank account had decreased.

Because. Do not misuse *because* for *for*. *Because* should be used to introduce a clause that tells the cause of some result rather than a clause that merely tells the reason for a belief or knowledge.

>WRONG: He has not been here today, *because* his tools are still in his locker.
>CORRECT: He has not been here today, *for* his tools are still in his locker.
>CORRECT: He stayed away *because* he was ill.

Beside, Besides. *Beside* means *by the side of*. *Besides* means *in addition to, except, moreover*.

>WRONG: *Beside* the reason mentioned, there are other reasons.
>CORRECT: *Besides* the reason mentioned, there are other reasons.

But what, But that. The word *but* in these expressions is meaningless, has no grammatical function, and should be omitted in formal writing.

>WRONG: We had no doubt *but what* (or *that*) he would be acquitted.
>CORRECT: We had no doubt *that* he would be acquitted.

Can, May. *Can* indicates ability. *May* indicates permission. In recent years, however, the objections to using *can* for *may* have grown less vigorous.

Can't hardly, Can't scarcely. These terms are illogical just as a double negative is illogical. Taken literally, either of them would cause a statement to mean the exact opposite of what it was intended to mean.

> WRONG: We *can't hardly* see it from here.
> CORRECT: We *can hardly* see it from here.

Can't help but. This expression cannot be defended from the grammatical standpoint and is avoided by careful writers except in extremely informal use. It is a distortion of the extremely formal expression, *cannot but.*

> WRONG: *I can't help but* agree with him.
> CORRECT: *I can't help* agreeing with him.
> CORRECT, BUT EXCESSIVELY FORMAL: I *cannot but* agree with him.

Come. Use of *come* for *came,* as in "We *come* here ten days ago," is highly illiterate. It persists in the oral language of many people who would not write it.

Considerable. *Considerable* is an adjective. It should not be used as a noun or adverb. As a noun it is colloquial; as an adverb, illiterate.

> WRONG: The company lost *considerable* during the first year.
> CORRECT: The company lost a *considerable amount* of money during the first year.
> WRONG: We were influenced *considerable* by what he said.
> CORRECT: We were influenced *considerably* by what he said.

Continual, Continuous. *Continual* means *constantly recurring* (as in *continual* interruptions). *Continuous* means *without cessation* (as in the *continuous* increase in pressure).

Could of, Would of, Should of, etc. In all such expressions, *of* is misused for *have* as the result of careless pronunciation in speech. The wrongness of these expressions is apparent when one realizes that they are comparable to "I *of* gone."

Data, Phenomena, Strata. These forms are all plurals. Their singulars are *datum, phenomenon, stratum.* There is a strong tendency, at present, to use *data* as a singular when referring to a mass of facts considered as a whole; but most authorities still insist that it should be treated as a plural.

Different from, Different than. *Different from* is preferred.

Differ from, Differ with. *Differ from* means *to be dissimilar. Differ with* means *to disagree.*

> Apple trees *differ from* pines in appearance.
> The manager *differed with* the president as to who should be promoted.

Enable.   *Enable* means *to make able, to give ability to.* It should not be used in such a sentence as, "This appropriation *enables* the road to be paved," for the road is not given any new ability. It would be preferable to say, "This appropriation *enables* the highway department to pave the road."

Farther, Further.   Careful writers prefer *farther* in reference to actual distance and *further* in reference to quantity or degree.

Fewer, Less.   *Fewer* should be used when numbers are referred to; *less*, when quantity is referred to.

> CORRECT: *Less* wheat was harvested this year than last; consequently *fewer* (not *less*) freight cars will be needed.

Good, Well.   See *Adjectives and Adverbs.*

Had of.   The *of* is meaningless, unnecessary, and wrong. *Had* is sufficient.

> WRONG: I wish I *had of* learned the news earlier.
> CORRECT: I wish I *had* learned the news earlier.

Had ought.   This is an undesirable form, mistakenly substituted for *ought* or *should.*

> WRONG: He'd (he had) *ought* to resign.
> CORRECT: He *should* resign.
> CORRECT: He *ought to* resign.

Hardly.   See *Can't hardly.*

Have got.   A redundant way of saying *have,* and an undesirable substitute for *must* or *have to* except on occasions of extreme informality.

> REDUNDANT: We *have got* many things to be thankful for.
> CORRECT: We *have* many things to be thankful for.
> UNDESIRABLE: *We've got* to reach the town before morning.
> CORRECT: We *have* to reach the town before morning.
> CORRECT: We *must* reach the town before morning.

If.   *If* is ambiguous when used in place of *whether* because it does not imply *or not* with sufficient emphasis.

> AMBIGUOUS: Let me know *if* you expect to visit us.
> CLEAR: Let me know *whether* you expect to visit us.

Imply, Infer.   A speaker or writer implies. A listener or reader infers. It is inaccurate to say, "I *implied,* as I read your letter, that. . . ." Similarly, it is wrong to say, "Did you mean to *infer,* by your remarks, that. . . ." The same distinction applies to *implication* and *inference.*

Irregardless. *Irregardless* is an illogical and unacceptable alternative for *regardless*. Superficially, it resembles *irresponsible* and *irrelevant,* in which the *ir* means *not,* whereas the *ir* in *irregardless* has no effect on meaning.

Is when, Is where, Is because. These phrases are often misused, as seen in the following sentences.

> WRONG: Insolvency *is where* one cannot pay one's debts.
> CORRECT: Insolvency is a lack of funds to make payments that are due.
> WRONG: Bankruptcy *is when* one's property is administered for the benefit of one's creditors.
> CORRECT: Bankruptcy is a state *in which* one's property is administered for the benefit of one's creditors.
> WRONG: The reason for his absence *is because* he is ill.
> CORRECT: The cause of his absence is *that he is ill.* (or *is illness.*)

Less. See *fewer.*

Lie, Lay. *Lie* is an intransitive verb meaning *recline. Lay* is a transitive verb meaning to *place,* or *cause to lie.*

> *Principal parts:* lie, lay, have lain
> lay, laid, have laid
> CORRECT: He lies on the bed. (present)
> CORRECT: He lay on the bed. (past)
> CORRECT: He *has lain* on the bed all afternoon.
> WRONG: He *lays* on the bed.
> WRONG: He *laid* (or *has laid*) on the bed all afternoon.
> CORRECT: He *lays* the newspaper on the bed.
> CORRECT: He *has laid* the newspaper on the bed.

Note. The confusion results primarily because the past tense of *lie* is identical with the present tense of *lay.* Some errors also result from the use of *laid* as the past tense or past participle of *lie.*

Like, As if. Careful writers do not use *like* as a conjunction, preferring *as if* or *as though. Like* is correct, however, as a preposition. In conversation, *like* as a conjunction slips into the language even of many literate people.

> CORRECT: He looked *like* a man of forty.
> CORRECT: He spoke *as if* (or *as though*) he had a cold.

Note. Being a preposition, *like* cannot introduce a clause; and the usual sign that the construction is a clause is the presence of the subject and verb. An easier way to avoid misuse of *like* is to try *as if* in its place, and use it if it sounds right. You may be sure that when *as if* is wrong it will sound wrong.

Most, Almost. These are two distinct words, with different meanings. Sometimes, however, *most* is used colloquially for *almost* as in "He made an error in *most* every copy." The simplest way to avoid this error is to remember that *most* is never correct when *almost* would convey the desired meaning.

> CORRECT: He made errors on *most* of the copies.
> CORRECT: It was a *most* unfortunate statement.
> COLLOQUIAL OR WRONG: Most everyone in the crowd saw the incident.

Of. Do not use *of* for *have* as in "He must *of* gone," for "He must *have* gone." Other examples of this error are *could of, should of,* and *might of.* The error is a carry-over from careless pronunciation.

Percent. This term should not be loosely used for *portion* or *part.* Ordinarily, it should be used only following a number. It should not be used for *percentage,* though to save space many writers disregard this distinction in tables and figures.

> CORRECT: The company earned a 10 *percent* profit.
> CORRECT: The *percentage* (not *percent*) of profit increased.
> WRONG: A large *percent* of the profit resulted from sales abroad.

Practicable, Practical. *Practicable* is used to indicate that a method *can* be used or that a result *can* or *could* be accomplished. *Practical* goes further; it indicates that whatever is being considered not only is a possibility but would actually give results that would justify action. For example, building a steam-powered automobile is *practicable* but has not proved to be *practical.* In most sentences either word makes sense, but they convey different meanings to well-informed readers.

Principal, Principle. *Principal* as an adjective means *main* or *chief,* and as a noun means *the main* or *controlling person,* as the *principal* of a school or one of the *principals* in a lawsuit. *Principle,* always a noun, means *fundamental truth, basic law.*

> CORRECT: The *principal* reason for its failure was defective workmanship.
> CORRECT: The *principles* of democracy must be preserved.

Reason is because, reason why. Both of these expressions are redundant, for neither *is because* nor *why* adds to the meaning already conveyed by *reason.* Although they might qualify as colloquialisms, careful writers avoid them even in informal writing.

Rise, Raise. The distinction between *rise* and *raise* is essentially the same as the distinction between *lie* and *lay*. *Rise* is an intransitive verb, *raise* is transitive.

> *Principal parts:* rise, rose, have risen
> raise, raised, have raised
>
> CORRECT: The prices *rise* when the supply becomes scarce.
> CORRECT: The dealers *raise* the prices when the supply becomes scarce.
> WRONG: The prices *raised* when the supply became scarce.
> CORRECT: The prices *rose* when the supply became scarce.

Seldom ever. Redundant; the full meaning is expressed by *seldom* alone.

Shall, Will. The careful discrimination between *shall* and *will* is not so common as in past times, and even literate writers frequently use *will* on some occasions when *shall* is called for by the rules. The main points to remember in order to make the traditional discriminations are as follows:

(a) To indicate simple future or expectation, *shall* is correct in the first person and *will* in the second and third persons.

> I *shall* be a senior next year.
> He *will* spend August in Omaha.

(b) *Will* is correct in the first person, and *shall* in the second and third person, to indicate determination or command.

> I *will* do it in spite of your objections.
> You *shall* change your itinerary.
> They *shall* not pass.

(c) *Will* is used in all persons to express willingness.

> I *will* go if no one else *will*.

Should, Would. Traditionally, the difference between *should* and *would* is the same as that between *shall* and *will*. That is, if we are meticulous we say, "I *should* enjoy reading the book," and "He *would* enjoy reading the book." Each of the two words has special uses, however, as the following comments show.

Should is used in all persons in the sense of *ought to* or in a conditional clause—which is usually a clause beginning with *if*.

> I *should* give him credit for working overtime.
> He *should* give me credit for working overtime.
> If I *should* refuse, no one could blame me.
> If he *should* refuse, no one could blame him.

Similarly, *would* is used in all persons to indicate customary action. It is also used whenever *should* would be misleading in that it might suggest *ought to.*

> *Customary action:* Long before closing time, I would begin preparing to go home.
>
> *Avoidance of the meaning of ought to:* Under the circumstances, I would have refused.

Sit, Set. The distinction between *sit* and *set* is essentially the same as the distinction between *rise* and *raise,* and *lie* and *lay. Sit* is an intransitive verb; *set* is transitive.

> *Principal parts:* sit, sat, have sat
> set, set, have set
>
> CORRECT: He *sits* at my desk, where he *sat* yesterday, and where he *has sat* so many days before.
>
> CORRECT: He *sets* the case on the desk, where yesterday he *set* it down so carelessly. He *has set* it there every time he has come in the office.

Some, Somewhat. *Some* should not be used as an adverb with the meaning of *a little.* The correct word for this use is *somewhat.*

> WRONG: I was feeling *some* better.
>
> CORRECT: I was feeling *somewhat* better.

These kind. This expression is wrong. The plural *these* should not modify the singular *kind.*

Try and. This expression should not be substituted for *try to. Try and* is just as illogical as "*I am able and*" for "*I am able to.*"

> WRONG: I will *try and* go.
>
> CORRECT: I will *try to* go.

Type. The common tendency among technical writers to use *type* as an adjective does not conform with the rules of good use. *Type* is a noun or verb, not an adjective. It should not directly modify a noun unless, in a hyphenated term, it becomes part of a compound adjective.

> WRONG: We installed the new *type* machinery.
>
> CORRECT: We installed the new *type of* machinery.
>
> CORRECT: The design calls for an *injector-type* condenser.

Very. It is not good idiom to use *very* to modify a past participle. Some word such as *much* or *well* should be used between *very* and the participle. It is a poor style to overwork *very* as an intensifier. In a phrase such as *very nauseating, very* is superfluous.

WRONG: The customers were *very pleased.*

CORRECT: The customers were *very much* (or *very well*) pleased.

While.   Care must be taken to prevent *while* from being ambiguous when it is used in the sense of *although;* its use with this meaning is often regarded as improper.

AMBIGUOUS: *While* his work was heavy, he did it well.

CORRECT: *Although* his work was heavy, as he did it well. (or: *As long as* his work was heavy, he did it well.)

AMBIGUOUS: The first floor was neat, *while* the second floor was cluttered.

CORRECT: The first floor was neat, *but* the second floor was cluttered.

With.   *With* should not be loosely used to establish some vague, undefined thought relationship.

WRONG: *With* the plans completed, the president of the company went on a vacation.

CORRECT: The plans having been completed, (or, When the plans had been completed,) the president of the company went on a vacation.

WRONG: The production was increasing, *with* the prices holding steady.

CORRECT: The production was increasing, *and* the prices were holding steady (or, *but* the prices were holding steady).

EXPLANATION OF GRAMMATICAL TERMS

Absolute.   An expression that is not grammatically connected with the sentence where it occurs; sometimes called a "nominative absolute."

The job being completed, the crew was laid off.

Adjective.   One of the parts of speech. A class of words used to modify (describe or limit) nouns or pronouns.

rough road; *this* year; *reasonable* profits

Adjective clause.   A clause used as an adjective to modify a noun or pronoun.

The tire *that blew out* was defective. (The clause *that blew out* modifies the noun *tire.*)

Everyone *who came* enjoyed himself. (The clause *who came* modifies the pronoun *everyone.*)

Adverb.   One of the parts of speech. A class of words used to modify verbs, adjectives, or other adverbs.

He drove *cautiously*. (*Cautiously* modifies the verb *drove*.)
The problem is *very* hard. (*Very* modifies the adjective *hard*.)
The frame was *very* strongly built. (*Very* modifies the adverb
  strongly.)

Adverb clause.   A clause used as an adverb to modify any part of speech
that an adverb might modify, usually a verb. Such a clause often tells
how, when, or where.

If the cost is excessive, we shall cancel the project. (The italicized
  clause modifies the verb *shall cancel*.)
They will burn the slashing *when the weather grows damper*. (The
  clause modifies the verb *will burn*.)

Agreement.   The necessary correspondence between a subject and verb
in person and number; between a pronoun and its antecedent in person,
number, and gender; and between a demonstrative adjective (*this, these*)
and the noun that it modifies, in number.

Antecedent.   The noun or the other pronoun to which a pronoun refers.

After reading the *report* he returned *it*. (*Report* is the antecedent
  of *it*.)

Appositive.   A noun or other substantive placed next to some other noun,
used in the same way grammatically, and referring to the same thing or
person.

The main plant, *a four-story building*, stood at the edge of town.
It was discovered by the watchman, *a reliable employee*.

Article.   Any of three words—*a, an,* or *the*.

Auxiliary.   A word used as part of a verb to assist in indicating the
tense, voice, mood, etc. of the main verb.

are working; *have* answered; *must have* heard; *shall* or *will* arrive

Clause.   A grammatical unit containing a subject and a finite verb. At
least one independent clause is necessary in any complete sentence.
(Sometimes part of the clause may be implied rather than expressed, in
which case it is called an elliptical clause.)

A main or independent clause is one that makes an independent
assertion.

A subordinate or dependent clause is one that is not self-sufficient
and does not of itself make an assertion, but is used as part of a main
clause. It is always used as if it were a single word—a noun, adjective,
or adverb.

Independent clause: The sky is blue.

Subordinate clause used as a noun: We heard *that the request was refused.*

Subordinate clause used as an adjective: I know of a hotel *that will be satisfactory.*

Subordinate clause used as an adverb: Please close the door *when you leave.*

Comparison. The change of form in an adjective or adverb, indicative of change of degree, as in *good, better, best; long, longer, longest; easily, more easily, most easily.*

Complement. An element that ordinarily follows a verb and completes the assertion made by the verb about the subject. Types of complements are shown in the following examples:

Direct object: The farmer raised *wheat.*

Indirect object: The manager gave *him* a check.

Predicate adjective as subjective complement: The tractor is *old.*

Predicate adjective as objective complement: The news made him *happy.*

Predicate noun as subjective complement: The tree is an *oak.*

Predicate noun as objective complement: The president made John Jones the *manager.*

Complex sentence. See *Sentence.*

Compound sentence. See *Sentence.*

Compound-complex sentence. See *Sentence.*

Conjunction. One of the parts of speech. A word used to connect words, phrases, or clauses. Coordinating conjunctions connect elements that are equal in grammatical rank. Subordinating conjunctions are used to connect subordinate clauses to main clauses.

Coordinating conjunctions: and, but, or, nor, for; sometimes *yet* and *so.* See also *Correlative Conjunctions.*

Subordinating conjunctions: after, as, because, since, when, where, etc. For example, in the sentence "He did not come because he was ill," *because* makes the clause it introduces subordinate.

Conjunctive adverb. An adverb that functions also as a conjunction to join main clauses (*also, however, moreover, consequently, furthermore,* etc.). Ability to discriminate between conjunctive adverbs and coordinating conjunctions is important as a means of avoiding the serious error known as the "comma splice." (See Unity of Structure under Unity and Coherence in the Sentence.)

Correlative conjunctions.   Conjunctions used in pairs to join elements of equal rank.

> either . . . or; neither . . . nor; not only . . . but also; both . . . and

Ellipsis.   The omission of words that are necessary for the grammatical completeness of the sentence, the words omitted being implied by what is expressed.

> While (I was) working out of doors, I gained weight.
> The new model is sturdier than the old (model is).
> (You) Send me the answer promptly.

The subject of an imperative sentence is practically always *you,* implied. Hence an imperative sentence is almost always elliptical.

Expletive.   A word that bears no real meaning but is used to fill out a sentence. An expletive at the beginning of a sentence is sometimes necessary, but sometimes causes postponement of the word that should be the subject.

> *It* is certain that production will increase.
> *It* is snowing.
> *It* is nine o'clock.

Gender.   The status of a noun or pronoun as masculine (*man, waiter, he*), feminine (*woman, she, waitress*), neuter (*tree, house, it*), or common (*mouse, person, you*).

Gerund.   A verb form used as a noun. A gerund plus its object is a gerund phrase.

> *Gerund: Walking* is good exercise.
> *Gerund phrase: Pruning a tree* demands judgment.

Idiom.   An expression that is accepted as correct even though it may not be in accordance with the normal patterns of the language, and that often conveys a meaning that would not be conveyed by the literal meaning of its component words.

> *The more the merrier* is a typical idiom. It conveys more meaning than can be accounted for by the words and is acceptable as a sentence even though it does not have the elements a sentence calls for. *Make off with* is an idiomatic expression meaning *take away. Of yours* (as in "that car *of yours*") is idiomatic, for it is acceptable even though the object of a preposition is in the possessive case. *Many a person is* is an idiom for it is accepted as correct even though *many,* a plural, modifies a singular word and though a subject with a plural meaning takes a singular verb.

When we refer to the *idiom of a language,* we refer to its particular manner of using words to convey ideas—a manner that cannot be entirely accounted for by rules or by the meaning of individual words. The use and meaning of prepositions, especially, depend upon the idiom of the language. When some expression in your speech or writing is corrected on the grounds that it is not idiomatic of the language, it is usually impossible to justify the correction by citing rules.

Infinitive.   The form of a verb that is used after *to,* though sometimes the *to* is implied rather than expressed. (He hoped *to be* present. He can *go.*) When it is part of the verb phrase, as in the examples above, an infinitive is called a complementary infinitive because it completes the verb.

An infinitive can also be used as a noun, an adjective, or an adverb.

> *Infinitive as a noun: To err* is human. (*To err* is used as the subject of the verb *is.*)
>
> *Infinitive as an adjective:* I have a confession *to make.* (*To make* modifies the noun *confession.*)
>
> *Infinitive as an adverb:* They were eager *to accept.* (*To accept* modifies the adjective *eager.*)

Inflection.   Variation in the form of words to indicate change of number, gender, person, tense, or mood. Inflection of nouns is referred to as declension; inflection of verbs as conjugation; inflection of adjectives and adverbs as comparison.

Interjection.   One of the parts of speech. A type of word placed at the beginning of a sentence or inserted into the sentence (whence *interjection—thrown in*) but not connected with the grammatical structure. Words such as *oh, alas,* and *well* (as an exclamation) are interjections.

Linking verb.   A verb that expresses neither action nor condition, but merely establishes a connection between its subject and a noun, pronoun, or adjective in the predicate. A noun or pronoun following a linking verb is in the nominative case. The commonest linking verb is the verb *to be.* Other characteristic linking verbs are *become, appear, seem, smell* (in the sense of "possess an odor"), *taste* (in the sense of "possess a flavor"), and *feel* (in the sense of "experience or provide a sensation").

> The seed *is* pure.
> The price *seems* right.
> The warmth *feels* good.

Modifier.   A word is a modifier when it is used to limit or change the meaning of some other word, or of a phrase or clause. A phrase or a clause, as well as a word, can be a modifier.

Word as a modifier: His *slightly* embarrassed manner was amusing. (*Slightly,* an adverb, modifies *embarrassed,* a participial adjective, which in turn modifies the noun *manner.*)

Phrases and clauses as modifiers: Construction *of the bridge* will be postponed *until the weather is warmer.* (*Of the bridge,* an adjective phrase, modifies *construction. Until . . . warmer,* an adverb clause, modifies the verb *will be postponed.*)

Mood. The form, in a verb, that indicates how the action or condition expressed by the verb is conceived by the writer or speaker. In the statement of a fact or the asking of a question, the indicative mood is used. In giving a command, the imperative mood is used. In expressing doubt, wish, or condition contrary to fact, the subjunctive mood is used.

Indicative: The rock *is* heavy. *Can* you *lift* it?

Imperative: Examine the surface carefully.

Subjunctive: If I *were* you, I *should accept* the offer.
I wish he *were* not so young.

Nonrestrictive modifier. A modifier that merely gives information about the term modified rather than limiting or identifying it.

The roof, *which was very old,* was damaged by the wind.

His oldest son, *who was more experienced,* handled the case efficiently.

Noun. One of the parts of speech. A word used to name a person, place, thing, quality, etc.

John; Aristotle; engineer; Atlanta; wood; courage

Many words can be used, of course, as more than one part of speech. Such words cannot be classified until they are actually used. For example, *green* is a noun in the sentence, "Green is its natural color" but an adjective in "The green grass is attractive."

Noun clause. A subordinate clause used as a noun.

I know *what you mean.* (Noun clause used as object of the verb *know.*)

How he had escaped was a puzzling question. (Noun clause used as subject of the verb *was.*)

Object. The word, phrase, or clause identifying the person or thing that receives the action indicated by a transitive verb, or the substantive referred to by a preposition.

Direct object of a verb: The company installed a *lathe.* (*Lathe* is the direct object of *installed.*)

Indirect object of a verb: He offered *me* a bribe. (*Me* is the indirect object of *offered,* for it indicates the person to whom the offer is made. *Bribe,* like *lathe* in the preceding example, is a direct object.)

Object of a preposition: He worked on a *farm.* (*Farm* is the object of *on.*)

Participle.   The form of a verb ending in *ing* (though this form may also be a gerund) or the form used after *have.* A participle may be used as the main element in a verb phrase or as an adjective.

Participles as verbs: I am *following* the route I have *followed* before.

Participles as adjectives: Running water purifies itself. The *rejected* casting was returned. The guard was *tired.*

Parts of speech.   The classifications under which all words are classified. The eight parts of speech in traditional grammar are nouns, pronouns, verbs, adjectives, adverbs, prepositions, conjunctions, and interjections. Each of these terms is defined in its proper alphabetical place in this Glossary.

A word qualifies as one or another part of speech only on the basis of its use in a sentence or phrase; it is possible for the same word to be used as more than one part of speech. Good usage, however, limits the *permissible* uses of most words. To determine whether it is correct to use a word as a certain part of speech, it is sometimes necessary to consult a dictionary.

Person.   Some pronouns and most verbs vary in form to indicate whether they refer to the person speaking or writing (first person), the person addressed (second person), or the person spoken or written about (third person).

First Person	Second Person	Third Person
I come	you come	he comes
I have	you have	he has
I shall find	you will find	he will find
We shall find	you will find	they will find
I am	you are	he is
I can	you can	he can

Phrase.   A group of related words that does not include a subject and predicate and that is usually used as a single part of speech—adjective, adverb, verb, or noun. Phrases may be prepositional, participial, gerund, infinitive, or verb.

Prepositional phrase: He discovered the address *of the owner.* (Adjective phrase, modifying the noun *address.*)

Prepositional phrase: The road will be closed *in December.* (Adverbial phrase, modifying the verb *will be closed.*)

Participial phrases: Entering the room, I noticed that the fan *standing in the corner* was noisy. (Adjective phrases, modifying, respectively, *I* and *fan.*)

Gerund phrase: Pitching hay is strenuous work. (Noun.)

Infinitive phrase: To predict the results is impossible. (Noun.)

Verb phrase: The method *has been found* successful.

Predicate.   The portion of a clause consisting of the verb, its complements, and the modifiers of both. Except for interjections and absolute phrases, neither of which appears frequently, the predicate consists of everything in a clause or sentence except the subject.

Preposition.   One of the parts of speech. A word used to introduce a noun or pronoun and establish its relationship to the sentence.

He will come *in* the morning *on* the plane *from* Chicago.

Principal parts.   The three forms of a verb from which, by help of auxiliary verbs, the various tenses are derived. These forms are the present infinitive, the past tense, and the past participle. (The past participle is the form used after *have.*)

Present Infinitive	Past Tense	Past Participle
work	worked	worked
sing	sang	sung
begin	began	begun
catch	caught	caught

Pronoun.   One of the parts of speech. A word used in place of a noun.

Personal pronouns: I, you, he, she, it, we, they

Interrogative pronouns: who, which, what

Relative pronouns: who, which, what, that

Demonstrative pronouns: this, that, these, those

Indefinite pronouns: each, either, neither, any, anyone, some, someone, one, no one, few, all, none

Reciprocal pronouns: each other, one another

Reflexive pronouns: myself, yourself, himself

Intensive pronouns: myself, yourself, himself

Relative clause.   A clause, always dependent, introduced by a relative pronoun. The clauses used as restrictive modifiers below are examples.

Restrictive modifier. A modifier that narrows, and thus restricts, the meaning of whatever it modifies.

> All persons *who drive recklessly* should be arrested.
> The cases *that were damaged* have been returned.

Sentence. A group of words expressing a thought and containing, either actually or by implication, a subject and a finite verb (predicate). Grammarians have never defined the term *sentence* to their own complete satisfaction, but this definition will serve our purpose. In grammatical form a sentence may be simple, compound, complex, or compound-complex.

A simple sentence consists of a single independent clause, though as indicated by the second example below, its subject, its verb, or both may be compound.

> *Simple sentence:* The building is old.
> *Simple sentence:* Both industry and labor support the new ruling and oppose the old.

A compound sentence consists of two or more independent clauses.

> *Compound sentence:* Sales were falling off, and inventories were increasing.

A complex sentence contains a single independent clause and one or more dependent clauses.

> *Complex sentence:* The police recovered the property that had been lost.

A compound-complex sentence contains at least two independent clauses, hence being compound, and at least one dependent clause, hence being complex.

> *Compound-complex sentence:* The wind was rising and the sky had clouded over when we finally reached the shelter.

Subject. The substantive that a verb or verbal makes an assertion about, asks a question about, or gives an order to.

> *Subject of a sentence: Birds* fly south in the winter.
> *Subject of an infinitive phrase (a verbal):* I know *him* to be honest. (*Him* is the subject of *to be.*)

Substantive. A noun, or any word or group of words used as a noun. Any of the following, in addition to a noun, may be a substantive: pronoun, infinitive phrase, gerund, or noun clause. (See separate listing for each of these terms.)

Tense. The distinctive form in a verb that indicates time.

Present tense: It *runs* well.

Past tense: It *ran* well.

Future tense: It *will run* well.

Present perfect tense: It *has run* well.

Past perfect tense: It *had run* well.

Future perfect tense: It *will have run* well.

Tenses appear in the progressive and emphatic forms of verbs, as well as in the simple forms.

Present progressive: I *am reading.*

Present emphatic: I *do read.*

Verb.   One of the parts of speech. A word or phrase that indicates action, being, or state of being. In a declarative sentence the verb is the word or phrase that actually makes the assertion.

A transitive verb takes an object.

He *bought* the *house.*

An intransitive verb cannot take an object.

The sun *rises* in the east.

Some verbs may be either transitive or intransitive.

Transitive: I *read* the letter

Intransitive: I *was reading.*

Verbal.   A word or phrase derived from a verb but not making an assertion. Infinitives, gerunds, and participles (participles used either in participial phrases or as adjectives) are the three types of verbals. Each is listed as a separate entry.

Voice.   The form of a verb that shows whether the subject of a verb or verbal performs the action indicated (active voice) or has the action performed upon it (passive voice). Only transitive verbs can be in passive voice.

Active voice: The government *collected* the taxes.

Passive voice: The taxes *were collected* by the government.

APPENDICES

A

ABBREVIATIONS FOR SCIENTIFIC AND ENGINEERING TERMS

The number of abbreviations in current use is so great and is expanding so rapidly that in a book like this only a small fraction of them can be listed. For those in positions that make it necessary for them to use long and highly specialized lists, such lists are usually available. Also it is worth noting that both collegiate and unabridged dictionaries provide abbreviations for hundreds of technical terms.

Since minor differences persist in actual use, the abbreviations in different lists are not always the same in form, but those in the list that follows are supported by well-regarded authorities.

As you use the list you will observe that some terms are abbreviated in one way when they appear alone and otherwise when they appear in the abbreviation for a phrase, especially if the phrase contains *per* or a diagonal line (/) standing for *per*. For example, *ft* is the abbreviation for *foot* or *feet* and *sec* for *second;* but *feet per second,* which includes both words, may be shortened to *fps.* Other examples of abbreviations that are used as combining forms are: *m* for *miles* and *h* for *hour* in *miles per hour* (*miles* standing alone is not abbreviated and *hour* is abbreviated as *hr* or *h*); *g* for *gallons* and *m* for *minute* in *gallons per minute* (the ordinary abbreviations for *gallons* and *minute* are *gal* and *min*); and *rf* for *radio frequency* (*radio* is not abbreviated standing alone, and *frequency* is abbreviated as *freq*). Even in abbreviating a phrase, however, it is better not to use the combining forms if there is any doubt about their being easily understood by the readers. For example, *ft per sec* is sometimes better than *fps.*

The abbreviations in this book do not include arbitrary signs and symbols. (Such signs and symbols are not actually abbreviations and should rarely be used in ordinary text.) They comply, so far as facts permit, with the sound generalization that capital letters are ordinarily used in abbreviations only when the terms for which the letters stand would be capitalized if written out. Periods are omitted except when omitting them might cause the abbreviation to be mistaken for a regular word, or when the term is abbreviated in general as well as in technical writing. You should not hesitate to use periods, however, if personal observation convinces you that it is customary to use them in the field that includes your subject.

abampere .... abamp
absolute .... abs
acoustic .... acst
acre .... (spell out)
acre-foot .... acre-ft
air horsepower .... ahp
air position indicator ... api
air speed indicator ..... asi
alternating-current (adj.) a-c
American Standard ..... Amer Std
American Wire Gage ... AWG
ammeter .... am.
ampere .... amp
ampere-hour .... amp hr
amplitude modulation ... am.
analysis of variance .... anova
angstrom .... A
approximate .... approx
atmosphere .... atm
atomic weight .... at wt
audio frequency .... a-f
auxiliary .... aux
azimuth .... az

balanced incomplete
 block designs .... bib designs
barometer .... bar
barrel .... bbl
Baumé .... Bé
best linear unbiased
 estimator .... blue
billion electron volts .... bev
biochemical oxygen
 demand .... bod
bits per inch .... bpi
board feet .... fbm
boiler horsepower ..... bhp
brake horsepower ..... bhp
brake horsepower-hour .. bhp-hr
Brinell hardness ...... Bh
Brinell hardness
 number .... Bhn
British thermal unit .... Btu

calorie .... cal
candle-hour .... c-hr
candlepower .... cp
carrier wave .... cw
cathode ray .... cr
Celsus C
centigrade C
centimeter .... cm
central processing unit .. cpu
chemically pure .... cp
circumference .... circ
cologarithm .... colog
continuous wave ...... cw
counts per minute ..... cpm
cubic .... cu
cubic centimeter cc
cubic feet per minute ... cfm
cubic feet per second .. cfs
cubic foot .... cu ft
cubic inch .... cu in.
cubic yard .... cu yd
cycles per minute ..... cpm
cycles per second ..... cps

data control block ...... dcb
decibel .... db
decimeter .... dm
degree .... deg*
dew point .... dp
direct-access storage
 device .... dasd
direct-connected .... dir-conn
direct-current (adj.) .... dc or d-c
distilled water .... dw
dry bulb .... db

effective horsepower .... ehp
efficiency .... eff
electromagnetic units ... emu
electromotive force ..... emf
electron volts .... ev
electronic data processing edp

* The symbol (°) is sometimes used in text as an abbreviation for degrees of latitude, longitude, or angles, but should not be used in place of the regular abbreviation for degrees of temperature. Since it is permissible, however, to use F, C, and K respectively for *degrees* Fahrenheit, Centigrade, and Kelvin, there is a strong tendency to do so and to omit *deg*.

elevation .... el
equation .... eq
extremely high
    frequency .... ehf

Fahrenheit (or degrees
    Fahrenheit) .... F
farad .... f
feet board measure
    (board feet) .... fbm
feet per minute .... fpm
feet per second .... fps
fluid .... fl
fluid ounce .... fl oz
foot .... ft
foot candle .... fc
foot per minute .... fpm
foot pound .... fp
fractional horsepower ... fhp
free on board .... fob
frequency .... freq.
frequency modulation .. fm
fusion point .... fp

gallon .... gal
gallons per day .... gpd
gallons per hour .... gph
gallons per minute ..... gpm
gallons per second ...... gps
gram .... gr
gravities (acceleration of
    gravity) .... g
ground position
    indicator .... gpi

hectogram .... hg
henry .... h
high explosive .... he
high frequency .... hf
high frequency current .. hfc
high-potential test ..... hipot
horsepower .... hp
horsepower hour .... hphr
hour .... hr or h
hundredweight .... cwt

inch .... in.
inch-pound .... in-lb
input/output .... i/o
inside diameter .... id
inside dimension .... id
intermediate frequency .. i.f.
International Critical
    Tables .... ICT
international unit ...... iu

jet propulsion .... jp
job control language .... jcl
joule .... j

Kelvin (or degrees
    Kelvin) .... K
kilo .... k
kilovolt .... kv
kilovolt-ampere .... kva
kilowatt .... kw
kilowatt-hour .... kwh
kinetic energy .... ke

lambert L
latitude .... lat
linear .... lin
linear foot .... lin ft
liquid .... lq
liquefied petroleum gas .. lpg
liter l
logarithm (common) .... log
logarithm (natural) ..... ln
longitude .... long.
low frequency .... lf
lumen .... lm
lumen hour .... lhr
lumens per watt .... lpw

man hour .... man hr
maximum .... max
meter .... m
miles per gallon .... mpg
miles per hour .... mph
millimeter .... mm
minute. .... min

modulator .... mod
modulus .... mod
molecular weight ...... mol wt
motor-generator .... mg
multiple programming
  with variable number
  of tasks .... mvt

National Aircraft
  Standards .... NAS
National Bureau of
  Standards .... NBS
National Electrical
  Safety Code .... NESC
National Electric Code
  Standards .... NEC
negative .... neg
net weight .... nt wt
nickel-silver .... ni-sil
normal temperature and
  pressure .... ntp
number .... no.

octane .... oct
optical .... opt.
optimum working
  frequency .... owf
oscillate .... osc
ounce .... oz
outside diameter .... od

parts per billion .... ppb
parts per million .... ppm
permeability .... perm
perpendicular .... perp
phase .... ph
phase modulation ...... pm
polar .... pol
potentiometer .... pot.
pound .... lb
pound-foot .... lb-ft
pound-force (avoirdupois) lbf
pound-mass (avoirdupois) lbm
pounds per cubic foot .. pcf
pounds per square foot .. psf

pounds per square inch .. psi
power amplifier .... pa
primary sampling unit .. psu
probable error .... pe
pulse-amplitude
  modulation .... pam

qualitative .... qual
quality .... qual

radio direction finder ... rdf
radio frequency .... rf
Rankine .... R
Réaumur .... Ré
reference line .... ref l
remote job entry .... rje
revolutions per minute .. rpm
revolutions per second .. rps
rheostat .... rheo
Rockwell hardness ..... Rh
roentgen .... r

scleroscope hardness .... sh
second .... sec
second-foot .... sec-ft
sequential access method sam
shaft horsepower .... shp
short wave .... sw
signal-to-noise ratio .... snr or s/n
socioeconomic status .... ses
solenoid .... sol
specific gravity .... sp gr
spherical candlepower .. scp
square .... sq
square meter .... sq m
square mile .... sq mi
standard metropolitan
  statistical area .... smsa

tachometer .... tach
tangent .... tan
temperature .... temp
template .... temp
tensile strength .... ts

tension . . . . tens.
thermal . . . . thrm
thermocouple . . . . tc
thermometer . . . . therm.
thousand . . . . m or M
thousand cubic feet . . . . mcf
time sharing option . . . . . tso
tolerance . . . . tol
transceiver . . . . xvr

ultrahigh frequency . . . . uhf
United States Standard . . USS

vacuum . . . . vac
vacuum tube . . . . vt
variable . . . . var

variable-frequency
  oscillator . . . . vfo
very high frequency . . . . vhf
video frequency . . . . vdf
volt v
volt-ampere . . . . va

watt . . . . w
watt-hour . . . . wh or whr
wavelength . . . . wl
wet bulb . . . . wb

yard . . . . yd
year . . . . yr
yield point . . . . yp
yield strength . . . . ys

B

HELPFUL PUBLICATIONS FOR TECHNICAL WRITERS

Technical Writing and Reports

BEAKLEY, GEORGE C., JR., and DONALD D. AUTORE, *Graphics for Design and Visualization.* New York: The Macmillan Company, 1973.

COMER, DAVID B., and RALPH R. SPILLMAN, *Modern Technical and Industrial Reports.* New York: G. P. Putnam's Sons, 1962.

FEAR, DAVID E., *Technical Writing.* New York: Random House, 1973.

GARLAND, KEN, *Graphics Handbook.* New York: Van Nos Reinhold, 1966.

GIESECKE, FREDERICK E., et al., *Technical Drawing,* 5th ed. New York: The Macmillan Company, 1967.

GRAVES, HAROLD F., and LYNE S. S. HOFFMAN, *Report Writing,* 4th ed. Englewood Cliffs, N.J.: Prentice-Hall, Inc., 1965.

HARWELL, GEORGE C., *Technical Communication.* New York: The Macmillan Company, 1960.

HOELSCHER, R. P., et al., *Graphics for Engineers: Visualizations, Communication and Design.* New York: John Wiley & Sons, 1968.

LUZADDER, WARREN J., *Basic Graphics,* 2nd ed. Englewood Cliffs, N.J.: Prentice-Hall, Inc., 1968.

MENZEL, DONALD H., HOWARD MUMFORD JONES, and LYLE G. BOYD, *Writing a Technical Paper.* New York: McGraw-Hill Book Company, 1961.

MILLS, GORDON H., and JOHN A. WALTER, *Technical Writing,* 3rd ed. New York: Holt, Rinehart & Winston, Inc., 1970.

SOUTHER, JAMES W., *Technical Report Writing.* New York: John Wiley & Sons, Inc., 1957.

SYPHERD, W. O., ALVIN M. FOUNTAIN, and V. E. GIBBENS, *Manual of Technical Writing.* Chicago: Scott, Foresman & Company, 1957.

TURNER, R. P., *Technical Report Writing,* 2nd ed. New York: Holt, Rinehart & Winston, Inc., 1971.

ULMAN, JOSEPH N., and JAY R. GOULD, *Technical Reporting,* 3rd ed. New York: Holt, Rinehart & Winston, Inc., 1972.

WEISS, HAROLD, J. B. MCGRATH, *Technically Speaking: Oral Communication for Engineers, Scientists, and Technical Personnel.* New York: McGraw-Hill Book Company, 1963.

ZETLER, ROBERT L., and ROBERT GEORGE CROUCH, *Successful Communication in Science and Industry.* New York: McGraw-Hill Book Company, 1961.

468

General English

CORBETT, EDWARD P., *Classical Rhetoric for the Modern Student,* 2nd ed. Fair Lawn, N.J.: Oxford University Press, 1971.

DEAN, HOWARD H., and KENNETH D. BRYSON, *Effective Communication,* 2nd ed. Englewood Cliffs, N.J.: Prentice-Hall, Inc., 1961.

KIERZEK, JOHN M., and WALKER GIBSON, *The Macmillan Handbook of English,* 5th ed. New York: The Macmillan Company, 1965.

LEGGETT, GLENN, C. DAVID MEAD, and WILLIAM CHARVAT, *Prentice-Hall Handbook for Writers,* rev. 6th ed. Englewood Cliffs, N.J.: Prentice-Hall, Inc., 1974.

A Manual of Style, rev. 12th ed. Chicago: University of Chicago Press, 1969.

PERRIN, PORTER GALE, and WILMA R. EBBITT, *Writer's Guide and Index to English,* 5th ed. Chicago: Scott, Foresman & Company, 1972.

STRUNK, W., JR., and E. B. WHITE, *The Elements of Style,* 2nd ed. New York: The Macmillan Company, 1972.

U.S. Government Printing Office Style Manual, rev. ed. Washington, D.C.: Government Printing Office, 1967.

Indexes

The Biological and Agricultural Index
Applied Science and Technology Index (1958–   )
Bibliography of Scientific and Industrial Reports
The Engineering Index
The Industrial Arts Index
International Index to Periodicals
The New York Times Index
Readers' Guide to Periodical Literature

Sources of General or Special Information

The Census Volumes (Department of Commerce, Bureau of Census)
 Agriculture
 Construction Industry
 Drainage of Agricultural Lands
 Irrigation of Agricultural Lands
 Manufacturers
 Retail Distributors
 Unemployment
 Wholesale Distribution
Encyclopedia Americana
Encyclopaedia Britannica
Minerals Yearbook (Department of the Interior, Bureau of Mines)
The Statesman's Yearbook
Technical Book Review Digest
U.S. Government Publications Monthly Catalog
Van Nostrand's Scientific Encyclopedia
The World Almanac

INDEX

471